SELECT
THERMODYNAMIC MODELS
FOR PROCESS SIMULATION

A PRACTICAL GUIDE USING
A THREE STEPS METHODOLOGY

Cover illustration : Esquif Communication
www.esquif.fr

Energies nouvelles **Publications**

▶ **Jean-Charles de HEMPTINNE**
Research Engineer, IFP Energies nouvelles
Professor, IFP School

SELECT THERMODYNAMIC MODELS FOR PROCESS SIMULATION

A Practical Guide using a Three Steps Methodology

2012

t Editions TECHNIP 5 av. de la République 75011 PARIS, FRANCE

Preface

In this "Practical Guide to Thermodynamics", the authors have provided a unique resource for practitioners, teachers and students for both understanding and use. The presentation has an orderly presentation of the developments from fundamentals to application case studies, as well as detailed guidance about specifying thermodynamic property problems and efficient, effective methods to solve them. The steps from the fundamentals of thermodynamics to properties, from properties to models, and from models to methods, are explicitly shown with rigorous derivation and solutions for realistic example situations. Their problem-solving procedure guides learners and users through the complex set of decisions to be made in order to achieve maximum accuracy and reliability of the desired results. This Guide's contents are far more than a manual; the perceptive user will also gain both pragmatic skills and valuable confidence to meet the challenges of obtaining thermodynamic and transport properties, as well as phase and reaction equilibria, in contemporary and future chemical technologies.

John P. O'Connell, University of Virginia

Foreword

This user guide **is not** designed as a *thermodynamics textbook*. Many very good thermodynamic handbooks exist for helping teachers in designing a course in a logical, linear fashion (Elliot & Lira [1], Prausnitz *et al.* [2], Smith & Van Ness [3], O'Connell [4], Vidal [5]...). Teachers can use this document as suggested below, but will be disappointed by the lack of demonstrations, and the non-linear conception of the logic.

This manual **is not** either *a review of the existing thermodynamic models*, that would help developers in understanding and code the most adapted model to their situation. Other authors have made a significant effort in summarising the models in a logical fashion, stressing their strengths and weaknesses and providing indications on how they should be coded, if needed (Poling *et al.* [6]; Kontogeorgis & Folas [7]; Mollerup & Michelsen [8]; Riazi [9]...). Thermodynamic experts will be disappointed by the lack of completeness, of numerical parameters and of algorithmic analysis.

Rather, our purpose is to provide a **vademecum** that should help the practicing engineer in finding the right questions to answer when faced with a novel type of thermodynamic problem: when the questions are correctly stated, the answer is half on its way.

The construction of the book, that may seem awkward at first, is designed for this purpose. It is constructed on three pillars (chapters 2, 3 and 4) that each represent a different point of view converging to the same goal: the development of an adequate thermodynamic set of models for an industrial problem. We believe that in order to analyse completely a physical modelling problem, these three pillars must be correctly mastered (figure 0.1):

1. understanding the fundamental principles,
2. use of the available mathematical models, and
3. knowledge of the system physical (phase) behaviour.

The domain of application that is aimed is petroleum and energy process design. Yet, readers working in other fields of chemical engineering may also find interesting topics.

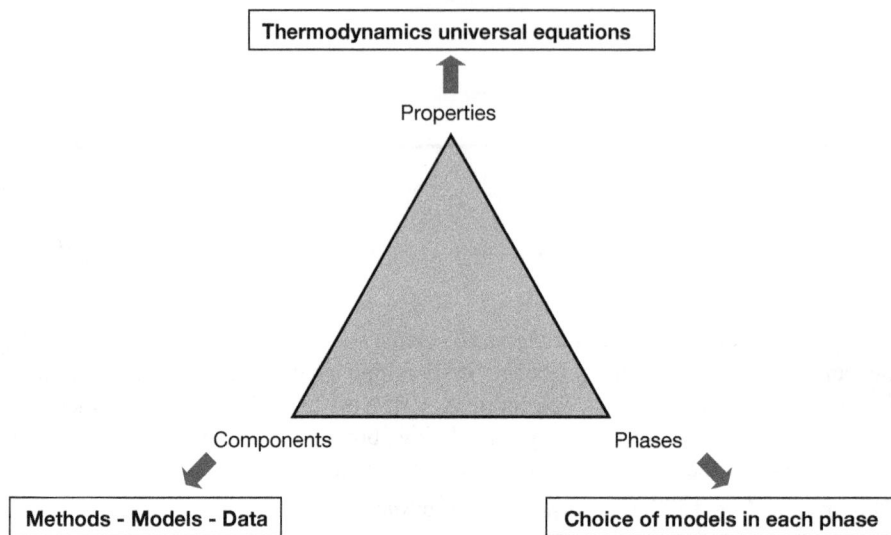

Figure 0.1

Triangular representation of the fundamental questions in thermodynamic problem-solving.

BOOK STRUCTURE

The **first chapter** goes into the details of the philosophy of the approach. The main messages we want to carry here is:

- A thermodynamic method may contain many different models (as in the image of the Russian doll); each of which has to be parameterised. The origin of the parameters is at least as important as the choice of the model.
- The three questions that will help the engineer solve his problem concern the relevant properties of his process, the type of mixture he has and the phases he might encounter.

The **second chapter** summarises some of the basic principles of thermodynamics (what O'Connell [4] calls 'Always True'). It is written so that most concepts and equations that the chemical engineer may need are provided in a very concise form. It contains two main sections:

- How to read and understand a phase diagram, using Gibbs phase rule?
- How are the fundamental thermodynamic principles used for calculating engineering properties? This results in a rather short presentation of residual properties, excess properties, and a few algorithmic aspects on phase equilibrium calculation. Chemical equilibrium is also touched upon.

The **third chapter** is the heart of this book. Rather than providing an exhaustive list of thermodynamic models, it emphasises the use of these models, starting from the concept of the fluid composition. In fact, this composition must be regarded as a set of parameters to be used with the chosen mathematical model. The link between this parameter + equation combination and the true physical behaviour is performed through a comparison with experimental data. This should explain the construction of the third chapter that has three parts:

- It starts with the description of experimental information: section 1 discusses pure components; section 2 discusses mixtures. The third section illustrates how this experimental information is used for parameter estimation using data regression.
- In section 4, the actual thermodynamic models are summarised. No effort was made to be exhaustive (the user is in principle limited to what his process simulator provides), but it was rather attempted to help the reader make the link between the molecular structure of the fluid phase and the significance of the parameters values.
- In a final, short but important section, the concept of key component is introduced. The sensitivity of the requested type of information to some parameters may be much larger than to others. Some guidelines are provided here to help the engineer focus his attention on the most important parameters (i.e. key components or key binaries).

The **fourth chapter** puts the modelling issue in the perspective of the true physical behaviour of a physical system. The phase behaviour depends greatly on the type of mixture that is considered. This is why a number of important industrial mixtures are discussed. For each system, some model recommendations are suggested. The discussion is organised in two main parts:

- First, the property behaviour is discussed through some thermodynamic diagrams. As phase properties are considered qualitatively independent of the fluid composition, the example of pure CO_2 is used.
- Second, the phase diagrams are analysed in some details. Here, the varying complexity of several types of mixtures is shown.

The true originality of our approach is presented in the final, **fifth chapter**. It both concludes the theoretical part of this endeavour (which is the book), and introduces the practical part that will be proposed on the web. The intention is to publish electronically a number of case studies that illustrate industrial examples that have been treated at IFP Energies nouvelles. The solution pattern will follow as closely as possible the approach suggested in this fifth chapter.

The same final chapter also lists some important industrial processes in order to guide the engineer in his problem-analysis.

HOW TO READ THE BOOK

As a result of the construction of the book, numerous cross-references are unavoidable. This may puzzle the reader who wants to start from the beginning and end at the final chapter. We apologise for this and we have made all our possible to make it readable in various modes. This has led us to re-state some items in different chapters.

Mode 1: The teacher

Even though this is not a textbook, the essential features for an advanced thermodynamics course are available. The main message of this course is condensed in the third chapter of the book. Nevertheless, several years experience have shown that some basics must be reviewed in order to make clear the links between the models (chapter 2), and that the physical picture must continually be kept in mind (chapter 4) in order to avoid that the student only looks at the mathematical relationships. This is why we suggest trying alternating physical understanding and modelling approach:

1. Discuss single phase properties:

    a. We suggest starting with reading diagrams. Section 2.1.3 provides some basics on the phase diagram, and section 4.1 discusses concretely how the various properties behave with pressure and temperature for a pure component.

    b. Discussion of the diagrams should lead the students to ask himself how these properties are calculated. This is discussed in section 2.2.2. The definition of the properties (section 2.1) and the fundamental relationships (section 2.2.1) are available if needed for detailed questions. At this first stage, it is good to stress the residual approach and the corresponding states principle, which are essential in most applications (as an example, we can refer to the Lee-Kesler method, section 3.4.3.3.).

    c. Now models have been introduced, it may be worth discussing the parameterisation of these models (i.e. how does one find the correct parameters considering the information available on the components). This is presented in section 3.1.

2. Discuss mixture vapour-liquid phase diagrams:

    a. Concerning phase diagrams, section 2.1.3 provides the basics of how an isothermal or an isobaric phase diagram should be read. The starting point should consist in illustrating Raoult's law, and, for mixtures with supercritical components, Henry's law. The theory of the asymmetric convention is discussed in some length in section 2.2.3.1, but it is probably of more use at this point to go to the Rachford-Rice equation (section 2.2.3.2) in order to allow some simple calculations.

    b. Non-ideality can be introduced next, with some example phase diagrams discussed in section 4.2.1 that goes more into the details of the various types of diagrams (understanding the significance of three-phase lines is generally not very simple).

3. Equations of state:

    a. Now the phase behaviour is understood, it may be of interest to introduce the calculation mode in a more general way ('always true') by defining fugacity and fugacity coefficient (sections 2.2.1.2 and 2.2.3.1). The link with the Rachford-Rice equation, previously seen, becomes then clear.

    b. It is now possible to move to the third chapter, and present in some detail the equations of state. The first step (because most often used) is to illustrate the cubic equations of state (section 3.4.3.4).

4. Activity coefficient models:

    a. The models can be introduced in two stages: first, discuss the significance of excess properties (section 2.2.2.2), then illustrate what is the impact of the activity coefficient on the phase diagrams (section 3.4.1).

    b. At this stage, the list of the activity coefficient models can be shown, as in section 3.4.2.

5. Now that all the basics are laid, the applications can be discussed one by one, with the model recommendations for each case (section 4.2). If more complex models are needed (as Huron-Vidal mixing rules or SAFT type equations), they should be introduced in good order, as referred to in chapter 3.

Mode 2: The student

The student obviously should follow the guidelines of his teacher, but he may find attractive the fact of finding the main concepts of chemical engineering thermodynamics separated in the three main chapters:

- Chapter 2 teaches what is 'always true'. Hence, all concepts and equations presented in that chapter can be applied in virtually all cases.
- Chapter 3 describes the most important thermodynamic models that can be found in commercial simulators. It will teach the approximate mathematical relationships that have been developed over the years for inter- or extrapolating the observed physical behaviour.
- Chapter 4 may seem of less interest from the student's point of view, as he is more often concerned with understanding the mathematical relationships than in viewing the actual complexity of the real world. Yet, this chapter illustrates that thermodynamics is in fact a physical science and that the numbers that can be generated through the model equations are to be compared with everyday observations.

Mode 3: The engineer

The book is really designed for the engineer. He should consider this as a working tool, which he can consult in all directions he wants using the many cross-references.

Most probably, he will start from the end, as he should recognise in the final (fifth) chapter some concerns he has when designing or upgrading process units. Chapter four should also help him identify the expected phase behaviour of his system and selecting the adequate model.

Unless he has been guided further in the other chapters through the cross-references, he could stop at this point. Yet, he may be really concerned with either accuracy or predictive power, in which case he will have to go to the third chapter which will teach him what experimental data are relevant to his problem and to which parameters it is most sensitive.

The second chapter will most likely be of less interest to him, unless he wants to better understand how the models are used for calculating the final properties, and what assumptions have gone into their choice.

CONCLUSION

As a conclusion, we hope that this book will be of use to many process engineers. It has been designed as a practical guide, and the reader is welcome to contact the main author for suggesting improvements and practical examples of use. An accompanying website is available where case studies will be proposed, which may be of help to the process engineer (http://books.ifpenergiesnouvelles.fr/ebooks/thermodynamics). New case studies will be added over time, so as to illustrate that the approach proposed in the book is applicable for a large range of conditions and processes.

Jean-Charles de Hemptinne

Jean-Marie Ledanois

Pascal Mougin

Alain Barreau

REFERENCE LIST

[1] Elliott, J.R. and Lira, C.T. "Introductory Chemical Engineering Thermodynamics", Prentice Hall PTR, Upper Saddle River, NJ, **1999**.

[2] Prausnitz, J.M., Lichtenthaler, R.N. and Gomes de Azevedo, E. "Molecular Thermodynamics of Fluid Phase Equilibria", 3rd Ed., Prentice Hall Int., **1999**.

[3] Smith, J.M., Van Ness, H.C. and Abbott, M.M. "Introduction to Chemical Engineering Thermodynamics", Sixth Edition, McGraw-Hill, Inc., New York, **2001**.

[4] O'Connell, J.P. and Haile, J.M. "Thermodynamics: Fundamentals for Applications", 1st Ed., Cambridge University Press, **2005**.

[5] Vidal, J. "Thermodynamics: Applications in Chemical Engineering and the Petroleum Industry", Editions Technip, Paris, **2003**.

[6] Poling, B.E., Prausnitz, J.M. and O'Connell, J.P. "The Properties of Gases and Liquids", 5th Ed., McGraw-Hill, New York, **2000**.

[7] Kontogeorgis, G.M. and Folas, G.K. "Thermodynamic Models for Industrial Applications: From Classical and Advanced Mixing Rules to Association Theories", Wiley, **2010**.

[8] Michelsen, M.L. and Mollerup, J. "Thermodynamic Models: Fundamental and Computational Aspects", 1st Ed., Tie-Line Publications, **2004**.

[9] Riazi, M.R. "Characterization and Properties of Petroleum Fluids", American Society for Testing and Materials, Philadelphia, **2005**.

Acknowledgements

This book is a step in a long-lasting pedagogical reflexion that is needed in the teaching of industrial thermodynamics at IFP School and IFP Training. We are indebted to the many generations of students, colleague teachers and industrial practitioners who have continuously challenged our vision on this complex material. We are conscious that this document is not an end on its own, but rather a tool that should still be improved through additional comments and critics.

The resulting monograph would not have seen light without the extremely careful reading and comments of Professor Jean-Noël Jaubert (ENSIC, Nancy) who, with his colleague Romain Privat, have spent their vacation time going through all the equations and nomenclature. We are similarly very much indebted to Dr. Marco Satyro, Associate Professor at Schulich School of Engineering and Chief Technology Officer at Virtual Materials Group, who brought his year-long experience of both educational and industrial practitioning to the service of improving this book.

Several other international colleagues, both from academia and from industry, have commented on the document, thus making it a true collaborative work. Olivier Baudouin, Process Manager at ProSim Company, has provided a number of examples that we have included. Claudio Olivera, Professor at Simon Bolivar University, made very detailed and incisive comments that we are sorry we could not all incorporate. Erich Müller, Professor at Imperial College, provided us with a very pertinent point of view that helped position this document in the wealth of existing thermodynamics handbooks. Last, but not least, John O'Connell, Professor at the University of Virginia and well-known author of the handbook "The Properties of Gases and Liquids", was so kind as to write a foreword for this book.

We could also mention many discussions related to the need for a document helping industrial users with this complex material. Dr. Eric Hendriks at Shell Global Solutions, Dr. Ralf Dohrn at Bayer Technology Services, Professor Georgios Kontogeorgis from the Technical University of Denmark and author of the highly recommended handbook on "Thermodynamic Models for Industrial Applications: From Classical and Advanced Mixing Rules to Association Theories", Professor John Shaw from Calgary University, and many others that we apologize for not mentioning.

We should also acknowledge the support we got from IFP Energies nouvelles, embodied by our Research Division Head Jacques Jarrin and the IFP School team headed by Christine Travers. We have received many help and suggestions from our IFP Group colleagues, as Raymond Bulle at IFP Training, Pierre Bichet and Jean-Luc Monsavoir at IFP School,

Vincent Coupard at IFP Energies nouvelles, and many other, in particular from the Thermo-dynamic and Molecular Simulation Department. Mireille Darthenay, under the supervision of Patrick Boisserpe, put a lot of useful work into the reading and correcting of the several proofs, and all the figures are the highly professional work of Dominique Allinquant.

Jean-Charles de Hemptinne
Jean-Marie Ledanois
Pascal Mougin
Alain Barreau

Table of Contents

Chapter 1
INTRODUCTION

Chapter 2
FROM FUNDAMENTALS TO PROPERTIES

Chapter 3
FROM COMPONENTS TO MODELS

Chapter 4
FROM PHASES TO METHOD (MODELS) SELECTION

Chapter 5
CASE STUDIES

List of authors

Jean-Charles de Hemptinne
Chemical Engineer, KUL (Katholieke Universiteit Leuven, Belgium)
PhD, MIT (Massachusetts Institute of Technology, USA)
National accreditation to direct research, Université Claude Bernard Lyon I (France)
Research Engineer, IFP Energies nouvelles
Professor, IFP School
Tuck Foundation Chair "Thermodynamics for biofuels"
IFP Energies nouvelles, 92852 Rueil-Malmaison Cedex
jean-charles.de-hemptinne@ifpen.fr

Jean-Marie Ledanois
Chemical Engineer, IFP School (France)
PhD, Renewable Energies, École Centrale de Paris (France)
Full Professor, Simon Bolivar University, Caracas (Venezuela)
Visiting Scientist, IFP Energies nouvelles (France)
jmledanois@gmail.com

Pascal Mougin
Chemical Engineer, ENSIC (Ecole Nationale Supérieure des Industries Chimiques, France)
PhD, INPL (Institut National Polytechnique de Lorraine, France)
National accreditation to direct research, INPL (Institut National Polytechnique de Lorraine, France)
Research Engineer, IFP Energies nouvelles
IFP Energies nouvelles, 92852 Rueil-Malmaison Cedex
pascal.mougin@ifpen.fr

Alain Barreau
Chemical Engineer, CNAM (Conservatoire National des Arts et Métiers, France)
Research Engineer, IFP Energies nouvelles
alb.barreau@wanadoo.fr

Nomenclature

SYMBOLS

Keyword	Description	SI unit
[A]	concentration (per unit volume) of A	mol m^{-3}
A	helmholtz energy	J
a	molar Helmholtz energy	J mol^{-1}
α	isobaric thermal expansion coefficient	K^{-1}
α	bunsen coefficient	
$\alpha(T)$	alpha function for cubic equations of state	
α_{ij}	relative volatility between components i and j	
α_{ij}	binary interaction parameter for the NRTL model	
A, B, C, D, E, F…	model parameters	
a_i	activity of component i	
a_i	a parameter for component i in a cubic equation of state	
B	second virial coefficient	
β	isochoric thermal pressure coefficient	K^{-1}
b_i	b parameter for component i in a cubic equation of state	
β_T	isothermal compressibility coefficient	Pa^{-1}
c	volume translation parameter (often used for cubic equations of state)	$\text{m}^3 \text{ mol}^{-1}$
C_{mi}	langmuir constant in equation (3.255)	
C_P	isobaric heat capacity	J K^{-1}
c_P	molar isobaric heat capacity	$\text{J K}^{-1} \text{ mol}^{-1}$
C_V	isochoric heat capacity	J K^{-1}
c_V	molar isochoric heat capacity	$\text{J K}^{-1} \text{ mol}^{-1}$
d	hard sphere diameter of a molecule or a segment	m
$\Delta^{A_iB_j}$	equilibrium constant for association between the site A on molecule i and site B on molecule j	m^{-3}
$\Delta c_{P,F}$	molar heat capacity of fusion (liquid – solid)	$\text{J K}^{-1} \text{ mol}^{-1}$
$\Delta c_{P,r0}$	molar isobaric heat capacity of reaction with all components in their reference conditions (products – reactants)	J mol^{-1}

$\Delta c_{Pf,i0}$	molar isobaric heat capacity of formation of component i in reference conditions (molecule – elements)	J mol^{-1}
$\Delta g_{f,i0}$	molar Gibbs energy of formation of component i in reference conditions (molecule – elements)	J mol^{-1}
Δh_F	molar enthalpy of fusion (liquid – solid)	J mol^{-1}
$\Delta h_{f,i0}$	molar enthalpy of formation of component i in reference conditions (molecule – elements)	J mol^{-1}
Δh_{r0}	molar enthalpy of reaction with all components in their reference conditions (products – reactant)	J mol^{-1}
Δh^{σ}	molar vapourisation enthalpy (vapour – liquid)	J mol^{-1}
δ_i	solubility parameter of component i	J$^{1/2}$ m$^{-3/2}$
Δv_F	molar volume change upon fusion (liquid – solid)	m^3 mol^{-1}
ε^{AiBj}	strength of the energy interaction between site A on molecule i and site B on molecule j	J mol^{-1}
ε_{ij}	strength of the dispersive (energy) interaction between components i and j	J mol^{-1}
$\mathfrak{J}$	degrees of freedom	
Φ	number of phases	
φ_i	fugacity coefficient	
f_i	fugacity	Pa
Φ_i	volume fraction of component i	
F_{Obj}	objective function for regression	
G	Gibbs energy	J
g	molar Gibbs energy	J mol^{-1}
$g(r)$	radial distribution function	
γ_i	activity coefficient	
g_{ij}	group interaction energy parameter	J mol^{-1}
Γ_k	activity coefficient of the group k (UNIFAC model)	
H	enthalpy	J
h	molar enthalpy	J mol^{-1}
H_i	Henry constant	Pa
I	ionic strength (ion concentration in solution)	
J	pseudo ionization energy	J
κ^{AiBj}	volume of interaction between the site A on molecule i and site B on molecule j, expressed as a fraction of the segment volume	
K_i	equilibrium or distribution coefficient	
k_B	Boltzmann constant	$1.3806488 \times 10^{-23}$ J K^{-1}
k_{ij}	binary interaction parameter for the dispersive parameters in equations of state	
K_W	Watson (or UOP) K factor	

l_{ij}	interaction parameter between components i and j, often referring to the mixing rule on the volumetric parameter	
μ	Joule Thomson coefficient	$K\ Pa^{-1}$
m	constant in the Soave function for the alpha function	
μ_i	chemical potential	$J\ mol^{-1}$
m_i	mass of component i	kg
M_w	molar mass	$kg\ mol^{-1}$
N	amount of matter	mol
$\mathcal{N}$	number of components	
n	refractive index	
N_{Av}	Avogadro number	$6.022141 \times 10^{23}\ mol^{-1}$
n_C	number of carbon atoms	
ν_i	stoichiometric coefficients	
ν_{mi}	number of groups of type m in molecule i	
P	pressure	Pa
P_c	critical pressure	Pa
$\wp_i$	Poynting correction	
P_i^σ	vapour pressure of component i	Pa
P_r	reduced pressure	
Q	amount of heat	J
$q_0,\ q_1,\ q_2,\ l$	additional parameters (EoS dependent) for G^E mixing rules	
θ	molar vapour fraction	
q_i	molecular surface area in the UNIQUAC/UNIFAC theory	$m^2\ mol^{-1}$
Q_k	group k surface area in the UNIFAC theory	$m^2\ mol^{-1}$
θ_i	surface fraction of component i (UNIQUAC and UNIFAC models)	
θ_{mi}	fraction of type m cavities occupied by component i in equation 3.255	
R	ideal gas constant	$8.314462\ J\ mol^{-1}\ K^{-1}$
ρ	molar density	$mol\ m^{-3}$
$\mathcal{R}$	number of additional constraints in Gibbs phase rule	
r_i	number of segments in a molecule, according to the Flory or Lattice-fluid theory, extended to molecular volume in the UNIQUAC theory	$m^3\ mol^{-1}$
R_k	group k volume in the UNIFAC theory	$m^3\ mol^{-1}$
S	entropy	$J\ K^{-1}$
s	molar entropy	$J\ K^{-1}mol^{-1}$
SG	Specific Gravity	

σ_{ij}	closest distance between the centers of the molecules i and j, represented as hard spheres (mean diameter)	m
T	temperature	K
T_b	Normal Boiling Temperature (NBP)	K
T_c	critical temperature	K
T_F	fusion (or melting) temperature	K
τ_{ij}	binary interaction parameter for the NRTL or UNIQUAC model	
T_r	reduced temperature	
U	internal energy	J
u	molar internal energy	J mol^{-1}
u	speed of sound	m s^{-1}
V	volume	m^3
v	molar volume	m^3 mol^{-1}
W	amount of work	J
ω	acentric factor	
W	number of possible combinations in the lattice	
Ω_a	constant for calculating the a parameter of a cubic equation of state using the corresponding states principle	
Ω_b	constant for calculating the a parameter of b cubic equation of state using the corresponding states principle	
w_i	mass fraction of component i	
W_i	weight corresponding to data point i in a regression	
X	any thermodynamic property	
ξ	extent of reaction	
x_i	molar fraction of component i in the liquid phase	
y_i	molar fraction of component i in the vapour phase	
Z	compressibility factor	
z	number of close neighbours	
z_i	molar fraction of component i in the feed (or in an unknown phase)	
ζ_l	reduced diameter (ζ_3 is the compacity or reduced density)	
Z_{RA}	Rackett compressibility factor	
W	number of distinguishable microstates	
$\underset{=}{\Delta}$	equals per definition	

SUB- OR SUPERSCRIPT

Keyword	Sub or superscript	Description
0	sub	refers to a reference state
#	super	refers to the ideal gas phase
(m)	super	molarity-based (for electrolyte properties, section 3.4.2.5)
*	super	refers to the pure component in the same pressure, temperature conditions and the same physical state as the mixture property
α, β	super	refers to a physical state
ass	super	refers to the association contribution
attr	super	attractive contribution (sum of all contributions, of which the dispersive contribution is most important, this is why they are often considered synonymous)
c	sub	critical property
cal	super	refers to calculated values
chain	super	chain contribution
corr	super	correction term
dev	super	deviation (difference between experimental and calculated)
disp	super	dispersive contribution
E	super	Excess property
eos	super	refers to the equation of state calculation
exp	super	refers to experimental values
F	super	refers to the feed (global)
Φ	super	phase counter
F	sub	fusion (crystallisation) property
f	sub	formation property
g	sub	gas
H	super	refers to the infinite dilution reference state, in the asymmetric convention
hs	super	hard sphere contribution
i	sub	component *i*
in	sub	incoming property
i,s	sub	component *i* in solvent *s*
id	super	ideal mixture property
ij	sub	refers to the binary interaction parameter between components *i* and *j*
L	super	refers to the Liquid phase
lr	super	long range contribution
m	sub	mixture property
$v+, v-$	super	cationic or anionic stoichiometric coefficients
out	sub	outgoing property

pol	super	polar contribution
r	sub	reduced property
r	sub	reaction property
ref	super	refers to a reference state
rep	super	repulsive contribution
res	super	refers to residual property
σ	super	property at saturation (vapour-liquid): for pure components, it means at the vapour pressure
S	super	refers to the solid phase
s	sub	solvent
sr	super	short range contribution
V	super	refers to the Vapour phase
$\bar{X}_i$		partial molar property for a property X
∞	super	infinite dilution property

ABBREVIATIONS

Keyword	Description	Section of this book
AAD	Absolute Average Deviation	3.3.1.3.B
AIChE	American Institute of Chemical Engineers	
API	American Petroleum Institute	
ASTM	American Standard for Testing of Materials	3.1.2.3.A
BIP	Binary Interaction Parameter (often k_{ij})	3.4.3.4.E
BWR	Benedict-Webb-Rubin EoS	3.4.3.3.C
COSMO	Conductor-Like Screening Model	3.4.2.2.A
CPA	Cubic Plus Association	3.4.3.5.B
DIPPR	Design Institute for Physical Properties (database of AIChE)	
EoS	Equation of State	
GCA	Group Contribution + Association EoS	3.4.3.4
GCLF	Group Contribution Lattice-Fluid EoS	3.4.3.6
G^E mixing rule	Mixing rules that are developed for cubic EoS, according to the Gibbs Excess model following Huron-Vidal principle	3.4.3.4.E
GPSA	Gas Processor's Suppliers Association	4.2.4.2.A
GS	Grayson and Streed model	3.4.4.4
LCVM	Linear Combination of Vidal and Michelsen mixing rules	3.4.3.4.E
LCST	Lower Critical Solution Temperature	4.2.1.1
LK	Lee and Kesler EoS	3.4.3.3.C
LLE	Liquid-Liquid Equilibrium	
LSE	Liquid-Solid Equilibrium	
MBWR	Modified Benedict, Webb and Rubin EoS	3.4.3.3.C
MHV	Modified Huron Vidal mixing rules (first and second order exist: MHV1 and MHV2)	3.4.3.4.E
NBP	Normal Boiling Point	3.1.1.1.B
NRHB	Non Random Hydrogen Bonding EoS	3.4.3.6
NRTL	Non Random Two Liquids model	3.4.2.2.C
PR	Peng and Robinson EoS	3.4.3.4
PSRK	Predictive SRK (UNIFAC introduced in the SRK EoS, using a G^E mixing rule)	3.4.3.4.E
RK	Redlich and Kwong EoS	3.4.3.4
RMSD	Root Mean Square Deviation	3.3.1.3.A
SAFT	Statistical Associating Fluid Theory EoS	3.4.3.5.A

SARA	Saturate - Aromatic - Resins - Asphaltene: this is a typical analytical information for petroleum heavy ends	
SBWR	Soave modification of Benedict-Web-Rubin EoS	3.4.3.3.C
SD	Simulated Distillation	3.1.2.3
SD	Standard Deviation	3.3.1.3.A
SRK	Soave modification of Redlich Kwong EoS	3.4.3.4
TBP	True Boiling Point curve (according to ASTM D2892)	3.1.2.3
TPT	Thermodynamic Perturbation Theory	3.4.3.2.A
UCST	Upper Critical Solution Temperature	4.2.1.1
UNIFAC	UNIversal quasi-chemical Functional group ACtivity model	3.4.2.4.C
UNIQUAC	UNIversal QUasi Chemical ACtivity model	3.4.2.4.B
VLE	Vapour-Liquid Equilibrium	
VSE	Vapour-Solid Equilibrium	
VTPR	Volume Translated modification of Peng Robinson EoS	3.4.3.4
WAT	Wax Appearance Temperature	4.2.2.3.B

1

Introduction

The chemical engineer today cannot be expected to deliver any new or improved project without using some kind of simulation tool. The number of such tools increases regularly, with an ever improved user interface, so that it becomes a true pleasure to "predict" the behaviour of a process.

In order to truly understand and interpret the numerical results that are thus generated, it is important to have a view of the way the simulator is constructed [1]. It contains numerical packages as well as models that translate into mathematical equations what is believed to be the physical behaviour of a unit operation [2]. For that purpose, thermophysical property calculation methods must be available. These calculation methods generally also come with a database that contains model parameters. Because of the importance of these properties, and the difficulty in obtaining sufficiently accurate results, a large number of options are available to the user so that he can "custom design" his property package.

The property packages available within commercial simulators generally contain many equations that can be found in well-known thermodynamics handbooks [3-5]. Today, using CAPE-OPEN interfaces, it is possible to use external property packages within these simulators [6], thus making it possible to rapidly take advantage of the numerous developments produced using the results of work by the thermodynamic research groups.

Yet, simultaneously, this wealth of possibilities may puzzle the practicing engineer who must make a final choice in order to go ahead with his process simulation. Some authors have discussed this problem in more or less depth [7, 8], but considering the importance of the challenge, it seemed to us important to write this book in order to help him or her in this choice.

In order to select the most appropriate thermodynamic method, it is important to analyse the problem correctly and understand how a thermodynamic model is constructed. This will enable the engineer to identify more precisely what choices are to be made, and how these choices do or do not affect the final result. That is the path proposed to readers throughout the book to help them determining the most accurate method to solve a given thermodynamic problem.

The first chapter summarises the philosophy of this book. After a first warning concerning the importance of identifying what is the limiting physical phenomenon, and a description of what is meant by a thermodynamic method, the three basic questions that are the foundation of this book are introduced. These questions should be kept in mind by any process engineer who has to define the most appropriate thermodynamic method for his purpose.

Chapters 2 to 4 discuss in more detail, illustrated with many practical examples, the same three questions.

Chapters 5 will then conclude by proposing a concrete procedure that is used as guide in the case studies.

1.1   IDENTIFY THE RIGHT PHYSICS IN PROCESS SIMULATION

Although we could say "thermodynamics is everywhere", it would be wrong to say "thermodynamics is responsible for everything". In fact, the thermodynamics discussed in this book deals with equilibrium properties. The transformations that the thermodynamic models can describe are those called "quasi-static" transformations. In other words, the models to be considered are intended to describe how physical systems would evolve if they were left under no restriction, for an infinite amount of time.

Obviously, process engineers do not want to let the systems free, but instead have them produce value-added products, if possible as easily and quickly as possible. Physical gradients will therefore be imposed to allow the fluid to flow, heat, react, etc. as desired.

It is also known that some physical processes are faster than others. It is the task of the engineer to split the global phenomenon into sequential sub-phenomena and to investigate which phenomenon is rate-limiting. All the other phenomena can then be considered at or close to equilibrium.

Unfortunately, the generalised use of process simulators tends to hide the physical phenomena they describe. As an example, if a pipe is represented by a connection between two units, the simulator will be unable to determine the pressure drop when a fluid flows through this pipe. It is the responsibility of the process engineer to evaluate whether or not this pressure drop is significant in reality and, if necessary, add a unit that describes the correct phenomenon.

Example 1.1 Mechanical entrainment

As a first example of the incorrect blame of a thermodynamic model, consider a stabilisation drum as shown in figure 1.1 that operates at 2 bar and 50 °C. Let assume that the feed composition can be represented by the mixture given in table 1.1:

Table 1.1   Data for example 1.1

Component	% (mol)
n-butane	50
n-pentane	30
n-hexane	20

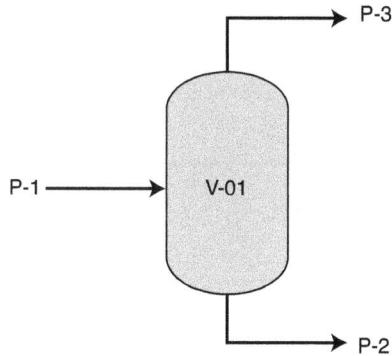

Figure 1.1

Illustration of a flash drum.

The flash calculation, under these conditions, yields the results shown in table 1.2 (using the Peng-Robinson model):

Table 1.2 Results for example 1.1

Component	Liquid P-2 % (mol)	Vapour P-3 % (mol)
n-butane	31.8	73.2
n-pentane	32.7	26.5
n-hexane	35.5	0.2

The operator, however, insists that he finds at least 10% hexane in the P-3 stream, which causes serious fouling problems down the line.

What happens?

Mechanical entrainment of the liquid could explain this observation. It may be caused by a defect in the deflector or any other internals. These phenomena are obviously not considered in the thermodynamic calculation.

Example 1.2 Bad mixing

In this second example, we consider two streams that merge into one. Using a process simulator, this is represented using a mixer. Assume (figure 1.2) that shortly after the mixer, the stream is split in two streams that go to different parts of the process. According to the simulator, the compositions of the streams leaving the splitter are identical. However, if the distance between the mixer and the splitter is not long enough, or if nothing is done to ensure complete mixing, there is a large probability that the compositions will be different, resulting in wrong simulation results.

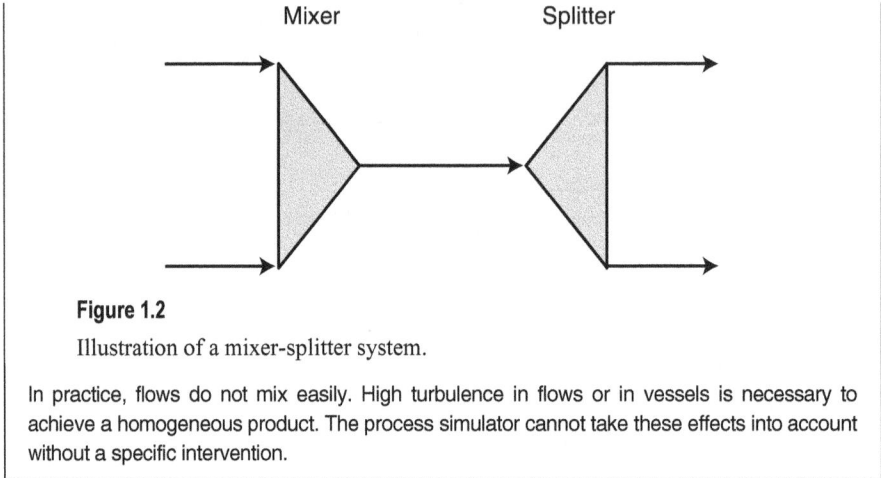

Figure 1.2

Illustration of a mixer-splitter system.

In practice, flows do not mix easily. High turbulence in flows or in vessels is necessary to achieve a homogeneous product. The process simulator cannot take these effects into account without a specific intervention.

1.2   WHAT IS A THERMODYNAMIC METHOD?

Thermodynamic methods are tools used to simulate the physical behaviour of a material system. They are used to calculate one or several physical properties. It can be viewed as a mathematical operation that starts from a list of input properties and produces one or several output properties (figure 1.3). The properties are generally fluid composition and two state variables. The definition of state input variables are given in section 2.1 (p. 23).

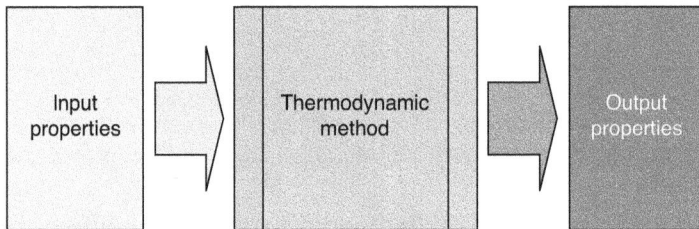

Figure 1.3

A conceptual view of a thermodynamic calculation.

In order to facilitate understanding, this concept will be illustrated with a small example:

Example 1.3 Simple Bubble temperature calculation

The input properties are:
• components in the mixture,
• compositions of the feed (z_i) and
• total pressure P.

The output properties are:
• temperature T and
• compositions of the vapour phase (y_i).

1.2.1 The physical model

Thermodynamic methods are constructed from one or more *physical models*. These models are essentially mathematical relationships that relate a physical property to one or more state variables. They are discussed in detail in section 3.4 (p. 160). Depending on the basic knowledge available, these models may be more or less empirical. They often contain adjustable parameters that are determined using experimental data, as further discussed in section 1.2.3 (p. 6).

A particularity of the thermodynamic equations is that the model may undergo some mathematical treatment (differentiation, integration, etc.) before they actually deliver the property that is asked for [9]. This is why such great care should be taken regarding the quality of the equations. This also helps us understand why it may be preferable to use a different model for different properties (the same model for both vapour pressure and enthalpy of vapourisation, for example, ensures thermodynamic compatibility between these properties, but reduces the accuracy for each single type of property).

In fact, in most situations encountered, several models will be combined in order to yield the expected result. The reason for this complexity is that the calculation of a given property may require the prior knowledge of one or several other properties. The combination of several models is a thermodynamic method. Figure 1.4 represents this visually.

The "russian dolls" image illustrates quite nicely this concept: several models are used together in order to reach the final results.

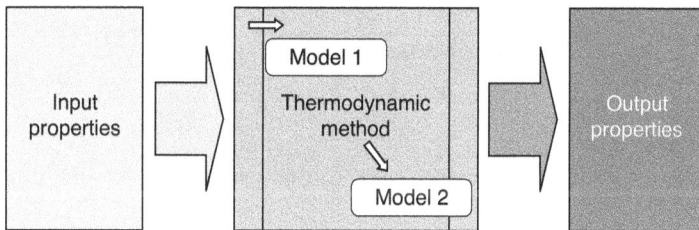

Figure 1.4

A conceptual view of a thermodynamic method using different models.

Example 1.4 Bubble temperature calculation for an ideal mixture

The solution to this example will require two different models:

1. The bubble pressure equation for an ideal mixture

$$P = \sum_i x_i P_i^\sigma (T) \tag{1.1}$$

2. The Antoine equation for calculating the pure component vapour pressures:

$$\log P_i^\sigma = A_i + \frac{B_i}{C_i + T} \tag{1.2}$$

1.2.2 The algorithm

As we can see in the simple example used here, the combination of equations 1.1 and 1.2 cannot yield the temperature as a direct answer. A solver will be needed to calculate it (figure 1.5, where the solver is represented in a more general way by the term "algorithm"). If a solver is used, the following must be provided:

- an initial estimate,
- a solution scheme,
- a convergence criterion.

Although users of thermodynamic tools are not often confronted with this problem directly, it is good they be aware of it, since it may help them understand that:

- The solver might not converge. In the best case it will provide a warning, but sometimes the simulator continues without warning.
- A bad solution may be caused by a local minimum in the solution scheme.
- If users compare the results of two simulators, using identical methods and parameters, there may still be differences.

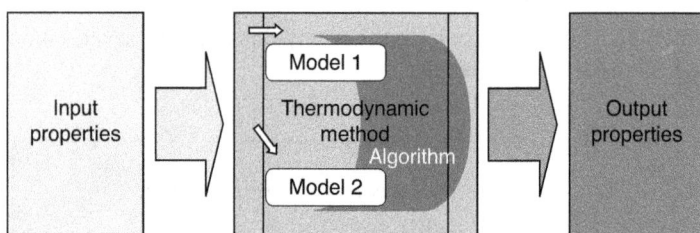

Figure 1.5

A conceptual view of a complex thermodynamic method with thermodynamic calculation algorithm.

1.2.3 The data: properties or parameters?

So far, only the **physico-chemical properties** or state variables that enter into the problem have been considered. State variables are physico-chemical properties that define the state of the system (typically: pressure, temperature, composition). Users should make sure that these basic properties are well known. This is further discussed in section 1.3.1.1 (p. 11) and later in section 2.1 (p. 23).

However, the calculation tools also employ **parameters**. These may or may not have a direct physical significance and are usually related to the model used. They are discussed in a more exhaustive manner in sections 3.1 (p. 102) and 3.2 (p. 142).

Example 1.5 Vapour pressure calculation using the Antoine equation

The Antoine equation is as follows:

$$Ln\left(P^{\sigma}\right) = A + \frac{B}{T+C} \tag{1.3}$$

The physico-chemical properties are: Temperature (T) and Vapour pressure (P^{σ});

The empirical parameters are A, B and C.

Validating the numerical parameters of the models is an essential part of the thermodynamic model evaluation. It may be worthwhile evaluating the sensitivity of the final result to the values of these parameters. As an example, Brulé *et al.* [10] show very nicely how sensitive the physical property calculations may be to a change of an input parameter. In figure 1.6, the effect of the uncertainty on the critical temperature on several property calculations is shown. Obviously, the sensitivity is very different depending on the property considered. Generally, the phase equilibrium calculation (in this case vapour pressure) is the property that is most sensitive to the energy parameter (in this case critical temperature). Note also that the minimum may be different depending on the property considered. Hence, if the parameter is adjusted on the liquid density, the optimal value will differ from that

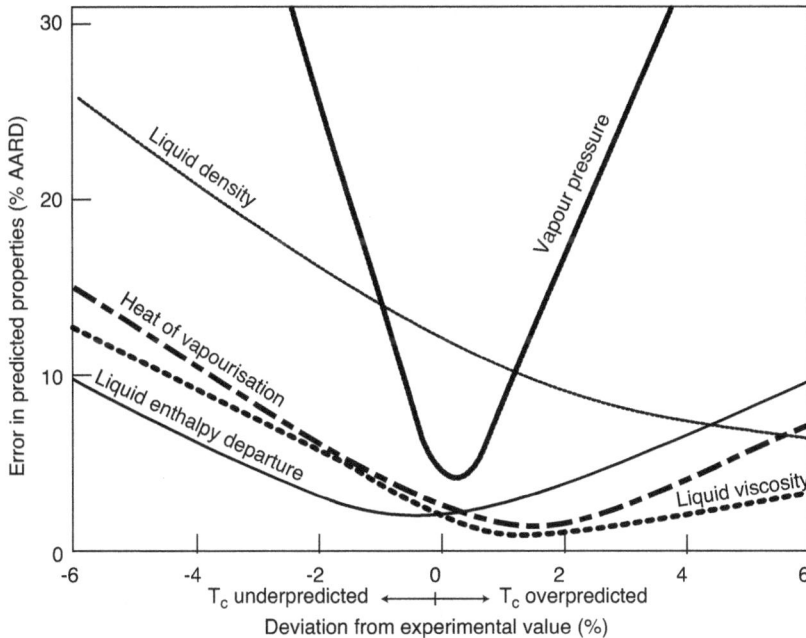

Figure 1.6

Sensitivity of the Peng-Robinson equation of state to the critical temperature of toluene, from [10].

found for the vapour pressure. In the example provided here the effect is small, but in other cases it may be more significant.

Dohrn [11] illustrates how such an uncertainty can have enormous consequences when it is amplified through the large number of thermodynamic calculations performed in a simulator. Dohrn considers a distillation column that separates an equimolar feed of toluene and chlorobenzene. Chlorobenzene is the heavier component, and will therefore be found at the bottom of the column. Using 40 theoretical stages, figure 1.7 shows how the calculated concentration of toluene in the residue varies as a function of the uncertainty on the critical temperature of toluene (theoretical value of 591.8 K). It is observed that the sensitivity is extremely large in this case, which means that if the toluene concentration is an issue, evaluation of this parameter will require extreme care.

A discussion on the key components, and therefore of the parameters, is proposed in section 3.5 (p. 225). At this stage, it is simply stated that parameters may come from different sources:

- The best method is to fit the parameters directly on experimental data. This is always the preferred choice if the expected properties must be calculated with great accuracy. Regression methods are discussed in detail in the book by Englezos and Kalogerakis [12]. They are also presented in section 3.3 (p. 148).

- They can be found in databases. This is essentially the case for parameters that have a physical significance [13]. However, all commercial simulators also provide parameter values so that the models proposed can be used with a large variety of components without any additional input from the user. This is very convenient but may be dangerous if the results are used in extrapolation outside the range for which the proposed values have been generated.

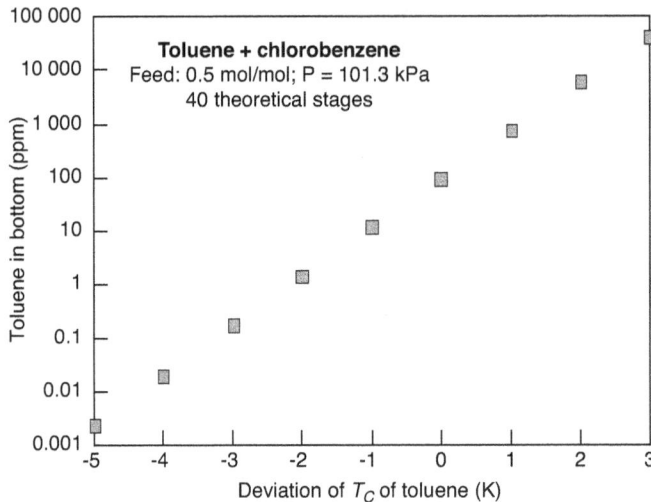

Figure 1.7

Effect of an error on the critical temperature of toluene on the calculated concentration at the bottom of the distillation column [11].

- If some information concerning the mixture (pseudo-)components is available (typically boiling temperature and density for distillation fractions or the molecular formula for complex molecules), it may be possible to use independent correlations [14] or group contribution methods that provide the required parameters (e.g. UNIFAC – [15] – for G^E models; Coniglio *et al.*, 2000 [16], for cubic equations of state). Very often, the accuracy of the results obtained in this way is not as good as when they are fitted on data, but they may be very useful as a first indication of trends.
- Today, molecular simulation tools exist that allow the use of, for example, quantum calculations for identifying model parameters. This issue is increasingly studied in the literature today and can in some cases lead to very interesting predictive results [17, 18] yet probably not as accurate.

1.2.4 Conclusions

A thermodynamic property calculation requires the combination of a good model, good parameters and a stable algorithm.

Three issues can be identified as a result of this observation:

- what intermediate properties are needed to compute the required property?
- what model is used for each of the intermediate property?
- what parameters are needed for each model?

These issues will guide the reflections throughout this book.

1.3   CRITERIA FOR PROBLEM ANALYSIS

Problem analysis will focus on three different issues. Each will give rise to a number of questions that must be clearly identified:

- what property(-ies) is(are) requested and what property(-ies) is(are) given?
- what is the fluid composition?
- where are the process pressure and temperature conditions located with respect to the phase behaviour?

Each of these issues is detailed below. In addition, a separate chapter will be devoted to each one in order to fully understand the consequences of the choices made. Table 1.3 summarises the questions and the possible answers.

Table 1.3   Checklist for problem analysis

Question	Type of answer	Discussed in section
Property – Chapter 2		
Given property	*PT, PΘ, TΘ, PH, TV, …*	Section 2.1.4
Requested property	Single phase: volume, enthalpy, entropy	Section 2.2.2
	Equilibrium: phase boundaries, solubilities	Section 2.2.3
	Chemical equilibrium	Section 2.2.4
Composition – Chapter 3		
Type of components	Simple	Section 3.1.2.1
	Light components	Section 3.1.2.1
	Heavy components	Section 3.1.2.2
	Pseudo-components	Section 3.1.2.3
Type of mixture	Non-polar (ideal)	Section 3.4.1
	Non-ideal	Section 3.4.2
	Electrolyte	Section 3.4.2.5
	Supercritical	Section 3.4
Key component Concentration range	Pure	Section 3.5
	Medium	Section 3.5
	High dilution	Section 3.5
Phase state and system – Chapter 4		
Physical state	Vapour	Section 4.1
	Liquid	
	Solid	
	Critical point	
PT conditions	List according to industrial systems	Section 4.2

1.3.1   What property is given/requested?

This question is further subdivided in two parts:
- what property is given?
- what property is requested?

Obviously, it makes no sense discussing the choice of model if the required property is not defined first. This is why it is a good idea to start the analysis of the problem by identifying it.

Various groups of properties will be identified, including:
- thermodynamic (single phase) properties,
- phase equilibrium properties,
- chemical equilibrium properties.

1.3.1.1 Thermodynamic properties

The definition of the thermodynamic properties discussed in this book is defined in more details in section 2.1.1 (p. 24). As an introduction, the list of the most important of these properties is given in table 1.4.

Table 1.4   List of the thermodynamic properties

Symbol	Meaning	Definition	SI unit
T	Temperature		K
P	Pressure		Pa
V	Volume		m^3
U	Internal energy		J
S	Entropy		$J K^{-1}$
H	Enthalpy	$= U + PV$	J
A	Helmholtz energy	$= U - TS$	J
G	Gibbs energy	$= U + PV - TS$	J

Many thermodynamic textbooks discuss the meaning of these fundamental quantities [4, 20, 21]. Simply note that for the internal energy, and therefore also for enthalpy, Helmholtz energy and Gibbs energy, only differences have a physical meaning. Hence, if a numerical value is to be calculated for one of these properties, a reference state must be defined. Although strictly (according to the third principle of thermodynamics also called Nernst's postulate), the entropy can be defined as an absolute value (available in databases – e.g. [13]), often only changes in entropy are important and consequently any reference state can be used for practical applications.

1.3.1.2 Phase equilibrium

In addition to the thermodynamic properties listed in table 1.4, phase equilibrium information is often required. This type of calculation is based on an algorithm that minimises the Gibbs energy of the system [21] for fixed pressure and temperature. This energy is computed from the chemical potentials or, equivalently, the fugacities. The frequently used Rachford-Rice approach (section 2.2.3.1.C) is based on the distribution coefficients. These are defined as the ratio of the molar fractions of each component in all phases. For vapour-liquid equilibrium, the equation is:

$$K_i = \frac{y_i}{x_i} \tag{1.4}$$

where y_i is the molar fraction of component i in the vapour phase, and x_i is its molar fraction in the liquid phase. The distribution coefficient is calculated from [1]

$$K_i = \frac{\varphi_i^L(x, P, T)}{\varphi_i^V(y, P, T)} \tag{1.5}$$

---

1. The origin of this expression and the significance of the fugacity coefficients φ_i are detailed in section 2.2.3 (p. 63).

The fugacity coefficients for each phase (liquid φ_i^L and vapour φ_i^V in this case, but other combinations may be considered), are calculated using the temperature, the pressure and the composition of each phase (which is unknown.) It therefore becomes clear that an algorithm will have to be used in order to solve for the phase compositions. Sometimes the terms equilibrium ratio or partition coefficient are used as synonyms of distribution coefficient.

It is essential to understand here that the fugacities must be computed separately for each phase that is potentially present in the system. In the case of fluid phases, equation of state (EoS) models are well adapted for use in all phases of the system. This will be called the homogeneous approach. However, for non-ideal systems, it may be recommended having separate methods for each phase. In particular, liquid phases are sometimes better described using activity coefficient models. Solid phases always have a separate model. Using a different model for each phase is called the "heterogeneous approach".

In difficult cases, a separate model may even be applied to different components within a single phase. A typical example of this is the aqueous phase where the solvent (water and possible co-solvents) and the solutes (dissolved hydrocarbons or ionic species) show very different behaviour and can therefore not be described with a single model. The resulting approach is called "asymmetric". These different approaches are discussed in more details in section 2.2.3.1 (p. 63). A discussion regarding the best combination is proposed in section 4.3 (p. 325).

1.3.1.3 Chemical equilibrium

In some processes, thermodynamic calculations are needed in order to determine how the equilibrium composition changes with pressure and temperature as a result of one or more chemical reactions. This requires an additional algorithm, also based on a Gibbs free energy minimisation. It can be shown that the chemical equilibrium equation can then be written as:

$$\prod_i \left(\frac{f_i}{f_i^o} \right)^{v_i} = K \qquad (1.6)$$

where f_i are the component fugacities, v_i are the stoichiometric coefficients and fugacities in the reference state. The chemical equilibrium constant is calculated from the formation properties of the components (see section 2.2.4, p. 86).

Note that in some cases, simultaneous phase and chemical equilibria must be calculated. Since the same type of property is used, no additional model need to be defined. Yet, the mathematical algorithm for solving such a system becomes rather complex.

1.3.2 What are the mixture components?

Chapter 3 will discuss how fluid composition affects the thermodynamic calculations. This issue relates to both the physical model to be used and the numerical values of the parameters. We will stress on several occasions that comparison with experimental data is essential in order to validate the choices.

Three subquestions are of interest at this level:

- the type of components,
- the type of mixture (how do the components mix?), and
- the composition range.

1.3.2.1   What type of components are considered?

It is essential that the process engineer has some idea of the chemical nature of his mixture. Pure component properties are required in order to develop and/or validate the model used: all models use pure component parameters. They may come from different sources. How the parameters are determined depends on the type of component. Four different types of components can be identified. In section 3.1.2 (p. 121), the two first types are both considered as database components:

1. **"Simple" components**: components whose properties have been investigated over large domains of pressure and temperature. Examples include hydrogen, methane, carbon dioxide, hydrogen sulphide, nitrogen, most refrigerants, etc. For these components, very precise equations of state have been developed (see the REFPROP package at NIST [22]), that can be used for calculating almost any property. These equations are extremely valuable when the "simple" components are used pure. When they are part of a mixture, however, they are described in the same way as "light" components.

2. **"Light" components**: Most hydrocarbons up to C_{10}, and some selected heavier components (*n*-alkanes) have been investigated extensively along their vapour pressure line. Hence, their characteristic parameters are generally well known, and most properties along the saturation lines (vapour pressure, but also sublimation pressure and crystallisation conditions) have been subject of many investigations. Critical reviews of the data exist and have resulted in correlations that provide these values with an accuracy that is close to the experimental uncertainty (DIPPR [13]).

It will be shown in chapter 3 that for both "simple" and "light" components, the parameters are often available in databases (section 3.1.2.1, p. 121).

3. **"Heavy" components**: Beyond C_{10}, the number of isomers becomes so large that for a given substance, few or no experimental data exist. Hence, no parameters are generally available. However, if the chemical structure of the component is known, group contribution methods can be used, as discussed in section 3.1.2.2 (p. 124).

4. **"Pseudo-"components**: A particularity of the oil industry is that the fluids contain far too many components (including large numbers of isomers) which, in addition, are generally very badly identified. Hence, the fluid is more usually described using assays and a number of representative experimental data [14, 23]. Correlations are then used in order to transform this information into a number of components that are treated as if they were pure components. A detailed description of the methods used is presented in section 3.1.2.3 (p. 129).

1.3.2.2   What is the mixture non-ideality?

A component behaviour is to a large extend influenced by its neighbours. A component in a mixture may behave very differently from its pure component behaviour. The deviation from the ideal mixture is used to quantify this effect.

14   *Chapter 1 • Introduction*

The ideal mixture is strictly defined using the fugacities (fugacity is proportional to the molar fraction, section 2.2.2.2, p. 60). Its bubble pressure curve in a *Pxy* diagram is a mere straight line. Its low-pressure behaviour can be expressed using Raoult's law:

$$y_i P = x_i P_i^\sigma \tag{1.7}$$

Even though a mixture of non-polar components is not exactly an ideal mixture, we may state that most hydrocarbon mixtures will follow closely this behaviour. More generally, this type of simple behaviour is encountered when the molecules in the mixtures are similar.

The deviation from this ideal behaviour is generally described using an activity coefficient, γ_i:

$$y_i P = \gamma_i x_i P_i^\sigma \tag{1.8}$$

The origin of non-ideal behaviour is to be found in the molecular interactions between the components. Two different types of interaction can be identified (section 3.4.2, p. 171):

A. The **interaction energy** between different molecules in the mixtures and among molecules of the same type may vary. Rare gases have almost no interactions, while water molecules interact very strongly with each other (figure 1.8). At the far right side of the figure, the interaction energy is so large that covalent bonds are formed.

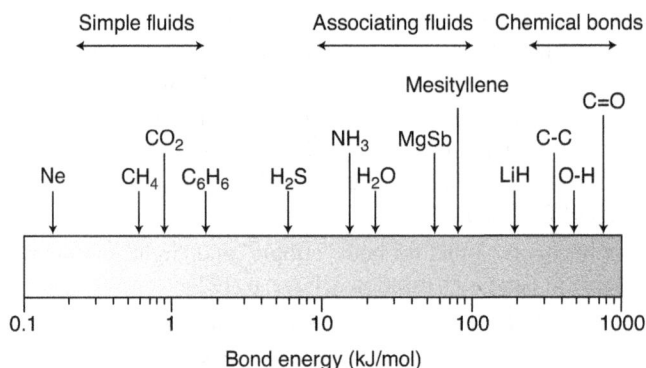

Figure 1.8
Continuous distribution of bond strengths showing the span from simple Van der Waals attractions to the formation of chemical bonds [24].

This criterion helps us understand why, if a mixture contains essentially non-polar hydrocarbons, the presence of heteroatoms leads to non-ideal behaviour.

If non-ideality is the result of strong differences in interaction energy, one may state that positive deviation from ideality (the bubble pressure is larger than that found using Raoult's law – eq. (1.7) or $\gamma > 1$ in (1.8)) is the result of larger attractions between molecules of the same kind in comparison with molecules of different kinds. This is most often the case in mixtures of interest in the petrochemical industry. In contrast, when, as a result of specific (for example hydrogen bonding) forces, attractive interactions between unlike molecules are stronger than between molecules of the same kind, negative deviations from ideality will result ($\gamma < 1$).

B. When the **sizes** or **shapes** of the molecules are different, non-ideal behaviour will also be observed. This is particularly the case when polymers are mixed with small solvent molecules. This effect is generally called "entropic".

The mixture non-ideality may have important consequences on the phase behaviour as will be further discussed in section 3.4.1 (p. 160). Yet, this non-ideality may also have an influence on the mixture single phase properties (mixing enthalpies and/or mixing volumes are observed). Section 2.2.2.2 (p. 60) will also discuss how the models used for activity coefficient calculations may also be used to calculate single phase mixture properties, in particular enthalpy, entropy and heat capacity.

1.3.2.3   What is the key component concentration range?

It is well known that the fluid mixtures in chemical and petrochemical industries contain generally a large number of components. In the above question, the use of the concept of "key component(s)" somehow indicates that not all components have the same importance in the final result. It is the responsibility of the process engineer to distinguish between majority components (which make up the main part of the feed), and key components, which may or may not be majority components. In the case of solubility calculations, or phase boundary calculations, this issue may be critical. For the calculation of the other properties discussed in the previous section, it can be considered that the key components are the majority components: impurities have no significant effect on the system properties.

It is possible to illustrate the importance of the key component concentration range using three extreme examples.

- In the case of **pure substances**, the question of key component is solved rather quickly. Here, the pure component properties are requested, which may or may not be well-known.
- In most cases, however, the fluid is a mixture, where only the majority components have a real significance. The other components are called impurities and are not looked at any further. The components of interest are within a concentration range of 5% to 100%. We will call this a **medium concentration range**. In this case, the data required to validate the model will cover the entire concentration range.
- It may occur that the volatility of a **minority component** becomes key. Such is the case, for example, when impurities must be removed with a very high degree of accuracy. Now, the process engineer expects his thermodynamic model to be accurate in the high dilution regime, which is much more difficult, as the component property strongly depends on both the key component and the solvent mixture. The data that should be used for determining the model parameters, and for model validation are the so-called infinite dilution data. Few such data exist, and hence few models have been tuned to yield accurate values in this domain.

A further analysis of how to identify the key components is provided in section 3.5 (p. 225).

1.3.3 Where are the process conditions located with respect to the phase envelope?

The very first issue that the process engineer will need to answer is whether his mixture is single phase (vapour, liquid or solid) or multiphase (vapour-liquid; liquid-liquid or liquid-solid) The reason for this is that a system property (the volume for example) is not the same whether the system contains two (or more) phases (in which case the volume is $V = V^L + V^V$) or whether it is single phase.

Few commercial simulators will identify complex phase behaviour unless it has been specifically requested. Hence, it is up to the process engineer to know what phases may appear in the process conditions.

Each phase comes with its own model. The choice of the model that should be used to describe a specific phase will strongly depend on its nature (solid, liquid or vapour). The description of critical phenomena is particularly delicate, because two distinct phases must be described, whose properties become identical at the critical point and whose compressibility behaviour is very peculiar, as pointed out by Levelt-Sengers [25].

Hence, the above question actually contains two distinct questions:

- What is the nature of the phases that may potentially be encountered?
- Where are the process conditions (pressure, temperature and composition) located with respect to this phase behaviour?

Chapter 4 discusses this issue by reviewing the types of phases observed for various mixtures of industrial interest.

1.3.3.1 Nature of the phases

Process engineers are most used to deal with liquid or vapour phases. A check whether vapour-liquid equilibrium is encountered in the pressure-temperature-composition domain of the process is always essential. Liquid-liquid equilibrium is generally not automatically performed by the simulator, but should be checked, as liquid phase split could occur unexpectedly (e.g. in low temperature, LNG production, the liquid methane may phase split with heavier hydrocarbons).

In some cases three or more phases may coexist. Vapour-liquid-liquid conditions are known to process engineers (heteroazeotropic extraction), but it may come as an unexpected feature. In fact, while only one vapour phase may exist in a given system, many different liquid phases can coexist (aqueous and organic are well known, but methane or CO_2 may also form their own phase). The heavy end of a hydrocarbon mixture may also decant as an asphaltene phase.

The situation becomes even more complex when solid phases may appear. This is the case with hydrates, ice, paraffins (or any pure compound), or scale (salt precipitation).

Depending on the situation, the models and the analysis of the simulation results will be different. Although fluid phases can all be described using the same model (homogeneous approach), it may be more accurate to use a different model for each phase (heterogeneous approach). Model recommendations for each case are made in sections 4.2 (p. 264) and 4.3 (p. 325).

1.3.3.2 Pressure- temperature conditions

Figure 1.9 [7] illustrates, for a regular vapour-liquid equilibrium, what model choice is best for single phase property calculations. The type of model employed largely depends on the compressibility of the phase considered. We will illustrate this with the volume. Since, according to the Maxwell relations (section 2.2.1.1.C, p. 46), all properties can be related to the volume, the same conclusions will be reached for the other thermodynamic properties, including the fugacity which is required for phase equilibrium calculations.

The **vapour phase** is a compressible phase. The volume strongly depends on pressure. The best known equation, that is valid for low pressure vapours (below 0.5 MPa) is the ideal gas equation of state [1]:

$$V = \frac{NRT}{P} \tag{1.9}$$

Unlike many other thermodynamic equations, the ideal gas equation has no empirical parameter. Its only parameter is R ($R = 8.314462$ J mol^{-1} K^{-1}), known as the universal gas constant. The pressure-volume-temperature relationship for an ideal gas is independent of its composition! This is so because the molecules are so far apart that their interaction may be neglected.

Unfortunately, nature rarely behaves in such a simple way, and as pressure increases, the interactions among molecules have a greater impact. Up to 1.5 MPa, the virial equation of state, truncated after the second term, is an appropriate approximation:

$$V = \frac{NRT}{P} + B(T) \tag{1.10}$$

The second virial coefficient, B, is a temperature-dependent parameter that may be found in databases as discussed in section 3.1.1.2.G (p. 120). It can also be calculated using the corresponding states principle [2] [26, 27]. It expresses the interactions between pairs of molecules.

At higher pressure, third and higher-order interactions become significant and a more complex equation of state is needed. The same equation as for high pressure liquids should be used here.

In the **liquid phases**, as the compressibility is low, the saturated properties can be used up to 1 MPa above the vapour pressure of the pure components. Saturated pure component properties are very often available in databases, as correlations that have been fitted on many experimental data. When mixtures are concerned, and especially for non-ideal mixtures [3],

---

1. The equations of state are further defined in section 3.4.3 (p. 189). It is a relationship between pressure, temperature and volume.
2. The corresponding states principle is a very powerful method for calculating properties of non-polar compounds (hydrocarbons). It is further discussed in section 2.2.2.1.C (p. 57).
3. The definition and characterisation of non-ideal mixtures is further discussed in chapter 3. To a first approximation, hydrocarbon mixtures may be considered ideal.

the excess properties, as defined in section 2.2.2.2 (p. 60), may have an effect on the final property. For example, the liquid molar volume can be written as:

$$V_i^L = \sum_i N_i v_i^{L\sigma} + V^E \tag{1.11}$$

where $v_i^{L\sigma}$ is the saturated molar volume of component i, N_i its mole numbers and V^E the excess volume. Note that the excess volume can most often be neglected, which is not the case for the excess enthalpy or the excess entropy. The excess properties for phase equilibrium calculations is the excess Gibbs energy whose expression is related to that of the activity coefficients.

The 1 MPa limit above the saturation pressure depends on the proximity to the critical point, as shown in figure 1.9. The reason for this is that the compressibility of the liquid becomes larger when this point is approached. Similarly, at higher pressure, the compressibility of the liquid can no longer be neglected, and a complete equation of state is needed for calculating the single phase properties. This type of equation is generally expressed as

$$P = f(T,V,\mathbf{N}) \tag{1.12}$$

The volume is found by reversing the equation at fixed pressure and temperature, while for the other thermodynamic functions, it requires a mathematical treatment yielding the so-called residual properties (section 2.2.2.1, p. 52 and section 3.4.3, p. 189).

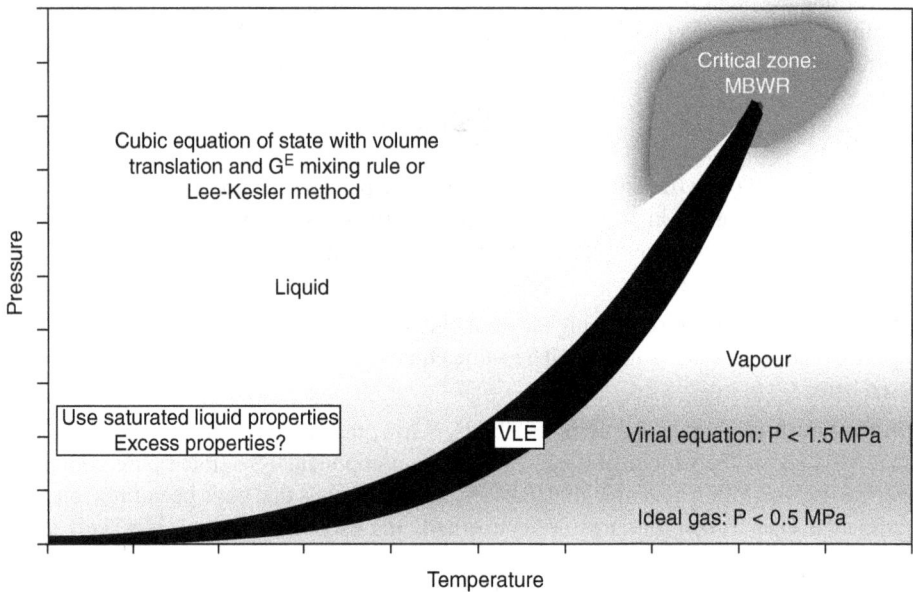

Figure 1.9

Model selection for single phase property calculations. The black area is the two-phase vapour-liquid equilibrium domain [7].

The Lee-Kesler method [28] or another BWR-type equation is adapted to hydrocarbon mixtures (SBWR) [29, 30]. It can be extended to mixtures containing hydrogen, CO_2 or H_2S. For more complex mixtures, where unlike interactions are essential, more complex equations will be needed (with volume translation and complex mixing rules) [1].

The situation is even more complex close to the critical point [25, 31] as here, the compressibility coefficient becomes infinite. As a matter of fact, it has been shown that no analytical equation can describe the true physical phenomena in the neighbourhood of this point. Yet, for engineering purposes, complex BWR-type equations (as, for example Kestin as a basis for the steam tables [32], or the IUPAC series, with the equations developed by Span and Wagner [33-35]) have been developed, with an acceptable accuracy. Other approaches, as the cross-over theory [36] allows for any equation of state to approach this point. This theory is rather complex, however, and is not available in commercial simulators.

1.4 CONCLUSIONS

In this first chapter, we have introduced the reader to the philosophy that will be further expanded in the three main chapters of this book. The main messages that have been delivered so far are:

- Thermodynamic methods can only be used where phenomena are at equilibrium.
- Thermodynamic calculation methods are most generally made out of a combination of models ("russian dolls") which each have a mathematical shape and numerical parameters.
- A three-step analysis is proposed in order to develop the thermodynamic model. Each of these steps is further detailed as a separate chapter of this book.

REFERENCE LIST

[1] Seider, W.D., Seader, J.D. and Lewin, D.R. "Product and Process Design Principles" 3rd; Wiley International, **2003**.

[2] McCabe, W., Smith, J. and Harriott, P. "Unit Operations of Chemical Engineering" 7; McGraw Hill, **2004**.

[3] Smith, J.M., Van Ness, H.C. and Abbott, M.M. "Introduction to Chemical Engineering Thermodynamics" Sixth Edition; Mc Graw Hill, Inc, **2001**.

[4] Prausnitz, J.M., Lichtenthaler, R.N. and Gomes de Azevedo, E. "Molecular Thermodynamics of Fluid Phase Equilibria" 3rd Ed.; Prentice Hall Int., **1999**.

[5] Sandler, K.E. "Chemical and Engineering Thermodynamics" 3rd edition, J. Wiley & Sons: New York, **1999**.

[6] Breil, M.P., Kontogeorgis, G.M., von Solms, N. and Stenby, E.H. "CAPE-Open: An International Standard for Process Simulation" *Chemical Engineering* **2007**, *114*, 13, 52-55.

---

1. Further discussed in chapter 3.

[7]   de Hemptinne, J.C. and Behar, E. "Thermodynamic Modelling of Petroleum Fluids" *Oil & Gas Science and Technology-Revue de l'Institut Français du Pétrole* **2006**, **61**, 3, 303-317.

[8]   O'Connell, J.P. Gani, R., Mathias, P.M., Maurer, G., Olson, J.D. and Crafts, P.A. "Thermodynamic Property Modeling for Chemical Process and Product Engineering: Some Perspectives" *Industrial & Engineering Chemistry Research* **2009**, **48**, 10, 4619-4637.

[9]   de Hemptinne, J.C., Barreau, A., Ungerer, P. and Behar, E. "Evaluation of Equations of State at High Pressure for Light Hydrocarbons" in *Thermodynamic Modeling and Materials Data Engineering*; Ed. Caliste, J-P, Truyol, A. and Westbrooks, J. H.; Data and Knowledge in a Changing World; Springer, **1998**.

[10] Brulé, M.R., Kumar, K.H. and Watanasiri, S. "Characterization Methods Improve Phase-Behaviour Predictions" *Oil & Gas Journal* **11-2-1985**, **83**, 6, 87-91.

[11] Dohrn, R. "Thermophysical Properties for Chemical Process Design", personal communication, **2008**.

[12] Englezos, P. and Kalogerakis, N. *"Applied Parameter Estimation for Chemical Engineers"* 1st Ed.; Marcel Dekker, Inc.: New York, Basel, **2001**.

[13] Rowley, R.L., Wilding, W.V., Oscarson J.L., Yang Y., Zundel, N.A., Daubert, T.E. and Danner, R.P. "DIPPR® Data Compiltation of Pure Compound Properties", Design Institute for Physical Properties; AIChE, New York, **2003**.

[14] Riazi, M.R. *"Characterization and Properties of Petroleum Fluids"*; American Society for Testing and Materials: Philadelphia, **2005**.

[15] Fredenslund, A., Jones, R.L. and Prausnitz, J.M. "Group Contribution Estimation of Activity Coefficients in Nonideal Liquid Mixtures" *AIChE Journal* **1975**, **21**, 6, 1089-1099.

[16] Coniglio, L., Tracy, L. and Rauzy, E. "Estimation of Thermophysical Properties of Heavy Hydrocarbons through a Group Contribution Based Equation of State" *Industrial & Engineering Chemistry Research* **2000**, 39, 5037-5048.

[17] Singh, M., Leonhard, K. and Lucas, K. "Making equation of state models predictive – Part 1: Quantum Chemical Computation of Molecular Properties" *Fluid Phase Equilibria* **2007**, **258**, 1, 16-28.

[18] Garrison, S.L. and Sandler, S.I. "On the Use of *ab initio* Interaction Energies for the Accurate Calculation of Thermodynamic Properties" *Journal of Chemical Physics* **2002**, **117**, 23, 10571-10580.

[19] Harvey, A.H. and Laesecke, A. "Fluid Properties and New Technologies: Connecting Design with Reality" *Chemical Engineering Progress* **2002**, **98**, 2, 34-41.

[20] O'Connell, J.P. and Haile, J.M. *"Thermodynamics: Fundamentals for Applications"* 1st Ed.; Cambridge University Press, **2005**.

[21] Vidal, J. "Thermodynamics: Applications in Chemical Engineering and the Petroleum Industry"; Editions Technip, Paris, **2003**.

[22] Lemmon, E. "NIST Reference Fluid Thermodynamic and Transport Properties Database (REFPROP): Version 7.0" **2006**.

[23] Riazi, M.R. "A Continuous Model for C7 + Fraction Characterization of Petroleum Fluids" *Industrial & Engineering Chemistry Research* **1997**, **36**, 10, 4299-4307.

[24] Muller, E.A. and Gubbins, K.E. "Molecular-Based Equations of State for Associating Fluids: A Review of SAFT and Related Approaches" *Industrial & Engineering Chemistry Research* **2001**, 40, 2193-2211.

[25] Levelt Sengers, J.M.H. "Scaling Predictions for Thermodynamic Anomalies near the Gas-Liquid Critical Point" *Industrial & Engineering Chemistry Fundamentals* **1970**, **9**, 3, 470-480.

[26] Tsonopoulos, C. "An Empirical Correlation of Second Virial Coefficients" *AIChE Journal* **1974**, 20, 263-272.

[27] Tsonopoulos, C. and Heidman, J.L. "High-Pressure Vapor-Liquid Equilibria with Cubic Equations of State" *Fluid Phase Equilibria* **1986**, 29, 391-414.

[28] Lee, B.I. and Kesler, M.G. "A Generalized Thermodynamic Correlation Based on Three-Parameter Corresponding States" *AIChE Journal* **1975**, **21**, 3, 510.

[29] Soave, G. "A Noncubic Equation of State for the Treatment of Hydrocarbon Fluids at Reservoir Conditions" *Industrial & Engineering Chemistry Research* **1995**, **34**, 11, 3981-3994.

[30] Soave, G. "An Effective Modification of the Benedict Webb Rubin Equation of State" *Fluid Phase Equilibria* **1999**, 164, 157-172.

[31] Kiran, E. and Levelt Sengers J.M.H. *"Supercritical Fluids; Fundamentals for Applications"*; Kluwer Academic Publishers, **1994**.

[32] Kestin, J. "Thermophysical Properties of Fluid H_2O" *Journal of Physics and Chemical Reference Data* **1984**, **13**, 1, 175-183.

[33] Angus, S., Armstrong, B., de Reuck, K.M., Altunin, V.V., Gadetskii, O.G., Chapela, G.A. and Rowlinson, J.S. Carbon Dioxide; International Thermodynamic Tables of the Fluid State; Pergamon Press, **1976**.

[34] Span, R. and Wagner, W. "A New Equation of State for Carbon Dioxide covering the Fluid Region from the Triple-Point Temperature to 1100 K at pressures up to 800 MPa" *Journal of Physical and Chemical Reference Data* **1996**, **25**, 6, 1509-1596.

[35] Wagner, W. and Span, R. "Special Equations of State for Methane, Argon, and Nitrogen for the Temperature-Range from 270-K to 350-K at Pressures Up to 30-MPa" *International Journal of Thermophysics* **1993**, **14**, 4, 699-725.

[36] Kiselev, S.B. "Cubic Crossover Equation of State" *Fluid Phase Equilibria* **1998**, 147, 7-23.

2

From Fundamentals to Properties

The first item to be investigated when solving a thermodynamic problem concerns the properties: what properties are known and what properties are to be calculated? The scope of this book is not to define these properties exhaustively, nor to derive the relationships among them (many excellent textbooks exist for this purpose [1-4]). However, in order to help the engineer understand how the simulation works, it is essential to summarise the most important relations.

> This chapter deals with the "Always True" in thermodynamic simulation: how the fundamental principles are used as tools for modelling concrete problems.

In a first part, a number of definitions will be provided. Gibbs phase rule will then be used to help the reader understand the general types of phase behaviour that can be encountered. Finally, Duhem's phase rule or theorem will help the reader explaining what is the information required in order to make a calculation possible.

A second part will focus on how these properties are computed. Fundamental thermodynamic relations between properties will be listed. The two types of fluid description (notion of residual property and excess approach) will be given to describe real fluid behaviour. The notion of activity coefficient will introduce the excess approach. The phase equilibrium will be associated with the equality of chemical potentials or fugacities. Distribution coefficients will be presented to solve various VLE problems like flash, bubble and dew points. The phase equilibrium problem will also be looked at from the point of view of the process, in order to provide a practical guide to the important processes (distillation, extraction, crystallisation, etc.). The presentation of chemical equilibrium will be based on knowledge of the thermodynamic properties.

2.1 PROPERTIES, STATES AND PHASES

The amount of information required for the system to be fully defined is provided by the so-called phase rules. These rules give a relation between the number of components, the number of phases and the number of physical variables or properties to define a thermodynamic state.

The meaning of the thermodynamic properties and of a thermodynamic state will be first defined. In a second point, the use of these rules will be presented in sections 2.1.3 (p. 27) and 2.1.4 (p. 41).

2.1.1 Properties

Properties are physical variables that describe a thermodynamic system qualitatively. A very short definition of these properties must be given for consistency. The physical dimension is provided based on the fundamental dimensions of the "Système International" (SI) (*i.e.* length [L], mass [M], time [T], amount of substance [N], temperature [θ]), and the SI units are given along with some of the most frequently used industrial units [5].

Extensive properties are properties proportional to the amount of material of the system. The most well-known examples are mass, volume, amount of matter, but enthalpy, internal energy, Gibbs energy, Helmholtz energy or entropy could also be included.

Intensive properties are properties that do not depend of the size of the system. Common examples are pressure, temperature, but also include extensive properties divided by the amount of substance, for example molar volume v or molar density ρ.

$$v = \frac{V}{N} = \frac{1}{\rho} \tag{2.1}$$

In thermodynamic applications, the molar properties are commonly used rather than the specific properties (per unit mass). The reason for this peculiarity will be made clear when presenting the residual approach (section 2.2.2.1, p. 52) and the excess approach (section 2.2.2.2, p. 60) for calculation of thermodynamic properties.

2.1.1.1 Volume

Volume is a simple physical dimension derived from the concept of length. For a parallelepiped the formula is the product of length, by width and by height:

$$V = l.w.h \tag{2.2}$$

The physical dimension of volume is $[L^3]$ and the SI unit is the cubic meter (m^3). Other units traditionally used, among others, are litre (L) and cubic inch (in^3).

2.1.1.2 Pressure

Pressure is defined as a normal force applied on a surface, even if the surface is very small:

$$P = \frac{dF}{dA} \tag{2.3}$$

The physical dimension is $[L^{-1}MT^2]$ and the SI unit is pascal (Pa). Industrial applications often use bar or psi. Pressure is measured with a manometer (U-tube with liquid), a Bourdon tube and more often with a strain gauge on a deformable surface [6].

Pressure measurements are often the result of a difference. One may distinguish:

- absolute pressure, with reference against a perfect vacuum. This is the value used in thermodynamic calculations,
- gauge pressure, with ambient pressure as reference,
- differential pressure, which is the difference between two measurements.

2.1.1.3 Temperature

Temperature is more complex to define. The thermodynamic temperature is defined from the zeroth law of thermodynamics that states: if two thermodynamic systems are separately in thermal equilibrium with a third, they are also in thermal equilibrium with each other. At microscopic scale, it is a measure of the average motion of particles in a gas; at macroscopic scale, it is a measure of the possibility of heat to be transfered.

The physical dimension of temperature is one of the fundamental dimensions, and its symbol is [θ]. The SI unit is kelvin (K), but in practice degree Celsius (°C) is used in non Anglo-Saxon countries where degree Fahrenheit (°F) is employed. In this case, absolute scale is expressed in degree Rankine (°R) for thermodynamic calculations. Thermodynamic calculations at all times use the absolute scale. Temperature is always measured in an indirect way: expansion of a liquid in conventional thermometers, expansion of a gas in earlier versions of thermometers, expansion of a solid in gauge, generation of a difference of electric potential in thermocouples, change of electrical resistivity in resistors. The most absolute measurement is made by observing the wavelength of radiation of a black body [6].

2.1.1.4 Entropy

Entropy is one of the most abstract basic concepts in thermodynamics. It is a measure of the organisation level or the information content of a system: a highly disordered system has a high entropy, while an ordered system has a low entropy. The second principle, briefly discussed below, teaches that entropy (degree of disorder) naturally always increases. It is also used to classify different forms of energy: some energy is "ordered", or "coherent" and can be used or transformed into work; other kinds of energy are "disordered", or "incoherent" as thermal energy. It is a measure of the impossibility of converting energy into work spontaneously. It can be seen as a way of classifying different forms of energy with a "quality" or "aptitude" for conversion into another kind of energy. Its first appearance is due to S. Carnot, in his book entitled "Réflexions sur la puissance motrice du feu" in 1824 who gave the basis of the second law. The actual concept was published by Clausius in 1850 and the formal definition was given by Boltzmann in 1896:

$$S = k_B \ln\left(W\right) \tag{2.4}$$

This statistic definition is based on the number W of distinguishable microstates that can be observed in a given macrostate and k_B is the Boltzmann constant. An interessant book of Ben-Naim named "Entropy demystified" [7] may help to understand its meaning.

The physical dimension [$L^2MT^2\theta^{-1}$] and the SI unit is the joule per kelvin (J K^{-1}).

2.1.1.5 Energies

Internal energy is directly related to the first law of thermodynamic. It includes potential energy contribution, kinetic energy contribution but also chemical, magnetic, electric or other forms of energy in some complex process. All material length scales may be concerned with the internal energy: sub-atomic (nuclear energy), intra-molecular (vibrations between atoms within a molecule), inter-molecular (interactions between molecules) and macroscopic (system position or speed). Since only differences are considered in classical thermodynamics, for all practical applications pertaining to chemical engineering (no nuclear reactions), only the intra- and inter-molecular internal energy are considered here.

Enthalpy, *Gibbs energy* and *Helmholtz energy* are different forms of energy, that are related to the internal energy, but include additional contributions that make them extremely valuable for thermodynamic calculations (see definitions in tables 1.4 and 2.9, p. 46), although their intuitive meaning is not straightforward. Enthalpy is a concept that was first introduced by Clapeyron and Clausius in 1827, discussed by Gibbs in 1875 and probably named by Kammerling Onnes from the Greek "enthalpos" that means "to put heat into" (see Howard, 2002 [8, 9]).

All kinds of energy have the same physical dimension $[L^2MT^2]$ and the SI unit is the joule (J). Other units such as the calorie (cal) and the British Thermal Unit (BTU) were traditionally used.

Entropy or energy content is proportional to the amount of substance, hence it is an extensive property. In the same way as for the volume in equation (2.1), the molar properties are often used:

$$h = \frac{H}{N} \tag{2.5}$$

$$s = \frac{S}{N} \tag{2.6}$$

Energy can not be measured directly, as only differences have a physical meaning. A *reference state* must be defined in order to provide a numerical value, and calorimetric techniques are necessary to obtain an indirect measurement.

2.1.2 Thermodynamic state

A state property is considered to be any property used to define the state of a system. Often, the *thermodynamic state* must be defined precisely (as for example in reference states). It is useful to remember the minimum amount of information needed to avoid ambiguity in this definition.

An unambiguous definition of a thermodynamic state may be given by:
- temperature
- pressure
- composition
- physical state (ideal gas, vapour, liquid or solid)

Very often the ideal gas physical state is used, even though it is not always physically stable, since it is convenient for thermodynamic calculations.

As an example, a "standard state" for the vapour phase may be used. However, there is no common agreement as to its exact definition as it can be appreciated in table 2.1. The table also shows the standard volume for each of these standards, using the ideal gas law.

Table 2.1 Principal definitions of the standard state for the vapour properties

Temperature (K)	Pressure (kPa)	Used by	Known as	$v(T,P)$ $(m^3 \; kmol^{-1})$
273.15 (0 °C)	101.325 (1 atm)	IUPAC (before 1982), NIST	Standard conditions, Normal cubic meter, DIN 1343	22.414
273.15 (0 °C)	100	IUPAC (since 1982)	Standard temperature and pressure (STP), Standard conditions, ISO 13443	22.711
288.15 (15 °C)	101.325 (1 atm)	European gas companies	International standard atmosphere (ISA), Normal cubic meter, ISO 2533	23.645
293.15 (20 °C)	101.325 (1 atm)	NIST [10]	Ambient, Room	24.055
298.056 (60 °F)	101.325 (14.696 psi)	American gas companies, SPE	Standard cubic feet	24.458
298.056 (60 °F)	101.5598 (14.73 psi)	American gas companies, OPEC	Standard cubic feet	24.401
298.15 (25 °C)	101.325 (1 atm)		Ambient, Room	24.465
298.15 (25 °C)	100	NBS [11]		24.790

2.1.3 Gibbs phase rule, or how to read a phase diagram

The Gibbs phase rule is of great help when drawing phase diagrams. In this case, no feed composition is required. It states that:

The number of intensive properties that must be fixed for a system to be entirely determined is given by:

$$\Im = \mathcal{N} - \phi + 2 - \mathcal{R} \tag{2.7}$$

Where $\mathcal{N}$ is the number of components in the system, ϕ is the number of phases and $\mathcal{R}$ are the number of additional relationships (constraints) as for example known chemical reactions, azeotropy, critical points or fixed compositions. Note that the "intensive properties" according to this definition are mole fractions within the phases along with temperature and pressure. This phase rule does not consider the relative amount of each phase and as a result it does not consider the so-called extensive properties which are proportional to the amount of substance. Examples below will illustrate the use of this rule.

2.1.3.1 Pure component application

For a pure component without additional relationships equation (2.7) reduces to:

$$\Im = 3 - \phi \qquad (2.8)$$

This is why on a pure component phase diagram, where pressure is shown as a function of temperature, the single phase ($\phi = 1$) region is a two-dimensional region, the two-phase ($\phi = 2$) regions are lines (vapour-liquid or **vapour pressure** line associated to evaporation/ condensation; vapour-solid or **sublimation/deposition** line and liquid-solid or **melting or fusion/crystallisation** line), and the three-phase "region" ($\phi = 3$) is a point. Figure 2.1 is an example of a pure component phase diagram.

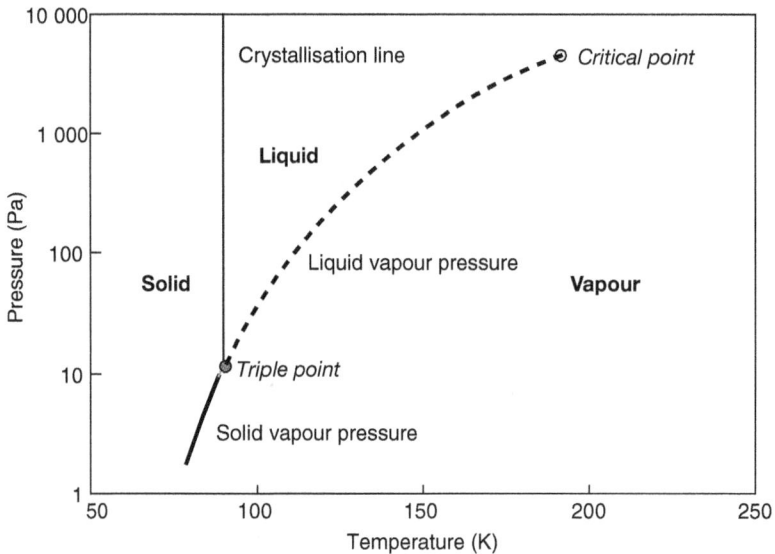

Figure 2.1

Behaviour of methane as a pure component (from DIPPR [12]).

Note that the vapour pressure line ends with a *critical point*. This point is defined as the conditions where the properties of the two phases merge, so that these phases become indistinguishable (same density, energy, etc.). This is a particularity of the fluid phases. When equation (2.7) is used, the critical point is considered as an additional constraint $\mathcal{R} = 1$ on top of the two-phase line, which explains why it is presented as a point (zero dimension).

Example 2.1 Refrigeration system

Propane is used to cool a process, as shown in the diagram of figure 2.2. The fluid that must be cooled in exchanger E103 should have an outlet temperature of -30 °C. At the inlet of the exchanger, propane is a bubbling liquid while it is a saturated vapour at the outlet. What pressure will the exchanger work at?

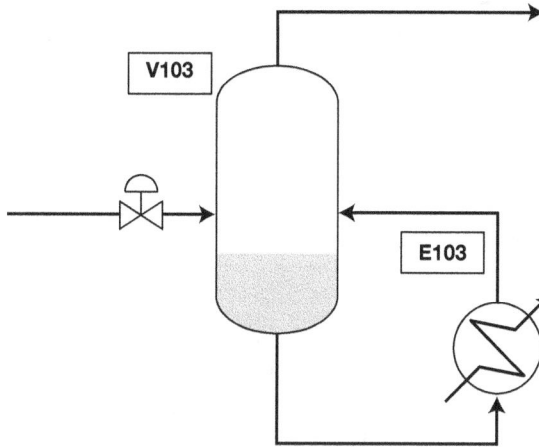

Figure 2.2

Sketch of the heat exchanger in a propane cooling example.

Analysis:

Only propane appears in the process. It is a well-known light hydrocarbon.

Propane will evaporate to cool the process. It is a vapour-liquid equilibrium of a pure component.

Only temperature is given. Saturation pressure (vapour pressure) must be calculated.

Model requirement:

A vapour pressure curve is enough to solve the question asked. As long as the temperature remains within the limits provided by the database, the Antoine [13] equation is sufficient (see chapter 3):

$$\ln\left(\frac{P^\sigma(T)}{P_c}\right) = A + \frac{B}{(T/T_c) + C}$$

For propane, the parameter values are given in table 2.2:

Table 2.2 Constants for Example 2.1

Parameter	T_c (K)	P_c (MPa)	A	B	C
Value	369.82	4.24	5.75442079	-5.48259492	-0.04723775

Solution:

The refrigeration system uses the heat of vapourisation of propane to cool the process fluid. The propane, a pure component ($\mathcal{N} = 1$), is at vapour-liquid equilibrium, *i.e.* a two-phase equilibrium ($\phi = 2$). The Gibbs phase rule (2.7) indicates that $\Im = 1$. If temperature is set to $-30\ ^\circ C$ (to ensure a reasonable driving force for heat exchange), then pressure is automatically fixed to the vapour pressure.

We therefore find, at $-30\ ^\circ C$, a pressure of 5.6 bar. With this information, the thermodynamic state is defined, indicating that temperature was a sufficient piece of information.

This example is discussed on the website:
http://books.ifpenergiesnouvelles.fr/ebooks/thermodynamics

The behaviour of the properties as a function of pressure and temperature in the single phase region is further discussed in the first section of chapter 4.

2.1.3.2 Binary systems

For a binary system, the single phase region of the phase diagram, according to equation (2.7), is a three-dimensional volume. It can be presented in a pressure-temperature-composition diagram as shown in figure 2.3 (top). In order to keep these figures readable, the solid regions are not represented (see for example [14]). The two-phase regions are two-dimensional surfaces on this plot, and the three-phase regions (not shown on the figure) are lines.

Due to the difficulty of reading this type of diagram, cuts or projections of isolines on one plane are preferred (figure 2.3, bottom). It is important to understand their meaning correctly.

A. Two phase temperature-composition or pressure-composition phase diagrams

When the phase diagram is cut at a given pressure (isobar), a temperature-composition, or *Txy* diagram appears (an example is shown in figure 2.4). Similarly, if figure 2.3 is cut at a given temperature (isotherm), a pressure-composition or *Pxy* plot is found, as shown in figure 2.5.

Two saturation lines are shown on both of these diagrams: they show the compositions of the phases at equilibrium (bubble and dew curves). They merge at the end of the diagram, for pure components (vapour pressure or boiling temperature are the same). For systems whose composition, pressure and temperature values are represented by a point located between the two lines, two phases (liquid and vapour) coexist.

A single liquid phase appears at higher pressures (for isothermal plots) or at lower temperatures (for isobaric plots). The bubble curve marks the boundary between the liquid phase and the two-phase region. It describes the **saturated liquid** composition. **Any liquid that coexists with a vapour phase is at its bubble point**.

A single phase vapour is observed at lower pressure (for isothermal plots), or higher temperatures (for isobaric plots). The dew curve is at the limit between the vapour region and the two-phase region. This curve describes the **saturated vapour** composition. **Any vapour that coexists with a liquid is at its dew point**.

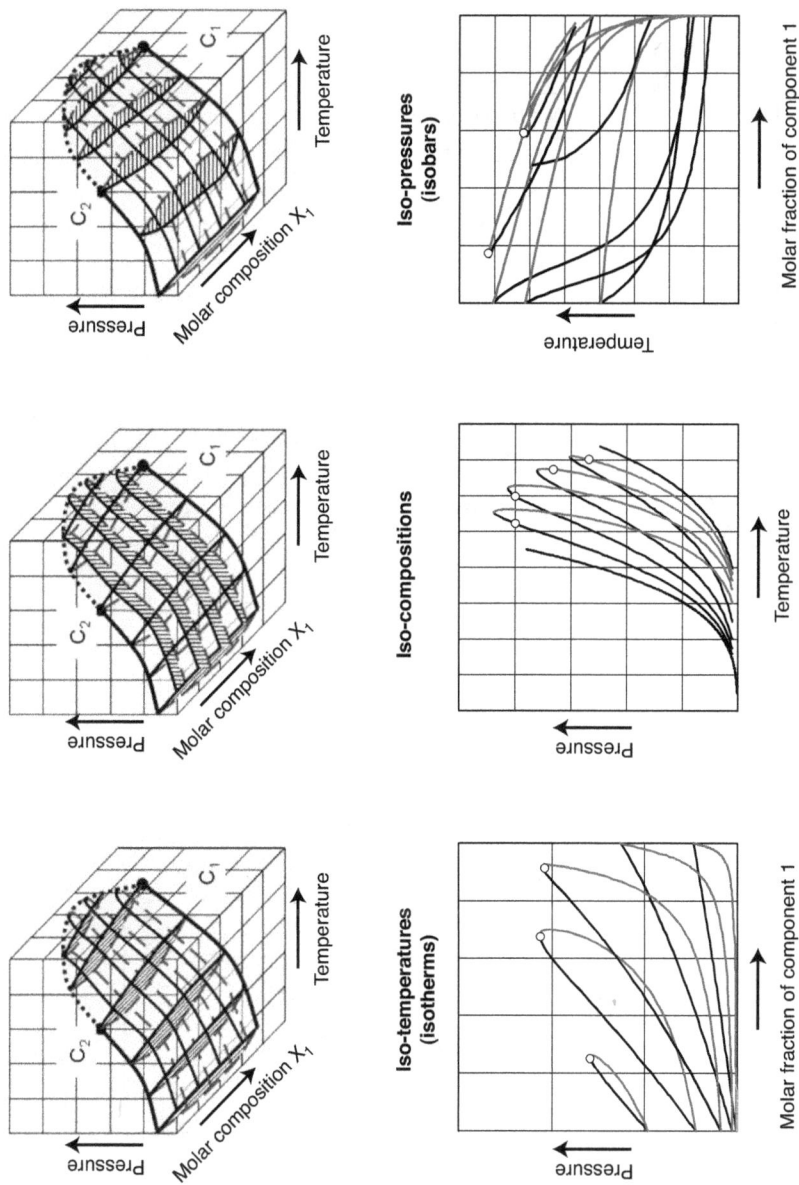

Figure 2.3

Three dimensional phase diagrams of a binary mixture and three types of cuts.

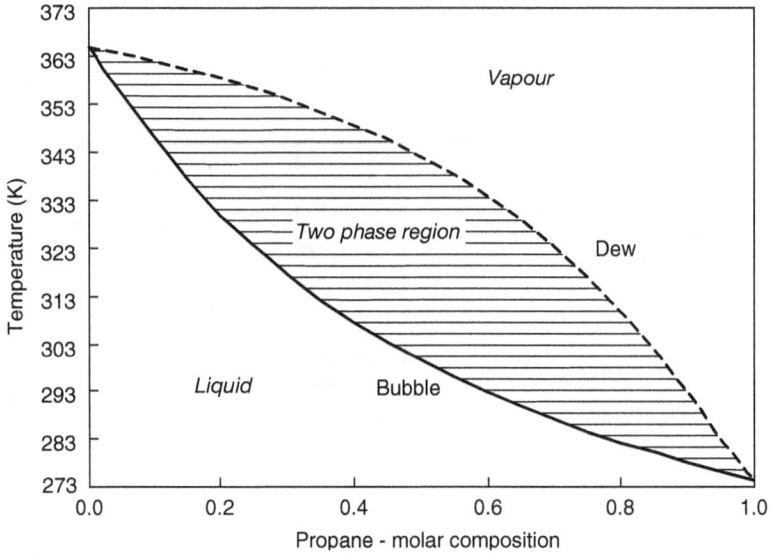

Figure 2.4

Example of a *Txy* vapour-liquid phase diagram (model of propane + *n*-pentane mixture at 0.5 MPa).

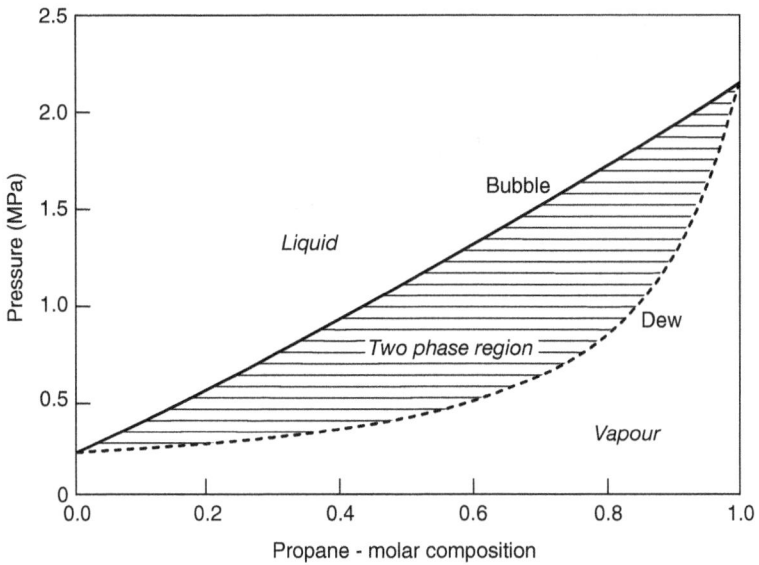

Figure 2.5

Example of a *Pxy* phase diagram (case of a nearly ideal mixture of propane + *n*-pentane at 333.15 K).

B. Pressure-temperature phase diagrams

Projections or cuts (iso-composition) are also provided on the *PT* plane. An example of such a projection is shown in figure 2.6. The vapour pressure curves of the pure components can be identified. These curves end with their critical point. The *PT* projection also shows a line that links these two points. It connects the critical points of all intermediate compositions and is called the ***critical point locus***.

For intermediate compositions (e.g. 80% ethane), a number of *PT* cuts are shown in figure 2.6. The two-phase region in this representation has become a surface, rather than a line as it was for a pure component. Application of the Gibbs phase rule (2.7), with $\mathcal{N} = 2$ and $\phi = 2$ ($\mathcal{R} = 0$) indicates that it is not sufficient to know that two phases coexist to have a single relationship between pressure and temperature. Yet, this two-phase region (or envelope) is limited by two distinct curves. One (the bubble curve, on the upper left) is the line where the liquid feed starts forming a first bubble of vapour, while the dew curve (on the lower right) is the line where a first drop of liquid condenses from the vapour feed. Both curves join at the critical point of the mixture (open circle). Points on the critical locus have an additional constraints (criticality: $\mathcal{R} = 1$), which explains why the critical locus is a line on this plot.

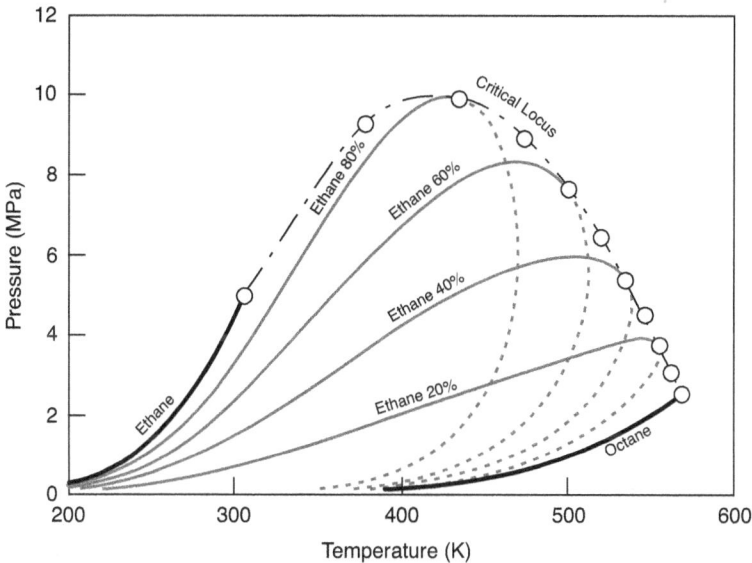

Figure 2.6

PT projection of the phase diagram of the ethane + octane binary mixture.

Bold lines are pure component vapour pressures; full lines are bubble points, dotted lines are dew points. The open circles represent the mixtures critical points and the dash-dot line the critical locus.

Example 2.2 VLE observation

A binary mixture of propane and n-pentane is available at 0.5 MPa. Using the information on figure 2.4, indicate the temperature range where vapour-liquid equilibrium can be observed.

If a system containing 40% (molar) propane is known to be at vapour-liquid equilibrium at 333.15 K, which is the proportion of each component in each phase? Can this reading be made on figure 2.5?

Analysis:

The properties to deal with are pressure, temperature and composition.

The mixture is a binary system with two light hydrocarbons.

Vapour and liquid are present together at equilibrium conditions.

Solution:

The Gibbs phase rule indicates that the degree of freedom of the system is $\Im = 4 - \phi$. If two phases are present, two intensive properties are necessary to specify the phase conditions. In a *Txy* diagram, pressure is fixed. Hence, a single additional property is required, which can be either temperature, liquid composition or vapour composition.

The corresponding extreme temperatures at 0.5 MPa are the boiling points of propane (275.15 K on the right of the figure) and of n-butane (367.15 K on the opposite side).

At 333.15 K, two phases can be observed with a liquid composition of propane of 0.18 and a vapour composition of 0.62. If composition of the overall mixture is less or greater than these two specified limits, only one phase will be observed. Any mixture with an overall composition included in this range will split into two phases. For example, the 40% mole propane mixture will lead to the following material balance:

$$F = L + V$$

$$Fz_1^F = Lx_1^L + Vy_1^V$$

Defining the vapour fraction as $\theta = V/F$ yields:

$$\theta = \frac{z_1^F - x_1^L}{y_1^V - x_1^L} = \frac{0.4 - 0.18}{0.62 - 0.18} = 0.5$$

This equation is known graphically as the **lever rule**.

When reading on the *Pxy* diagram, the compositions obtained are exactly the same.

This example is discussed on the website:
http://books.ifpenergiesnouvelles.fr/ebooks/thermodynamics

2.1.3.3 Three phases *Txy* or *Pxy* diagrams

When three phases are at equilibrium in a binary mixture, the Gibbs phase rule indicates that a single degree of freedom remains. When represented on a *Txy* or *Pxy* diagram, this means that either pressure or temperature is fixed and hence that no degree of freedom remains.

Figure 2.7 shows an example of three-phase equilibrium. It is represented by a straight, horizontal line to describe the three-phase conditions: temperature, as well as the composition of the three phases are fixed. On the left of the horizontal line, composition of the heavy liquid phase (aqueous) is found (liquid 1 on the figure with 0.035 isobutanol mole fraction). On the opposite side the composition of the light liquid phase (organic) is shown as liquid 2 (0.4 isobutanol mole fraction). The third point in the middle of the line indicates the composition of the vapour (0.34 isobutanol mole fraction) in equilibrium with the two liquids. At this temperature (362.65 K or 89.5 °C), cooling a system that contains a vapour will result in condensation of the vapour and formation of two liquids, while the temperature remains constant, as with a pure component.

In between the two almost vertical lines and below the rigorously horizontal three-phase line, two liquids coexist. Outside this zone, at temperatures below the three-phase temperature (89.5 °C) only liquid 1 or liquid 2 can be found. At the top of the diagram, there are two different vapour-liquid equilibrium zones. On the left, the dew line starts from the normal boiling point of water and ends at the vapour composition on the three-phase line. On the right side, the dew line starts from the normal boiling point of isobutanol and reaches the same point. Bubble lines join the normal boiling points to the liquid compositions on the three-phase line.

This type of diagram, where the vapour composition on the three phase line lies between the two liquid compositions is called a ***heteroazeotrope***. However, other types of three-phase diagrams may exist, as will be illustrated in chapter 4 of this book.

In all cases for binary systems, the three phase conditions are represented by a straight horizontal line.

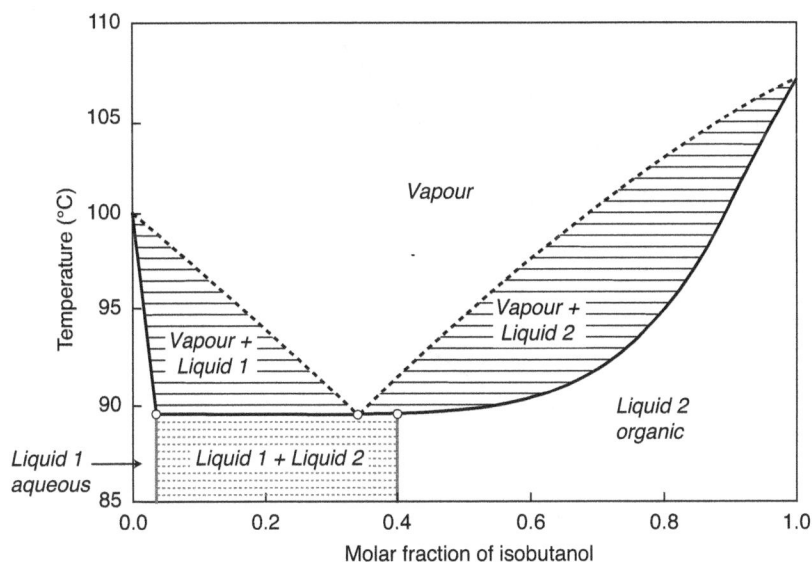

Figure 2.7
Atmospheric vapour-liquid-liquid diagram of the water-isobutanol system.

2.1.3.4 Ternary mixtures

The study of ternary mixtures is of great interest for a number of process applications, as it helps us understand the role of solvents or co-solvents. The Gibbs phase rule indicates that there are 3 degrees of freedom for a two-phase equilibrium (vapour-liquid or liquid-liquid). The two-phase zone can therefore be represented by a volume and the single phase zone by a four-dimensional hyper-volume. Very often, triangular cuts at fixed pressure and temperature are used (see, for example, figures 2.8, 2.9 and 2.10).

A. Vapour-liquid equilibrium

For ternary mixtures, if either pressure or temperature is fixed, a three-dimensional diagram is found again. This technique can be used to visualise the aspect of two-phase vapour-liquid equilibrium. A triangle is used as a basis to describe the composition of each component

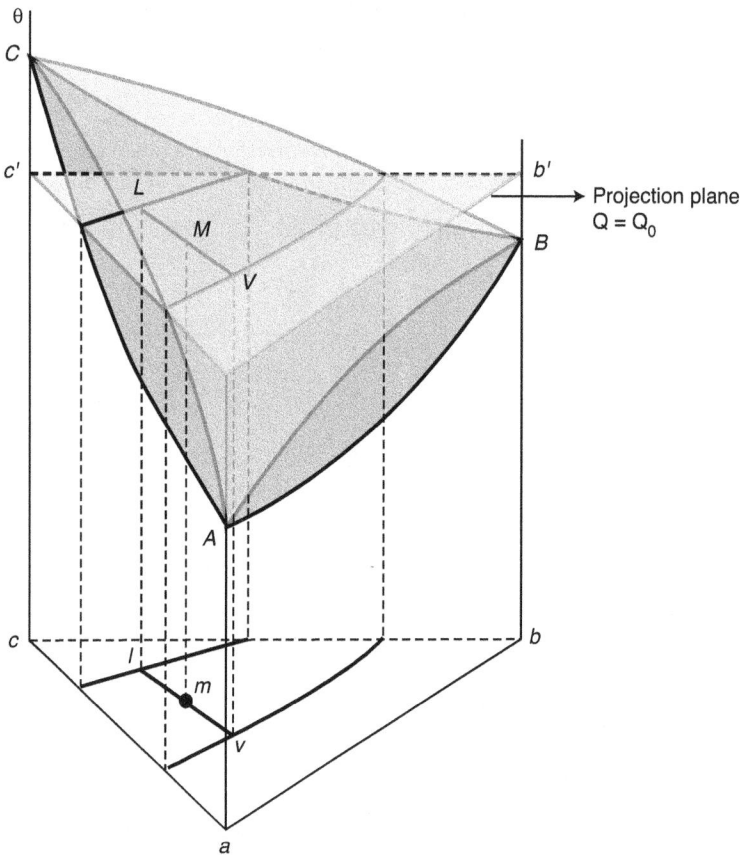

Figure 2.8

Isobaric vapour-liquid equilibrium representation for a ternary mixture.

(note that the sum of the heights is constant in a triangle, which is why it is used to describe composition). On this diagram bubble and dew locus are surfaces. In the constant pressure diagram as shown in figure 2.8, the vapour lies on top, at high temperature, above the dew surface. The liquid zone is found below the bubble surface.

B. Liquid-liquid equilibrium

Ternary mixtures of partially miscible products are commonly used in industry to purify one of the products by extraction. Pressure and temperature are usually constant in such processes and the Gibbs phase rule shows that the two remaining degrees of freedom can describe the system. Two molar fractions are sufficient to describe a single-phase system. If two phases are present, the phase compositions are located on the saturation lines, and only one molar fraction is necessary. Triangular phase diagrams are used, as in figure 2.9. The two-phase zones are bordered by *saturation lines* (similar to bubble and dew lines except that, since both describe a liquid phase, this terminology can no longer be used), and hatched by the *tie-lines*, that are straight lines connecting the compositions of the two liquids in equilibrium with each other. Often, a *plait point* or *critical point* can be seen, at the limit where the two liquid phases become identical (the tie-line is reduced to a point).

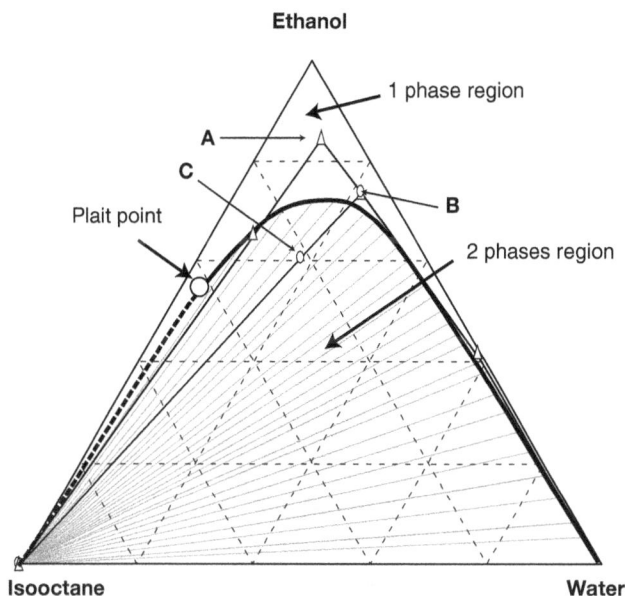

Figure 2.9

Liquid-liquid equilibrium for the ternary mixture of isooctane + ethanol + water at 298.15 K and 101.32 kPa (kind I). The A, B, C labels refer to example 2.3.

   Different kinds of phase diagram can be encountered (figure 2.10). Kind I corresponds to two partially miscible components and one fully miscible with both others, as in figure 2.9. In kind II, two components are fully miscible and both are partially miscible with the third. Kind III implies that no two components are fully miscible: in this case, a three-phase equilibrium can be encountered. Note that three phase equilibrium in a ternary mixture, at fixed pressure and temperature, leaves no degrees of freedom. Hence, the compositions of the three phases are fixed. They are represented by the summits of the internal triangle.

   It is quite common to observe that the size of the two-phase zone decreases with a temperature increase. This effect is represented in figure 2.10 for the different types of diagrams. Diagrams of type II may have one or two two-phase zones. For type III, the three different zones that may exist at lower temperature may merge, yielding a single three-phase zone.

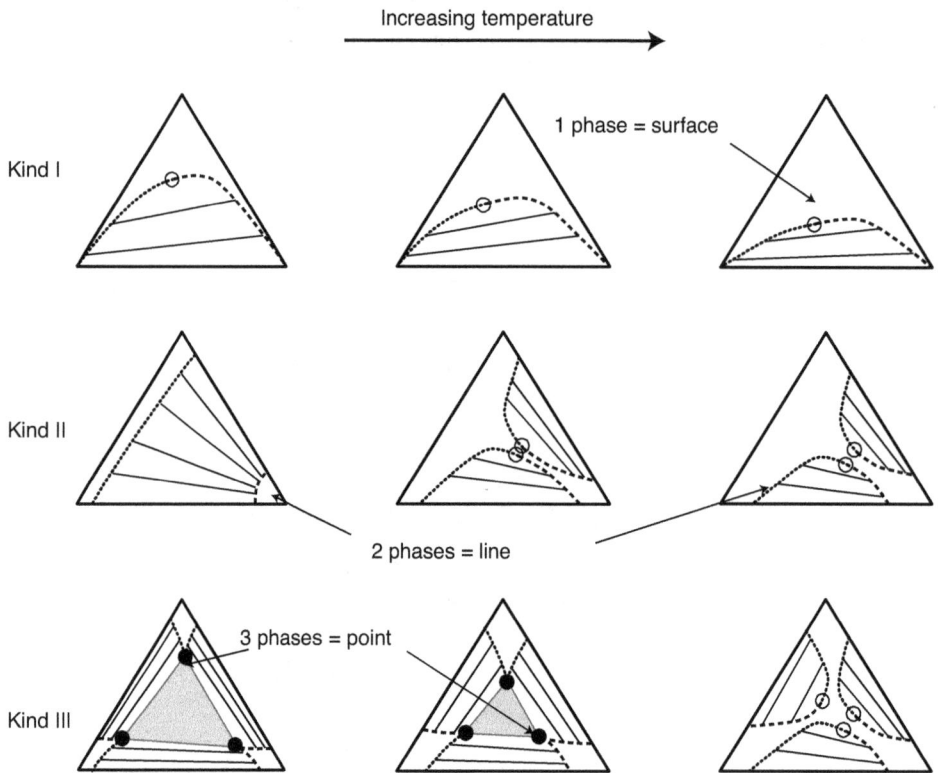

Figure 2.10

Different kinds of liquid-liquid equilibrium for a ternary mixture.

Example 2.3 Flexfuel model

E85 gasoline may be represented in a simplified way using a mixture of isooctane for the hydrocarbon (15% volume) and ethanol (85% volume). For the sake of the example we consider that the alcohol used for the mixing has been obtained by simple distillation: its composition in ethanol is very close to the binary azeotrope between water and ethanol. A 0.965 volume fraction composition will be used. The mixture is made at 298.15 K (25 °C) and the molar volumes are available at this temperature for each component.

Locate this E85 gasoline on figure 2.9.

In a distillery region, a driver adds 5 litres of a 70% volume alcohol to the 25 litres of E85 in the car fuel tank. What happens?

To repair the "mistake", the driver decides to fill up the tank with only gasoline (isooctane) since the flex motor is designed to work with any of these fuels. Suppose he fills the contaminated tank with 20 litres of isooctane. What happens? Is the phenomenon due to the "mistake"? Some data related to the components are given in table 2.3:

Table 2.3   Constants for example 2.3

Component	Molar mass (kg kmol^{-1})	Molar density (mol L^{-1}) at 298.15 K
Isooctane	114.2	6.0205
Ethanol	46.1	17.04
Water	18.0	55.384

Analysis:

Pressure and temperature are near ambient. Only compositions and phase behaviour are of interest.

Components are hydrocarbon, alcohol and water. Water is known not to mix with hydrocarbons. Ethanol is a solvent for both other fluids. A two-phase liquid-liquid zone will be observed. The diagram is kind I.

Solution:

First of all, equilibrium data are given in molar composition and gasoline measurements are made on a volume basis. Conversion must be carried out using molar volume.

Location of the point describing the E85 is known exactly (table 2.4). It is made from 15% isooctane and 85% of the ethanol-water mixture. If the "Ethanol" mixture contains 0.965 of pure ethanol then the net ethanol volume fraction will be 0.82025. With a total volume of 25 L, the molar fractions are calculated in table 2.4.

Table 2.4   Original E85 gas for example 2.3

Component	x (vol/vol)	Vol (L)	Amount (mol)	x (mol/mol)
Isooctane	0.1500	3.75	22.58	0.0546
Ethanol	0.82025	20.51	349.42	0.8457
Water	0.02975	0.74	41.19	0.0997
Total		25.00	413.19	

With 0.8457 of ethanol in the mixture, the E85 is very near the top of the triangle (point A on figure 2.9). Water composition is almost on the 10% isoline. This point belongs to the single-phase zone, so the fuel mix is OK.

5 litres of raw alcohol from a retort is now added to the fuel. The same table starting with a volume composition of {0,0.7,0.3} is created. The mixture is obtained adding the moles of each component, resulting in the composition given in table 2.5:

Table 2.5   First mixture of gas for example 2.3

Component	Vol (L)	x (vol/vol)	Amount (mol)	x (mol/mol)
Isooctane	3.75	0.1250	22.58	0.0406
Ethanol	24.01	0.8002	409.07	0.7359
Water	2.24	0.0748	124.27	0.2235
Total	30.00		555.91	

The resulting point is slightly below the former (point B on figure 2.9). On the 323.15 K diagram, the point is still in the one-phase area, but on the 298.15 K diagram the point is located very near the two-phase limit. If the temperature drops a few degrees, the liquid will split into two phases. It can be considered a dangerous mistake.

Will adding 20 litres of pure isooctane correct this mistake? Another balance of the same kind is made to obtain the composition shown in table 2.6:

Table 2.6   Second mixture of gas for example 2.3

Component	Vol (L)	x (vol/vol)	Amount (mol)	x (mol/mol)
Isooctane	23.75	0.4750	142.99	0.2114
Ethanol	24.01	0.4801	409.06	0.6049
Water	2.25	0.0449	124.27	0.1837
Total	50.00		676.32	

Now the mixture contains close to 60% ethanol and a similar amount of water and isooctane (point C on figure 2.9). This point is located in the two-phase zones indicating a separation into two layers: at the bottom, the aqueous phase (with 66% ethanol and 20% water at 323.15 K) and an organic phase with very little ethanol and essentially isooctane on top. The cause is the excessive presence of water. Industrial E85 is made of very pure (99.8% volume) ethanol. Water is removed using a special dehydration column or molecular sieves. Check the result in this case.

This example is discussed on the website:
http://books.ifpenergiesnouvelles.fr/ebooks/thermodynamics

2.1.3.5 Multicomponent systems

Equation (2.7) is valid for any number of components. It is clear, however, that a multidimensional plot is almost impossible to use. However, the *PT* type diagram discussed in the previous section remains identical: whatever the number of components, the bubble curve and the dew curve keep their significance and the mixture critical point is located at the point where both curves merge.

Note however that if a three-phase region exists, it is located by a *PT* curve if the mixture is binary (the projection of the three-phase line seen in figure 2.7). When more components are present in the mixture, the three-phase region becomes a surface in the same way as the two-phase regions.

Whatever the number of components, equation (2.7) indicates that if the mixture composition is known, the number of supplementary constraints is equal to $\mathcal{R} = \mathcal{N} - 1$, and the number of remaining degrees of freedom is

$$\Im = \left(\mathcal{N} - \phi + 2\right) - \left(\mathcal{N} - 1\right) = 3 - \phi \tag{2.9}$$

Then, in a two-phase system for example, if the pressure is given, the temperature is fixed as well as the phase compositions (bubble or dew lines). This case is also discussed below and will be called a *T*θ or *P*θ calculation.

2.1.4 Duhem phase rule (theorem)

In process simulation calculations, the feed composition is generally provided, yet it is not known how many phases may be present. In this case Duhem's phase rule states that:

> If the amount of substance of all components in the feed is given, any combination of two state properties are sufficient to fully define the system.

As a result, a system may be defined according the classification provided in table 2.7. It will be seen later that in the case of phase equilibrium calculations, they will be called "Flash types".

The abbreviations used in table 2.7 will be kept throughout this book in order to simplify the notations.

Table 2.7   Most frequently encountered combinations of data types given

Flash type	Meaning	Example applications
PT	Pressure and temperature given	Basic case Used in all calculations
*T*θ or *P*θ	Temperature or pressure and vapour fraction given	Bubble point Dew point Partially vapourised flash
TV	Temperature and volume given	Closed vessel at known temperature
PH	Pressure and enthalpy given	Adiabatic distillation columns Adiabatic expansions
PS	Pressure and entropy given	Ideal adiabatic compressors Pumps Turbines

Example 2.4 Phase envelope of a natural gas with retrograde condensation

A natural gas reservoir is characterised by the following molar composition (table 2.8, compositions under 10^{-5} are omitted). The corresponding phase envelope has been generated (figure 2.11). Conditions in the reservoir are 333.15 K and 15 MPa (A). In a separator, the gas is first cooled to 298.15 K (B) and expanded at this temperature down to 9 MPa (C). Next, the fluid is further expanded at the same temperature down to 1 MPa (D). Describe the behaviour of the fluid.

Table 2.8   Sample composition of a natural gas

Component	molar fraction
Methane	0.74610
Nitrogen	0.00630
CO_2	0.00010
Ethane	0.12526
Propane	0.08043
Butanes	0.03081
Pentanes	0.00811
Hexanes	0.00212
C_7+	0.00056
C_8+	0.00021

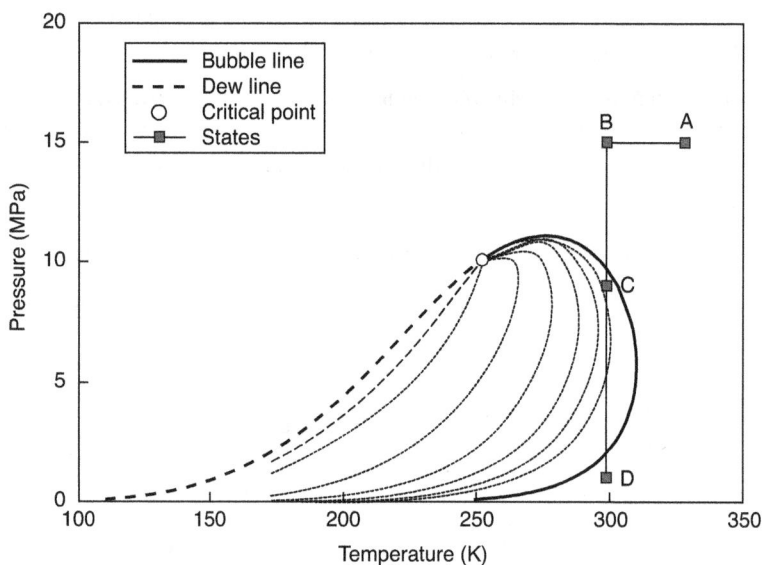

Figure 2.11

Retrograde condensation in a natural gas PT envelope diagram.

Analysis:

The only properties involved are pressure and temperature. Components include hydrocarbons, nitrogen and carbon dioxide. Only the phase behaviour is investigated. States are supercritical and in the vapour-liquid zone. The choice of the model is very important but in this example the figure is used.

Solution:

Natural gas is fully described by its molar composition. Duhem's phase rule therefore indicates that two properties are sufficient to describe the system. In this case, each state is described by two intensive properties P and T. The different states can be plotted on the phase envelope (figure 2.11). This figure also shows the lines of constant liquid fraction.

The first two points (A and B) are clearly located in the supercritical zone: no phase change appears. When the pressure is lowered at constant temperature, the **dew line** is crossed, and a denser (liquid) phase appears. At the temperature considered, a liquid phase surprisingly appears when the pressure drops below 9.7 MPa. This can be understood physically from the observation that the denser a gas is, the better a solvent it is for the heavy fraction of the fluid. Consequently, decreasing the pressure means decreasing its solvent power. The liquid dropout will continue to increase until the pressure reaches 6.5 MPa. Further pressure reduction will re-vapourise most of the liquid phase, which is the normal behaviour. The pressure range of a liquid deposit is called the retrograde region and the process is called **retrograde condensation**. At the normal dew pressure, on the bottom of the figure, the fluid is entirely vapourised (D).

This example is discussed on the website:
http://books.ifpenergiesnouvelles.fr/ebooks/thermodynamics

2.2 PROPERTY COMPUTATION

Obviously, the property or properties to be computed must be well identified. In the remainder of this document, we will consider that pressure, temperature and phase composition are the fundamental state variables. That means that the basis for calculation of any other property is the knowledge of these fundamental state variables. Most models are in fact constructed on these variables. In other cases, for example the equations of state, pressure is expressed as a function of temperature, volume and composition. Consequently, the equations of state will need an associated "solver" that provides the phase volume at the given pressure (see section 3.4.3.1, p. 189).

Note that in practice, other state variables may be used. The cases presented in table 2.7, other than TP, $T\theta$ or $P\theta$, will require a model for the additional state variable (e.g. PH will require a model for enthalpy, TV will require a model for volume, TS will require a model for entropy, etc.).

In principle, any thermodynamic calculation should start with a full (both phase and chemical) equilibrium calculation, since they will provide the true compositions of the individual phases required for further property calculations. In practice, it is up to the user to evaluate whether there is a risk of phase or chemical change. Even if this risk truly exists, there may be no need to consider it when the kinetics of the phase split or of the chemical equilibrium are slow.

Often, in process simulation, a vapour-liquid equilibrium calculation is performed, but liquid-liquid equilibrium and chemical equilibrium is rarely checked.

Even though there is a tendency to develop single models that can describe all the properties (a good test for equations of state is whether they can correctly describe all the properties by derivation or integration, in particular heat capacities), **it is often recommended for the practicing engineer to use different methods for different physical properties**. In particular, it is recommended to use different methods for the calculation of "single phase properties" (volume as well as enthalpy, entropy and heat capacity) on the one hand, and of phase equilibrium on the other hand.

In this section, we first quickly present the fundamental relationships of thermodynamics. They are essential in order to understand the mechanism that is used for computing single phase properties: it is shown that two main families of models exist, the ones based on the residual approach and the ones based on the excess approach.

The fundamental property needed for equilibrium calculations is the chemical potential, which is often expressed as a fugacity. Using this concept, and knowing the method for calculating these quantities (using excess or residual approach), the methods for phase equilibrium calculation and for chemical equilibrium calculations will be discussed.

2.2.1 Some fundamental relationships

Before going ahead and providing the methods required to calculate thermodynamic properties, it is useful to recall some basic principles and definitions.

2.2.1.1 System properties

A system is uniquely defined by a feed composition (amount of substance *i.e.* the number of moles) and two thermodynamic properties called state properties (see table 2.7). Most often, these are temperature and pressure, but other thermodynamic properties can be used for this purpose, in particular volume V, or energetic properties such as internal energy U, enthalpy H and entropy S. These properties are related to each other, as summarised in table 2.9. The basic quantities used are the volume V, pressure P, absolute temperature T and the entropy S. The fundamental relations can be easily obtained from the basic principles of thermodynamics. They are summarised here, but a more detailed description is available in thermodynamic textbooks [4, 15].

A. First principle of thermodynamics:

First of all, remember the differential version of the first principle applied to a closed system [1]:

$$dU = \delta Q + \delta W \tag{2.10}$$

and reversible mechanical work is given by:

$$\delta W = -PdV \tag{2.11}$$

The sign '−' indicates that the mechanical energy of the system increases as the volume is reduced. Heat flow (δQ) is defined using entropy, as shown below in equation (2.15).

For an open and unsteady system (*i.e.* a system where material flows in and/or out: this is most often the case in process conditions), this equation is complemented with the in- and out-flow of energy carried by each of the material streams, s:

$$dU = \delta Q + \delta W + \sum_s h_s dN_s \tag{2.12}$$

where dN_S is positive for inflow and negative for outflow.

The fact that the energy flow is expressed as enthalpy ($H = Nh = U + PV$) rather than energy can be understood considering that the stream "pushes" the system as a result of its pressure. Equation (2.12) is often written for stationary systems (*i.e.* no change with time, meaning that the internal energy does not change [2]):

$$\dot{H}_{out} - \dot{H}_{in} = \dot{Q} + \dot{W} \tag{2.13}$$

B. Second principle of thermodynamics

The second law uses the concept of entropy (S) whose change can have two distinct causes:

$$dS = dS_e + dS_i \tag{2.14}$$

the index e refers here to "external" and i to "internal". The change in entropy from external origin is related to the heat flux between the system and its environment:

$$dS_e = \frac{\delta Q}{T} \tag{2.15}$$

where T is the absolute temperature (not °C nor °F).

The change of entropy from internal origin (which is related to the dissipation of the coherent energy into heat) is always positive, according to the second law:

$$dS_i \geq 0 \tag{2.16}$$

using the equations (2.10), (2.11), (2.14), (2.15) and (2.16), it is possible to conclude:

$$-TdS_i = dU - TdS + PdV \leq 0 \tag{2.17}$$

---

1. The symbol d refers to an infinitesimal **change** (*i.e.* a difference between the state property value before and after the transformation); the symbol δ rather refers to an infinitesimal **quantity** (there is no difference involved).
2. The dot on top of the symbols refers to flow rate (per unit time).

This last equation informs, in very general terms, how a system changes spontaneously. In particular, it will teach us how components should be distributed between the phases, considering a system under some constraints. If the constraints are fixed temperature and volume, equation (2.17) becomes:

$$d(U - TS) = dA \leq 0 \qquad (2.18)$$

This is how A, the Helmholtz energy, is introduced, as defined below in equation (2.22).

Yet, most often, the systems are taken at fixed temperature and pressure, and equation (2.17) is written as:

$$d(U - TS + PV) = dG \leq 0 \qquad (2.19)$$

This is how G, the Gibbs energy, is introduced, as defined below in equation (2.23). Hence, the second principle, as we will use it in the rest of the text, can be rephrased as:

> A system that is left to itself, at fixed pressure and temperature, will behave so as to minimise its Gibbs energy.

C. The fundamental thermodynamic relations

When a system is at equilibrium, the inequalities in the above equations become equalities. The first of these equations is directly obtained from (2.10), (2.11) and (2.15):

$$dU = TdS - PdV \qquad (2.20)$$

This expression indicates that the fundamental dependence of internal energy is with entropy and volume, so the basic expression for internal energy is of the form $U = U(S, V)$.

Table 2.9　Definition of the fundamental energy properties [a]

$H \triangleq U + PV$	(2.21)		$dH = TdS + VdP$	(2.24)
$A \triangleq U - TS$	(2.22)		$dA = -SdT - PdV$	(2.25)
$G \triangleq U + PV - TS$	(2.23)		$dG = -SdT + VdP$	(2.26)

[a] The $\triangleq$ symbol indicates that the relation is a definition.

> The variables H, A and G are only Legendre transformations of variables V, U and S. There is no "new information" contained in these variables.

Note that both the internal energy and the entropy must be given a reference state (see section 2.1.2, p. 26) if a numeric value is expected.

Very often, a reference state is given for the enthalpy H and that for the internal energy is inferred using (2.21). For the entropy, the third principle of thermodynamic states that the pure component entropies are zero at the absolute zero temperature. This definition is not directly useable, but makes it possible to provide the absolute value of the entropy under reference conditions. It can be found in most databases.

Equations (2.24) to (2.26) are direct consequences of (2.20) and the definition equations (2.21) to (2.23). They show how the various properties are related to each other. Through additional mathematical manipulations on these exact differential equations, the equations of table 2.10, known as the **Maxwell relations**, can be found. The manipulations are very simple and correspond to:

$$df = \left(\frac{\partial f}{\partial x}\right)_y dx + \left(\frac{\partial f}{\partial y}\right)_x dy = \alpha dx + \beta dy$$

$$\Rightarrow \left(\frac{\partial \alpha}{\partial y}\right)_x = \left(\frac{\partial}{\partial y}\left(\frac{\partial f}{\partial x}\right)_y\right)_x = \left(\frac{\partial}{\partial x}\left(\frac{\partial f}{\partial y}\right)_x\right)_y = \left(\frac{\partial \beta}{\partial x}\right)_y$$

(2.27)

Table 2.10   The Maxwell relations between the fundamental properties

$\left.\dfrac{\partial U}{\partial S}\right\|_V = \left.\dfrac{\partial H}{\partial S}\right\|_P = T$ (2.28)	$\left.\dfrac{\partial H}{\partial P}\right\|_S = \left.\dfrac{\partial G}{\partial P}\right\|_T = V$ (2.30)
$\left.\dfrac{\partial U}{\partial V}\right\|_S = \left.\dfrac{\partial A}{\partial V}\right\|_T = -P$ (2.29)	$\left.\dfrac{\partial A}{\partial T}\right\|_V = \left.\dfrac{\partial G}{\partial T}\right\|_P = -S$ (2.31)
$\left.\dfrac{\partial T}{\partial V}\right\|_S = -\left.\dfrac{\partial P}{\partial S}\right\|_V$ (2.32)	$\left.\dfrac{\partial S}{\partial V}\right\|_T = \left.\dfrac{\partial P}{\partial T}\right\|_V$ (2.34)
$\left.\dfrac{\partial T}{\partial P}\right\|_S = \left.\dfrac{\partial V}{\partial S}\right\|_P$ (2.33)	$\left.\dfrac{\partial S}{\partial P}\right\|_T = -\left.\dfrac{\partial V}{\partial T}\right\|_P$ (2.35)

A final equation that is of great use to calculate the enthalpy is the Gibbs-Helmholtz equation:

$$\left.\frac{\partial(G/T)}{\partial(1/T)}\right|_P = H \quad \text{or} \quad \left.\frac{\partial(G/T)}{\partial(T)}\right|_P = \frac{-H}{T^2}$$

(2.36)

Other properties are defined as derivatives of the above fundamental properties:

• the isochoric or constant volume heat capacity (very often, the heat capacity is expressed in intensive units: $c_V = \dfrac{C_V}{N}$ or $c_P = \dfrac{C_P}{N}$.

$$C_V = \left.\frac{\partial U}{\partial T}\right|_V$$

(2.37)

• the isobaric or constant pressure heat capacity:

$$C_P = \left.\frac{\partial H}{\partial T}\right|_P$$

(2.38)

- the isobaric thermal expansion coefficient:

$$\alpha = \frac{1}{V}\frac{\partial V}{\partial T}\bigg|_P \tag{2.39}$$

- the isothermal compressibility coefficient:

$$\beta_T = -\frac{1}{V}\frac{\partial V}{\partial P}\bigg|_T = -\frac{1}{\kappa_T} \tag{2.40}$$

- the isochoric thermal pressure coefficient:

$$\beta = \frac{1}{P}\frac{\partial P}{\partial T}\bigg|_V \tag{2.41}$$

- the Joule-Thomson coefficient:

$$\mu = \frac{\partial T}{\partial P}\bigg|_H \tag{2.42}$$

- and finally the Mayer relation:

$$C_P - C_V = T\left(\frac{\partial P}{\partial T}\bigg|_V\right)\left(\frac{\partial V}{\partial T}\bigg|_P\right) = -T\left(\frac{\partial P}{\partial T}\bigg|_V\right)^2\left(\frac{\partial V}{\partial P}\bigg|_T\right) \tag{2.43}$$

With these basic properties definitions and the Maxwell equations, all property differentials can be expressed as a function of only T and P or T and V, as shown in tables 2.11 and 2.12.

Table 2.11 Relations of thermodynamic properties as a function of T and P

$dU =$	$\left(C_P - P\frac{\partial V}{\partial T}\bigg|_P\right)$	dT	$-\left(T\frac{\partial V}{\partial T}\bigg|_P + P\frac{\partial V}{\partial P}\bigg|_T\right)$	dP
$dH =$	C_P	dT	$+\left(V - T\frac{\partial V}{\partial T}\bigg|_P\right)$	dP
$dA =$	$\left(-S - P\frac{\partial V}{\partial T}\bigg|_P\right)$	dT	$-P\frac{\partial V}{\partial P}\bigg|_T$	dP
$dG =$	$-S$	dT	$+V$	dP
$dS =$	$\frac{C_P}{T}$	dT	$-\frac{\partial V}{\partial T}\bigg|_P$	dP

Table 2.12 Relations of thermodynamic properties as a function of T and V

$$dU = C_V \, dT + \left(T \frac{\partial P}{\partial T}\bigg|_V - P \right) dV$$

$$dH = \left(C_V + V \frac{\partial P}{\partial T}\bigg|_V \right) dT + \left(V \frac{\partial P}{\partial V}\bigg|_T + T \frac{\partial P}{\partial T}\bigg|_V \right) dV$$

$$dA = -S \, dT - P \, dV$$

$$dG = \left(-S + V \frac{\partial P}{\partial T}\bigg|_V \right) dT + V \frac{\partial P}{\partial V}\bigg|_T dV$$

$$dS = \frac{C_V}{T} \, dT + \frac{\partial P}{\partial T}\bigg|_V dV$$

2.2.1.2 Phase properties

A phase is a part of a system that is such that the intensive properties (temperature, pressure, molar or specific volume, molar or specific enthalpy, molar or specific entropy, etc.) are identical in any point of the phase. The intensive properties are generally different in distinct phases. A system can contain several phases (vapour, liquid or solid). Only a single vapour phase can exist in a given system, but several liquids or solids can coexist. The exponents (V, L or S) will be used to identify them.

All property calculations will refer to phase properties. The molar properties are expressed using a lower case symbol. For example, the liquid molar volume is written as:

$$v^L = \frac{V^L}{N^L} \tag{2.44}$$

For the extensive properties (proportional to the mole number as V, U, H, A, G, S), a balance equation can be written to find the system property knowing the phase properties:

$$V(system) = \sum_{\phi} N^\phi v^\phi \tag{2.45}$$

where the sum is made over all phases ϕ.

2.2.1.3 Mixture

A mixture is the aggregation of $\mathcal{N}$ different pure components and characterised by the amount of substance of each in the mixture.

$$N = \left\{ N_1, ..., N_{\mathcal{N}} \right\} \tag{2.46}$$

The molar composition of component i is therefore given by:

$$z_i = \frac{N_i}{\displaystyle\sum_{j=1}^{\mathcal{N}} N_j} \tag{2.47}$$

Usually the variable z_i is used when the phase of the mixture is unknown. For a vapour, y_i is preferred and x_i is reserved for liquids. If different liquids are present a superscript is used to differentiate them.

The fundamental relationships (equations (2.10) to (2.42), as well as tables 2.11 and 2.12) used in thermodynamics have been presented as independent of composition. They are valid for pure components as well as mixtures. They need to be generalised, however, since, in case of mixtures, the number of independent variables is no longer 2, but $\mathcal{N}+2$.

The fundamental thermodynamic properties should be expressed as a function of all independent variables (for any property expressed here with the letter X, which can represent the volume V, the internal energy U, the enthalpy H, the Gibbs energy G, the Helmholtz energy, the entropy S or the heat capacities):

$$\mathbf{X} = X\left(T, P, N_1, ..., N_{\mathcal{N}}\right) \tag{2.48}$$

From these definitions, it is clear that the derivatives of the thermodynamic properties (as in table 2.9) must include these new elements that capture the compositional dependence. Thus, the differential of G, for example, will be written as:

$$dG = \frac{\partial G}{\partial T}\bigg|_{P,N} dT + \frac{\partial G}{\partial P}\bigg|_{T,N} dP + \frac{\partial G}{\partial N_1}\bigg|_{T,P,N_{j\neq 1}} dN_1 + ... + \frac{\partial G}{\partial N_{\mathcal{N}}}\bigg|_{T,P,N_{j\neq\mathcal{N}}} dN_{\mathcal{N}}$$

$$= -SdT + VdP + \sum_{i=1}^{\mathcal{N}} \frac{\partial G}{\partial N_i}\bigg|_{T,P,N_{j\neq i}} dN_i \tag{2.49}$$

2.2.1.4 Property of a pure component in a phase

The phase properties can be compared with the same property of the components taken individually, at the same pressure and temperature (comparing properties require the states to be unambiguously defined: see section 2.1.2, p. 26). It will be found more meaningful, however, to define the partial molar properties that describe the effect of adding an infinitesimal amount of a given component to the property of the phase.

A. The pure component at the same pressure, temperature and physical state as the mixture

The properties of an individual component in a mixture, at the same pressure and temperature as the mixture, and in the same physical state, are written using the index of the component. A lower case indicates molar properties and an asterisk will refer to the pure component properties. For example, the molar volume of component i when it is pure, is written as v_i^*. If the mixture is a liquid, this molar volume will refer to the liquid state, if the mixture is a vapour, it will refer to the vapour state, irrespective of the true physical state of component i at the given pressure and temperature. It can be considered that this property is the limiting value for the phase property when the composition becomes equal to that pure component.

As an example, let us consider a mixture of propane and pentane at 500 kPa and 300 K. It is clear from figure 2.4 that the propane as a pure component will be found as a vapour. In contrast, pure pentane is a liquid. Nevertheless, if a mixture containing 80% of propane is maintained under these conditions, a vapour-liquid equilibrium will be observed: the liquid will have approximately 49% propane while the vapour will contain 91% of propane. This means that propane (light component) also exists in the liquid phase and pentane (heavy component) also exists in the vapour phase. This is why it is important to be able to describe properties of the individual components in a phase in which they are "naturally" unstable. In the example, $v_i^{*,L}$ and $v_i^{*,V}$ for either pentane or propane, will be two distinct values.

When the pure component can be encountered in the same physical state as that of the mixture, its properties under these conditions can be measured or calculated readily. However, when the pure component does not exist in the same physical state, this value cannot be measured directly. Its value should then be considered as a mere mathematical, intermediate value for the mixture property calculation.

B. Partial molar properties

The partial molar properties express the true contribution of the component i to the mixture property. It is defined as the partial derivative:

$$\bar{x}_i \overset{\Delta}{=} \left(\frac{\partial X}{\partial N_i} \right)_{T,P,N_{j\neq i}} \tag{2.50}$$

where

$\bar{x}_i$ is the partial molar property,

X is the extensive property,

N_i is the amount of substance (expressed as the number of moles) of component i.

According to Euler's theorem for homogeneous first order functions, the extensive property can be written as:

$$X = \sum_{i=1}^{\mathcal{N}} N_i \bar{x}_i \tag{2.51}$$

C. The chemical potential and the fugacity

Since the Gibbs energy is the basic property for calculating phase equilibrium, the partial molar Gibbs energy will be of particular importance in the calculation procedures. It is more commonly known as the *chemical potential* μ_i:

$$\bar{g}_i = \mu_i = \left(\frac{\partial G}{\partial N_i} \right)_{T,P,N_{j\neq i}} \tag{2.52}$$

The chemical potential is an essential quantity in fluid phase thermodynamics, but it is inconvenient to use. The reason for this is that it tends to minus infinity at infinite dilution, requires the definition of a reference state and is difficult to grasp intuitively. Another concept has therefore been developed, which describes the same physical quantity, yet is more often used. The *fugacity* f_i is defined from:

$$d\mu_i\big|_T \stackrel{\Delta}{=} RT \, d \ln f_i\big|_T \qquad (2.53)$$

which results in an equivalence between chemical potential difference and fugacity ratio (on the condition that both thermodynamic states, α and β, are at the same temperature):

$$\mu_i^\alpha - \mu_i^\beta = RT \ln \frac{f_i^\alpha}{f_i^\beta} \qquad (2.54)$$

2.2.2 Calculation of single phase "thermodynamic" properties

The properties concerned here are any of the energy properties (U, H, G, A, S) or volume V. Two thermodynamic frameworks have been developed over time, whose philosophy is to calculate the properties as a sum using some kind of ideal behaviour and a deviation from ideality. The deviation is then calculated using the fundamental equations that can be derived mathematically from tables 2.11 and 2.12.

The first framework, called the "residual approach", uses the ideal gas as ideal behaviour. The deviations from this behaviour are called *residual properties*. They are calculated using equations of state. This approach allows the computation of properties in both liquid or vapour phases, for any composition. It is nowadays increasingly used, and will be presented first.

The second framework is based on the definition of an ideal mixture. Hence, this framework proposes a way to find mixture properties assuming that the pure component properties are known. These pure component properties can be calculated either using correlations (as those presented in section 3.1, p. 102) or using the residual approach. The mixture properties are then calculated using the ideal mixture and the so-called *excess properties*. Although this approach is mainly used for liquid phases, it can also be used for solid phases. It is very powerful for strongly non-ideal mixtures.

2.2.2.1 Residual approach (equations of state)

Any thermodynamic property X can be calculated using:

$$\begin{aligned} X(P,T) = X^{\#}(P_0,T_0) + \left[X^{\#}(P,T_0) - X^{\#}(P_0,T_0) \right] \\ + \left[X^{\#}(P,T) - X^{\#}(P,T_0) \right] + \left[X(P,T) - X^{\#}(P,T) \right] \end{aligned} \qquad (2.55)$$

where the # indicates the ideal gas property. The term $X^{\#}(P_0,T_0)$ is the reference state that must be defined in order to calculate any energy property. Note that a thermodynamic state is fully defined when pressure (here P_0), temperature (here T_0), composition (here the same as the mixture composition for which the property should be calculated) and physical state (here ideal gas) are given.

The terms in the first two brackets are fully determined knowing the ideal gas behaviour (section A below), the last bracketed term is defined as the residual property (section B below). Table 2.14 explicits the above equation for all properties.

A. The ideal gas

The ideal gas [1] was first described in independent studies of Boyle (1660 [16]) and Mariotte (1679 [17], works compiled in 1717 in Leiden) who observed $Pv|_T = k$. Other works of Charles (unpublished, but cited by Gay-Lussac about 1787) observed the relation $T/v|_P = k$ for many different light gases. Finally the last historical relation $P/T|_v = k$ was introduced by Gay-Lussac in 1802 [18]. The symbol # is further used to indicate ideal gas properties.

The ideal gas equation of state, as written by Clapeyron (1834 [19]), is well known:

$$PV^\# = NRT \tag{2.56}$$

The ideal gas is an extremely important concept in thermodynamic calculations, as all residual models (*i.e.* equations of state) calculate real fluid properties as a departure from the ideal gas behaviour. Following observations must be made regarding equation (2.56):

- It is expressed using the mole numbers N. If another basis had been used (mass, for example), the value of R would depend on the molar mass. This is one of the two reasons why thermodynamic calculations always require molar properties: all deviations to this law will have to be expressed in terms of molar properties.
- It introduces a universal constant, R, called the universal gas constant. Its value is provided by NIST in SI units as: $R = 8.314462$ J mol^{-1} K^{-1}.
- Temperature T is an absolute temperature: the unit is K, not °C.

To a first approximation, low pressure vapours (below 500 kPa) can generally be described as ideal gases.

The use of the fundamental relationships presented in tables 2.11 and 2.12 using equation (2.56) yields:

$$dU^\# = C_v^\# dT \tag{2.57}$$

$$dH^\# = C_p^\# dT \tag{2.58}$$

$$dS^\# = C_v^\# \frac{dT}{T} + NR \frac{dV}{V} = C_p^\# \frac{dT}{T} - NR \frac{dP}{P} \tag{2.59}$$

$$dA^\# = -\frac{NRT}{V} dV - S^\# dT \tag{2.60}$$

---

1. The term *gas* was coined by Jan Baptista van Helmont [1579-1644] to describe matter which was invisible but with mass, formerly called "spirit". Its use in the general literature is sometimes ambiguous, as it may refer to a physical state (here called vapour) or a component (generally supercritical). In this work, it is only used for referring to the ideal gas.

$$dG^{\#} = \frac{NRT}{P}dP - S^{\#}dT \tag{2.61}$$

$$C_p^{\#} - C_V^{\#} = NR \tag{2.62}$$

Any of the energy properties of the ideal gas can be calculated using equations (2.57) to (2.62) provided that a reference state is defined and that the ideal gas isobaric heat capacity $C_P^{\#}$ is known as a function of temperature. It is important to note that $C_P^{\#}$ and $C_V^{\#}$ depend only on temperature and not on pressure. Consequently, the ideal gas internal energy or enthalpy are also independent of pressure.

Numerical values for $C_P^{\#}(T)$ (and therefore $C_V^{\#}(T)$, using (2.62)), are important in order to perform thermodynamic calculations. Section 3.1.1.2.E, p. 116 discusses this topic.

Example 2.5 Entropy rise in a ideal gas expansion

Air, available at 0.5 MPa and 300 K, is expanded continuously in an adiabatic process through a valve. The outlet pressure is atmospheric. What is the state of the gas at the outlet and what is the rate of entropy increase if flow is 1 mol/s.

Analysis:

No work is done during expansion, nor is heat exchanged (lack of area for heat transfer). Hence, according to the first principle this is an isenthalpic expansion ($h_{in}(T_{in},P_{in}) = h_{out}(T_{out},P_{out})$). The properties to be calculated are h (*PH* calculation) and s (explicitly asked for) at the inlet and the outlet of the valve (the differences are looked for, which means that no reference state is required). The exit state must be defined using the first law of thermodynamics.

Air can be considered to behave as an ideal gas (pressure below 500 kPa)

Solution:

At the inlet, T and P are known, so enthalpy is calculated. At the outlet, P and h are known, so temperature is obtained.

$$h_{out} - h_{in} = \int_{T_{in}}^{T_{out}} C_p^{\#}dT = 0 \Rightarrow T_{in} = T_{out}$$

With an ideal gas, there is no change in temperature during expansion (the Joule Thomson coefficient is zero). In their experiment of 1856, P. Joule and W. Thomson observed a small decrease in temperature using air, indicating clearly that it is not exactly an ideal gas.

For the change of entropy, the same reasoning applies:

$$s_{out} - s_{in} = \int_{T_{in}}^{T_{out}} \frac{C_p^{\#}}{T}dT - R\int_{P_{in}}^{P_{out}} \frac{dP}{P} = 0 - R\ln\left(\frac{P_{out}}{P_{in}}\right) = -R\ln\left(\frac{101.325}{500}\right) = 13.27\,\text{J mol}^{-1}\text{K}^{-1}$$

Hence, the rate of increase of entropy will be:

$$\dot{S}_{out} - \dot{S}_{in} = \dot{N}\left(s_{out} - s_{in}\right) = 13.27\,\text{W K}^{-1}$$

This example is discussed on the website:
http://books.ifpenergiesnouvelles.fr/ebooks/thermodynamics

B. The residual properties

Residual properties are defined as the difference between the true fluid property and the fluid property in a virtual ideal gas state. The calculation of these properties is based on the assumption that when pressure is lowered to zero, the true fluid behaviour converges to the ideal gas behaviour. Furthermore, the Maxwell relations are used in order to obtain the derivative properties. For any property X, except volume, the residual property is therefore calculated as:

$$X^{res}(T,P) = X(T,P) - X^{\#}(T,P) = \int_0^P \left(\frac{\partial X}{\partial P}\bigg|_T - \frac{\partial X^{\#}}{\partial P}\bigg|_T \right) dP$$

$$= \int_\infty^V \left(\frac{\partial X}{\partial V}\bigg|_T - \frac{\partial X^{\#}}{\partial V}\bigg|_T \right) dV + \int_V^{V^{\#}} \left(\frac{\partial X^{\#}}{\partial V}\bigg|_T \right) dV$$

(2.63)

The expression of volume is:

$$V^{res}(T,P) = V(T,P) - \frac{NRT}{P}$$

(2.64)

Using the fundamental equations (tables 2.11 and 2.12), following relations can then be found (table 2.13):

Table 2.13 Molar residual properties as a function of temperature and pressure

$$u^{res}(T,P) = \int_0^P \left(v - T \frac{\partial v}{\partial T}\bigg|_P \right) dP + RT - Pv = \int_\infty^v \left(T \frac{\partial P}{\partial T}\bigg|_v - P \right) dv$$ (2.65)

$$h^{res}(T,P) = \int_0^P \left(v - T \frac{\partial v}{\partial T}\bigg|_P \right) dP = \int_\infty^v \left(T \frac{\partial P}{\partial T}\bigg|_v - P \right) dv + Pv - RT$$ (2.66)

$$a^{res}(T,P) = \int_0^P \left(v - \frac{RT}{P} \right) dP + RT - Pv = \int_\infty^v \left(-P + \frac{RT}{v} \right) dv - RT \ln \frac{Pv}{RT}$$ (2.67)

$$g^{res}(T,P) = \int_0^P \left(v - \frac{RT}{P} \right) dP = \int_\infty^v \left(-P + \frac{RT}{v} \right) dv - RT \ln \frac{Pv}{RT} + Pv - RT$$ (2.68)

$$s^{res}(T,P) = \int_0^P \left(-\frac{\partial v}{\partial T}\bigg|_P + \frac{R}{P} \right) dP = \int_\infty^v \left(\frac{\partial P}{\partial T}\bigg|_v - \frac{R}{v} \right) dv + R \ln \frac{Pv}{RT}$$ (2.69)

Note that only the relationship between pressure, volume and temperature is needed to calculate all the above equations. Equations of state (EoS) provide such relationships. This is why an equation of state can be used to calculate any or all of the residual energy properties. It provides a coherent framework for all these properties. Nevertheless, the precision of the results may not be the same for all of them. They are discussed in section 3.4.3 (p. 189).

Also note that because of the importance of the chemical potential in thermodynamic calculations, the residual partial molar Gibbs energy is also expressed using the fugacity coefficient φ_i and defined as (equation (2.54) where the thermodynamic state β is the ideal gas):

$$\mu_i(T,P,N) - \mu_i^\#(T,P,N) = RT \ln \varphi_i = RT \ln \frac{f_i}{Pz_i} \tag{2.70}$$

where z_i is the mole fraction of the mixture considered (single phase). Using the residual approach, the fugacity coefficient can be calculated from a pressure-volume-temperature relationship, using either (for volume-explicit equations):

$$RT \ln \varphi_i = \int_0^P \left(\bar{v}_i - \frac{RT}{P} \right) dP \tag{2.71}$$

where $\bar{v}_i$ is the partial molar volume as defined by equation 2.50, or (for pressure-explicit equations):

$$RT \ln \varphi_i = \int_\infty^V \left(-\frac{\partial P}{\partial N_i}\bigg|_{T,V,N_j} + \frac{RT}{V} \right) dV - RT \ln \left(\frac{PV}{NRT} \right) \tag{2.72}$$

We can appreciate that the knowledge of a mathematical expression for an EoS is sufficient not only to fully describe all the residual properties (table 2.13), but also the chemical potential and the fugacity (equations 2.70 through 2.72).

Table 2.14 illustrates how the total phase properties are computed using the residual approach (equation 2.55).

Table 2.14   Total properties from a residual approach

		Real mixture		
		Ideal mixture		
	Ideal gas state (P_0, T_0)	Ideal gas state change to (P, T_0)	Ideal gas state change to (P, T)	Residual property at (P, T)
$U =$ $U^\#(T_0, P_0, N)$ +	0	+	$\int_{T_0}^T C_V^\#(T,N)dT$	+ $U^{res}(T,P,N)$ (2.73)
$H =$ $H^\#(T_0, P_0, N)$ +	0	+	$\int_{T_0}^T C_P^\#(T,N)dT$	+ $H^{res}(T,P,N)$ (2.74)
$A =$ $A^\#(T_0, P_0, N)$ +	$NRT_0 \ln\left(\dfrac{P}{P_0}\right)$	$- \int_{T_0}^T \left(S^\#(T,P,N)+NR\right)dT$	+ $A^{res}(T,P,N)$ (2.75)	
$G =$ $G^\#(T_0, P_0, N)$ +	$NRT_0 \ln\left(\dfrac{P}{P_0}\right)$	$- \int_{T_0}^T S^\#(T,P,N)dT$	+ $G^{res}(T,P,N)$ (2.76)	
$S =$ $S^\#(T_0, P_0, N)$ -	$NR \ln\left(\dfrac{P}{P_0}\right)$	$- \int_{T_0}^T \dfrac{C_P^\#(T,N)}{T}dT$	+ $S^{res}(T,P,N)$ (2.77)	

C. The corresponding states principle

From the beginning of the investigation of equations of state, the notion of corresponding states has been proposed. This is the case of Van der Waals (1873, 1910 [20, 21]), Clausius (1881 [22]) and Berthelot (1903 [23]). The most general expression of the two parameters corresponding states principle indicates that it is possible to find reduced (dimensionless) thermodynamic variables such that all reduced residual properties behave identically: a universal function can be found to describe all such properties.

In practice, the most conventional reduction parameters are critical pressure and critical temperature. This is why the principle is stated here as:

Any dimensionless residual property can be expressed using a universal relationship that uses only reduced temperature $T_r = T/T_c$ and reduced pressure $P_r = P/P_c$.

In other words, the reduced (*i.e.* dimensionless) residual property of any component at given T_r and P_r is identical to that of any other component (in particular for a simple fluid with exponent ^{0}) at the same values of T_r and P_r:

$$X^{res}\left(T_r, P_r\right) = X^{res,0}\left(T_r, P_r\right) \qquad (2.78)$$

where X^{res} may equal

- the reduced residual internal energy $u^{res}(T_r, P_r)/RT_c$,
- the reduced residual enthalpy $h^{res}(T_r, P_r)/RT_c$,
- the reduced residual entropy $s^{res}(T_r, P_r)/R$,
- the reduced residual heat capacity $c_P^{res}(T_r, P_r)/R$.

The corresponding state principles is very often also used for the ***compressibility factor*** defined as:

$$Z\left(T_r, P_r\right) = \frac{Pv}{RT} \qquad (2.79)$$

which is the ratio (and not the difference) of the volume and an ideal gas volume and therefore not truly a residual propriety.

This principle is very interesting but not sufficiently accurate for most practical purposes (in particular calculation of vapour pressures). Pitzer *et al.* (1955 [24]) therefore defined the ***acentric factor*** ω, using the reduced vapour pressure at the reduced temperature of 0.7:

$$\omega \overset{\Delta}{=} - \log_{10}\left(\left. P_r^\sigma \right|_{T_r=0.7} \right) - 1 \qquad (2.80)$$

The new model (2.78) is now expressed as:

$$X^{res}\left(T_r, P_r\right) = X^{res,1}\left(T_r, P_r\right) + \left(\omega - \omega^{(1)}\right)\left[\frac{X^{res,2}\left(T_r, P_r\right) - X^{res,1}\left(T_r, P_r\right)}{\omega^{(2)} - \omega^{(1)}} \right] \qquad (2.81)$$

where exponents 1 and 2 refer to the properties of two reference fluids.

This principle is used by Lee and Kesler [25] in order to predict single phase thermodynamic properties (see section 3.4.3.3.C, p. 202) with methane as fluid 1 $(\omega^{(1)} \cong 0)$ and n octane as fluid 2. An example of the graph of the compressibility factor (figure 2.12) and non-dimensional residual enthalpy (figure 2.13) is shown below. It is calculated for methane, but is as a first approximation valid for any non-polar fluid.

As an extension of this principle, the (cubic) equation of state (see section 3.4.3.4, p. 204) parameters are often calculated using the critical coordinates T_c, P_c and the acentric factor ω. The same principle can be applied to other properties, as for example transport properties.

Although the corresponding states principle is best adapted to pure component properties, it is also applied to mixtures. In this case, mixing rules for the critical coordinates are needed. The resulting properties are called pseudo-critical in order to differentiate them from the true critical point of the mixture. They can be calculated using various empirical mixing rules, the most well-known of which are those of Kay [26]:

$$T_{cm} = \sum x_i T_{ci} \tag{2.82}$$

$$P_{cm} = \sum x_i P_{ci} \tag{2.83}$$

$$\omega_m = \sum x_i \omega_i \tag{2.84}$$

Chapter 3 will present some other such mixing rules to be used with the Lee-Kesler method (section 3.4.3.3.C, p. 202) or in order to lump components into single pseudo-components in section 3.1.2.3.E, p. 137.

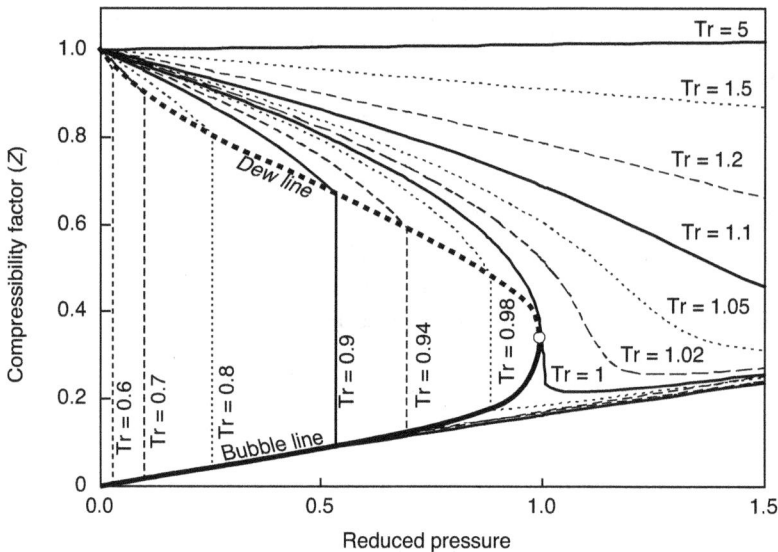

Figure 2.12

Compressibility factor of methane (simple fluid in corresponding states).

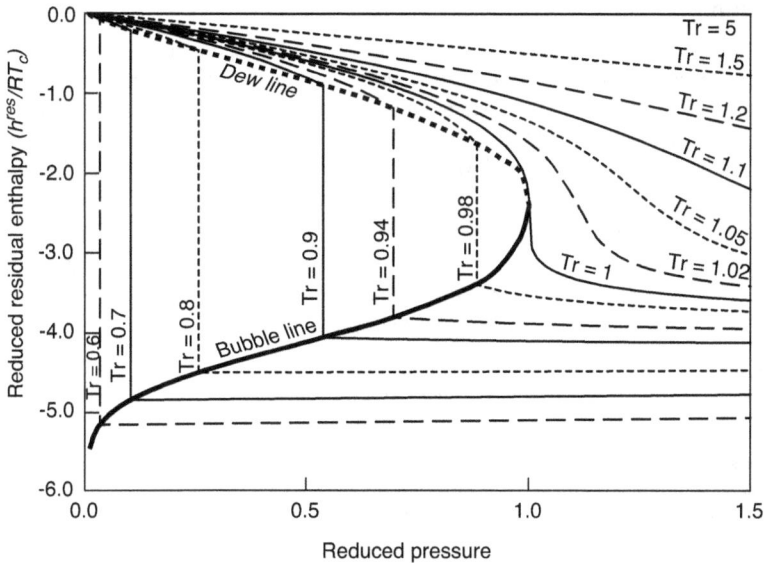

Figure 2.13

Residual enthalpy of methane (simple fluid in corresponding states).

It must be kept in mind that the quality of the results obtained using the corresponding states principle greatly depends on the similarity between the property of the fluids used as "template" and the mixture to be investigated. It is very often used for non-polar mixtures (petroleum-like mixtures), but should be used with caution for either high molecular weight material or polar components.

Example 2.6 Cryogenic plant

Methane must be condensed for intercontinental transport or when no pipeline is available. If 10 000 (metric) tons a day is to be transported, how much energy is required per year to accomplish this task? Pressure is atmospheric.

The data provided are the critical pressure (4600.1 kPa), the critical temperature (190.55 K), the normal boiling temperature (111.66 K) and the molar mass (16.04 kg kmol^{-1}). Use figure 2.13 for methane.

Analysis:

To condense methane from vapour to liquid, the heat of vapourisation (vapourisation enthalpy) has to be eliminated (supposing gas at saturation).

Pressure, temperature and enthalpy are required in this problem.

Methane is the only compound and is transported at its bubble point. The corresponding states principle can be applied, and the compressibility factor or residual enthalpy values can be read from figures 2.12 and 2.13.

Solution:

The normal boiling point is very low ($T_r = 0.59$). Working at higher pressure would make it possible to condense at a higher temperature, thus limiting the amount of cooling. If pressure is 1 MPa ($P_r = 0.22$) temperature of the liquid methane would be near $T_r = 0.8$, which corresponds to slightly over 150 K. Thermal losses against ambient temperature are very high and insulation is the only way to prevent excessive evaporation from the tanker (up to 0.9 m thick).

The energy required to condense methane can be read (in its non-dimensional form) from figure 2.13. The 0.6 isotherm is so low that it is barely visible on the graph. At this level, the width of the two-phase zone (difference of height between the dew line on top and the bubble line at the bottom) measures 4.8. The following value is obtained:

$$\frac{\Delta H}{RT_c} = \left(\frac{H-H^\#}{RT_c}\right)_L - \left(\frac{H-H^\#}{RT_c}\right)_V = -4.8 \Rightarrow \Delta H^{cond} = -4.8 \times R \times 190.55 = -7605 \text{ J mol}^{-1}$$

An experimental value of 8250 J/mol is reported in DIPPR (more accurate than the graphical reading). The molar flow rate of methane and the annual energy necessary for condensation will be:

$$\dot{N} = \frac{\dot{mt}}{M_w} = \frac{10000 \times 10^6 \times 365}{16.04} = 228 \times 10^9 \text{ mol/year} \Rightarrow \dot{Q} = \dot{N}\Delta H^{cond} = -1.7 \times 10^{12} \text{ kJ year}^{-1}$$

This value is equivalent to 55 MW. As liquefaction represents almost 40% of the cost of LNG production, we can see why the price of liquid methane is so high in comparison with gaseous production.

This example is discussed on the website:
http://books.ifpenergiesnouvelles.fr/ebooks/thermodynamics

2.2.2.2 The excess approach (activity coefficients)

This framework is generally used for the dense phases, as it is most adapted where molecular interactions are strong. This is the case in liquid or solid phases. It is less general than the equation of state approach because it implies that the pure component properties are known. These must be calculated through some other method (correlation or equation of state). The models proposed here focus on the effect of mixing. We first define the ideal mixture, and then show the properties used for describing departure from ideality.

A. The ideal mixture

The ideal mixture is defined using the fugacity (equation (2.53)), based on Lewis' relationship:

$$f_i(T,P,x) \triangleq f_i^*(T,P)x_i \tag{2.85}$$

Note that $x = (x_1,...,x_{\mathcal{N}})$ is the molar composition of the mixture.

The asterisk denotes the pure component property at the same pressure, temperature and physical state (generally liquid) as the mixture property (see section 2.2.1.4, p. 50). Note that in some cases, the pure component may not exist in the same physical state as the mixture. In that case, we deal with a virtual property that can be considered as a calculation intermediate, but has no physical meaning.

Using equation (2.85) along with the fundamental relationships, it is possible to demonstrate [3, 15] that the mixture properties for the ideal mixture behave as follows:

$$V^{id} = \sum N_i v_i^*$$ (2.86)

$$H^{id} = \sum N_i h_i^*$$ (2.87)

$$G^{id} = \sum N_i g_i^* + RT \sum N_i \ln x_i$$ (2.88)

$$A^{id} = \sum N_i a_i^* + RT \sum N_i \ln x_i$$ (2.89)

$$S^{id} = \sum N_i s_i^* - R \sum N_i \ln x_i$$ (2.90)

B. The excess properties

The excess properties are defined as the difference between the true mixture properties and those of the ideal mixture. Using Euler's theorem (2.51), they can be written as:

$$V^E = V - V^{id} = \sum N_i \left(\bar{v}_i - v_i^* \right)$$ (2.91)

$$H^E = H - H^{id} = \sum N_i \left(\bar{h}_i - h_i^* \right)$$ (2.92)

$$G^E = G - G^{id} = \sum N_i \left(\bar{g}_i - g_i^* \right) - RT \sum N_i \ln x_i$$ (2.93)

$$A^E = A - A^{id} = \sum N_i \left(\bar{a}_i - a_i^* \right) - RT \sum N_i \ln x_i$$ (2.94)

$$S^E = S - S^{id} = \sum N_i \left(\bar{s}_i - s_i^* \right) + R \sum N_i \ln x_i$$ (2.95)

Note that the excess volume and the excess enthalpy are equal to the volume of mixing or the enthalpy of mixing. In the case of the ideal mixture, the mixing properties are zero.

The excess property models are based on the excess Gibbs energy, which, using the definition of the fugacity (2.54) can also be written as ($\bar{g}_i$ is nothing but a chemical potential):

$$G^E = G - G^{id} = \sum N_i \left(\bar{g}_i - \bar{g}_i^{id} \right) = RT \sum N_i \ln \frac{f_i}{f_i^{id}}$$ (2.96)

If the activity coefficient is defined as

$$\gamma_i = \frac{f_i}{f_i^{id}} = \frac{f_i}{f_i^* x_i}$$ (2.97)

It is easy to see that the excess Gibbs energy can be calculated from:

$$G^E = RT \sum_i N_i \ln \gamma_i \tag{2.98}$$

Inversely, the activity coefficients can be calculated from the excess Gibbs energy using:

$$RT \ln \gamma_i = \left(\frac{\partial G^E}{\partial N_i} \right)_{T,P,N_{j \neq i}} \tag{2.99}$$

It is further possible to obtain all excess properties from an expression of the excess Gibbs energy. Using the fundamental equation (2.30):

$$\left. \frac{\partial G^E}{\partial P} \right|_T = V^E \tag{2.100}$$

The Gibbs-Helmholtz equation (2.36) provides the excess enthalpy:

$$H^E = \left. \frac{\partial \left(G^E / T \right)}{\partial \left(1/T \right)} \right|_P \quad \text{or} \quad \frac{-H^E}{T^2} = \left. \frac{\partial \left(G^E / T \right)}{\partial \left(T \right)} \right|_P \tag{2.101}$$

The excess entropy can then be computed using (2.23):

$$S^E = \frac{H^E - G^E}{T} \tag{2.102}$$

And similarly the excess internal energy:

$$U^E = G^E - PV^E + TS^E \tag{2.103}$$

In principle, the excess Gibbs energy is a function of temperature, pressure and the mixture composition. Yet, its dependence with pressure (2.100), is usually neglected, so that the expressions are given solely as a function of temperature and composition. A summary of the property calculations using the excess approach is given in table 2.15.

It is true that V^E is generally small and that consequently equation (2.100) can be assumed close to zero. Under high pressure, however, even a small value of V^E may have a significant impact on the numerical value of the Gibbs free energy. Using an excess Gibbs energy model above 1.5 MPa is therefore not recommended, unless the parameters have been fitted at high pressure.

Table 2.15   Total properties from an excess approach

		Real mixture		
		Ideal mixture		
		Pure component	**Mixing contribution**	**Excess property**
V	$=$	$\sum_{i=1}^{\mathcal{N}} N_i v_i^*$ $+$	0 $+$	$\sum_{i=1}^{\mathcal{N}} N_i \left(\bar{v}_i - v_i^* \right)$ (2.104)
U	$=$	$\sum_{i=1}^{\mathcal{N}} N_i u_i^*$ $+$	0 $+$	$\sum_{i=1}^{\mathcal{N}} N_i \left(\bar{u}_i - u_i^* \right)$ (2.105)
H	$=$	$\sum_{i=1}^{\mathcal{N}} N_i h_i^*$ $+$	0 $+$	$\sum_{i=1}^{\mathcal{N}} N_i \left(\bar{h}_i - h_i^* \right)$ (2.106)
A	$=$	$\sum_{i=1}^{\mathcal{N}} N_i a_i^*$ $+$	$RT\sum_{i=1}^{\mathcal{N}} N_i \ln\left(x_i \right)$ $+$	$\sum_{i=1}^{\mathcal{N}} N_i \left(\bar{a}_i - a_i^* \right)$ (2.107)
G	$=$	$\sum_{i=1}^{\mathcal{N}} N_i g_i^*$ $+$	$RT\sum_{i=1}^{\mathcal{N}} N_i \ln\left(x_i \right)$ $+$	$\sum_{i=1}^{\mathcal{N}} N_i \left(\bar{g}_i - g_i^* \right)$ (2.108)
S	$=$	$\sum_{i=1}^{\mathcal{N}} N_i s_i^*$ $-$	$R\sum_{i=1}^{\mathcal{N}} N_i \ln\left(x_i \right)$ $+$	$\sum_{i=1}^{\mathcal{N}} N_i \left(\bar{s}_i - s_i^* \right)$ (2.109)

2.2.3 Phase equilibrium

2.2.3.1 Some basic principles for phase calculations

A. The phase equilibrium condition

According to the thermodynamic principles, equilibrium is reached, considering the constraints on the system and at given pressure and temperature, when the Gibbs energy is lowest (see equation (2.19)). It can be shown that this minimum leads to the statement that the chemical potential is identical in all phases, for any component *i*. For two phases α and β, this is written as:

$$\mu_i^{\alpha} = \mu_i^{\beta} \qquad (2.110)$$

For a two-phase equilibrium, equation (2.110) provides as many relationships as components in the mixture ($\mathcal{N}$). For an equilibrium with φ phases, $\mathcal{N}(\phi-1)$ relationships can be written.

Using the definition of the fugacity (2.53), the same rule can be written as:

$$f_i^{\alpha} = f_i^{\beta} \qquad (2.111)$$

a. The vapour phase fugacity

The vapour phase fugacity is always expressed using the residual approach.

$$f_i^V = P y_i \varphi_i^V \tag{2.112}$$

where

P total pressure;

y_i vapour mole fraction of component i;

φ_i^V vapour phase fugacity coefficient of component i. This property is computed with an equation of state using (2.72). In the limit of low pressure (below 500 kPa), the ideal gas approximation can be used, which states: $\varphi_i^V = 1$.

b. The liquid phase fugacity

The fugacities in the liquid phase can be computed using either the residual or the excess approach (see also the FFF – Famous Fugacity Formulae – of O'Connell [4, 27]).

Using the **residual approach**, i.e. with an equation of state, the same expressions can be written as for a vapour phase:

$$f_i^L = P x_i \varphi_i^L \tag{2.113}$$

with

x_i liquid mole fraction of component i;

φ_i^L liquid phase fugacity coefficient of component i. This property is computed with an equation of state using (2.72).

Using the **excess approach**, the fugacity is calculated using the definition of the activity coefficient (2.97):

$$f_i^L = f_i^{L*}(T,P)\gamma_i x_i \tag{2.114}$$

The most general expression for the liquid phase fugacity is given by:

$$f_i^L = f_i^{L*}\gamma_i x_i = P_i^\sigma \varphi_i^\sigma \wp_i \gamma_i x_i \tag{2.115}$$

where

- $\wp_i = \exp\left(\dfrac{1}{RT}\int_{P_i^\sigma}^{P} v_i^L dP\right) \approx \exp\left(\dfrac{v_i^L\left(P - P_i^\sigma\right)}{RT}\right)$ is the Poynting correction that uses

  v_i^L, the liquid molar volume of component i at T and assumed to be independent of P;
- φ_i^σ is the fugacity coefficient of component i at saturation calculated using the vapour phase (equation (2.72);
- γ_i is the activity coefficient of component i in the liquid.

When the vapour pressure of component i, P_i^σ is lower than 500 kPa (i.e. most often, except for very light components), the pure component liquid fugacity can be approximated as:

$$f_i^{L*} = P_i^\sigma \tag{2.116}$$

c. Excess approach with the asymmetric convention:

The use of expression (2.114) assumes that component i exists as a pure liquid in the pressure and temperature conditions of the mixture (or at least that its properties can be calculated in these conditions). This may be a strong restriction (e.g. supercritical gases or ionic species). This is why the use of the excess approach is extended by defining a generalised reference state.

$$f_i^L = x_i f_i^{L,ref} \gamma_i^{(ref)} \qquad (2.117)$$

This reference state can be the pure component, in which case equation (2.114) is recovered, but it also can be user-defined (often at infinite dilution). The exponent *(ref)* is added to the activity coefficient in order to indicate that its value depends on the reference state chosen.

Figure 2.14a shows, for the example of a binary mixture, how the fugacity varies with composition. Each component has its own fugacity (in the same way as each component has its own chemical potential). In the limit of zero concentration of component *1*, its fugacity tends to zero, while the fugacity of the other component is f_2^* (and vice-versa). If the plot had been linear between these two end-points, the mixture would have been ideal according to Raoult's law ($\gamma_i = 1$ in equation (2.114) yielding equation (2.85)).

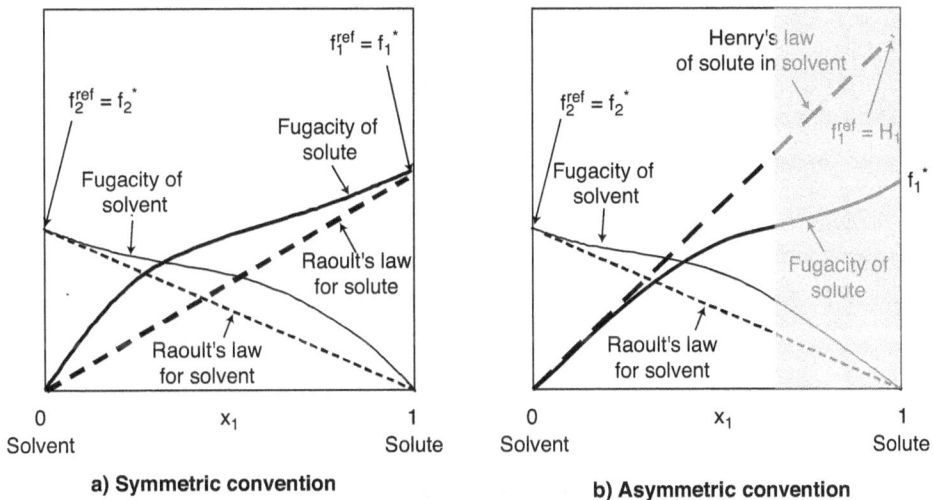

Figure 2.14

Fugacities for symmetric and asymmetric conventions, at fixed temperature and pressure, and variable composition for a binary mixture.

In case one of the components is supercritical (say component 1), the right side of the diagram becomes unphysical (figure 2.14b). Only the left side of the diagram has a physical significance: component 2 is a liquid, and is called the **solvent**. Component 1, that does not exist in the liquid state at the given temperature and pressure conditions, is called the **solute**. As a result, it is impossible to determine f_1^{L*} in equation (2.114). Equation (2.117) needs to be used, and a specific reference state needs to be defined for this solute. An alternative method, based on the slope at infinite dilution, may be used to calculate its fugacity: the **Henry constant**, defined as:

$$H_{i,s}^{\sigma}(T) \triangleq \lim_{\substack{x_i \to 0 \\ P \to P_s^{\sigma}}} \left(\frac{f_i^L(T,P)}{x_i} \right)$$

(2.118)

In other words, the reference state used for defining the component's properties is taken at infinite dilution (as shown in table 2.16). Note that the Henry constant is necessarily defined at the solvent's vapour pressure (this is where the solute concentration is zero) and is therefore not a function of pressure. In order to account for pressure, a Poynting correction must be considered. This is why the liquid fugacity for the solute is now calculated as:

$$f_i^L = H_{i,s}^{\sigma} \wp_i^{\infty} \gamma_i^H x_i$$

(2.119)

where, in addition to the Henry constant, the following factors are used:

- the dilute Poynting correction:

$$\wp_i^{\infty}(T,P) = \exp\frac{\overline{v}_i^{\infty}\left(P - P_s^{\sigma}\right)}{RT}$$

(2.120)

It requires the infinite dilution partial molar volume of component i in the solution ($\overline{v}_i^{\infty}$). This factor is close to one if the pressure is less than 2 MPa above the solvent vapour pressure (P_s^{σ}),

- the "asymmetric" activity coefficient, $\gamma_i^H(T,\mathbf{x})$, that is computed using the same type of models as the symmetric activity coefficient. It is related to the symmetric activity coefficient in such a way that at infinite dilution in the solvent ($x_i = 0$), the asymmetric activity coefficient becomes one:

$$\gamma_i^H(T,\mathbf{x}) = \frac{\gamma_i(T,\mathbf{x})}{\gamma_i^{\infty}(T, x_i = 0)}$$

(2.121)

where $\gamma_i^{\infty}(T, x_i = 0)$ is the activity coefficient of the solute at infinite dilution of component i in the solute given by the symmetric activity convention.

Expression (2.119) is only used for the solutes. For the solvents, equation (2.115) remains valid. This is why, in a solvent + solute mixture, this approach is called **asymmetric convention** (the definition of the reference state, as used in (2.117), is different depending on the type of component).

In fact, the use of the asymmetric convention can be extended to all cases where the solutes are in low concentration in a phase:

- gases dissolved in a liquid (they do not exist as pure liquids under the pressure and temperature conditions of the system) [28];
- all solutes (including liquids) in an aqueous phase (their properties are very different from those of the pure component) [29, 30];
- ions in an aqueous phase (they do not exist as ionic species in a pure state).

Table 2.16 Possible reference states of components

Convention	Component	Temperature	Pressure	Composition	Phase
Symmetric	Solute	T_{System}	P_{System}	Pure	Liquid
	Solvent	T_{System}	P_{System}	Pure	Liquid
Asymmetric	Solvent	T_{System}	P_{System}	Pure	Liquid
	Solute	T_{System}	$P_{Solvent}^{\sigma}$	Infinite dilution in the solvent	Liquid

It is important to note that as a result of its definition (2.118), the Henry constant is not a pure component property, but rather a binary, as it depends both on the solute (i) and on the solvent (in which i is diluted). This is why in mixed solvents, an exact definition as given by (2.118) becomes quite difficult, and a mixing rule is required. Carroll [31] suggests the use of

$$\ln H_{i,m} = \overset{solvents}{\underset{j}{\sum}} x_j \ln H_{i,j} + \overset{solvents}{\underset{j}{\sum}} \overset{solvents}{\underset{k \neq j}{\sum}} a_{jk} x_j x_k \qquad (2.122)$$

where $H_{i,j}$ is the Henry constant of solute i in solvent j. The a_{jk} parameters must be fitted on experimental values. It becomes very simple when $a_{jk} = 0$.

A useful summary of the use and mis-use of Henry's law is given by Carroll [32]. Some additional comments on this issue are provided in section 3.4.3 (p. 189).

d. The solid phase fugacity

Calculating this fugacity requires a specific approach which depends on the phase being looked for. The most general is the excess approach:

$$f_i^S = x_i^S f_i^{S*} \gamma_i^S \qquad (2.123)$$

In cases where the solid phase is a pure component, no activity coefficient γ_i^S is required. However, the solid phase is sometimes a mixture, in which case this coefficient must be taken into account. The models for the solid phase activity coefficient are generally identical to that for the liquid phase (Coutinho [33]).

The pure component solid fugacities are calculated using the fact that at their crystallisation temperature, the phase equilibrium condition holds (2.111):

$$f_i^{L*}\left(T_{F,i}, P_{F,i}\right) = f_i^{S*}\left(T_{F,i}, P_{F,i}\right) \qquad (2.124)$$

where the subscript F stands for the crystallisation or fusion conditions. In order to calculate $f_i^{L*}(T,P)$ at the system pressure and temperature, the fundamental equations (2.30) and (2.36) are used. The derivation can be found in numerous thermodynamics handbooks (e.g. Vidal [34]). The resulting expression is:

$$\ln\left(\frac{f_i^{S*}(T,P)}{f_i^{L*}\left(T_{F,i}, P_{F,i}\right)}\right) = \frac{\Delta h_{F,i}}{RT_{F,i}}\left(1 - \frac{T_{F,i}}{T}\right) + \frac{\Delta c_{P,F,i}}{R}\left[\frac{T_{F,i}}{T} - 1 - \ln\left(\frac{T_{F,i}}{T}\right)\right] + \frac{\Delta v_{F,i}\left(P - P_{F,i}\right)}{RT_{F,i}} \qquad (2.125)$$

where $\Delta h_{F,i} = h_i^L - h_i^S$ is the enthalpy of fusion of component i (taken at $T_{F,i}$ and $P_{F,i}$);

$\Delta c_{P,F,i} = c_{P,i}^L - c_{P,i}^S$ is the difference in molar isobaric heat capacity between the liquid and the solid phase.

$\Delta v_{F,i} = v_i^L - v_i^S$ is the molar volume difference upon fusion, taken at $T_{F,i}$ and $P_{F,i}$. Once again, this property is considered constant with respect to pressure and temperature. This volume is generally small and has little influence on the final result when $P\text{-}P_{F,i}$ is not too large. Often, it is considered independent of temperature.

B. The distribution or partition coefficient

In practice, **vapour-liquid** phase equilibria are often calculated using the so-called *distribution coefficient*, "equilibrium coefficient" or "equilibrium ratio", which describes, for each component i, the ratio of molar fraction in the vapour phase and in the liquid phase:

$$K_i \triangleq \frac{y_i}{x_i} \qquad (2.126)$$

According to the phase equilibrium relationship (2.111), the distribution coefficient can be computed using either a similar (residual) approach in both phases, or a different (residual and excess) approach in each phase. The former case is called a homogeneous method (or phi-phi), the latter a heterogeneous method (or gamma-phi). This is summarised in table 2.17.

Table 2.17 Nomenclature of the thermodynamic methods

Approach	Vapour fugacity calculation	Liquid fugacity calculation
homogeneous ($\varphi - \varphi$)		Residual approach
heterogeneous, symmetric ($\gamma - \varphi$)	Residual approach	Excess approach, symmetric convention
heterogeneous, asymmetric ($\gamma - \varphi$)		Excess approach, asymmetric convention

Using equations (2.112) and (2.113), the *homogeneous* method results in:

$$K_i = \frac{\varphi_i^L\left(T,P,\mathbf{x}\right)}{\varphi_i^V\left(T,P,\mathbf{y}\right)} \qquad (2.127)$$

In the *heterogeneous* approach, equation (2.112) is combined with (2.115) (symmetric convention), to yield:

$$K_i = \frac{P_i^\sigma\left(T\right)\wp_i\left(T,P\right)\,\varphi_i^\sigma\left(T,P_i^\sigma\right)\gamma_i\left(T,\mathbf{x}\right)}{P} \frac{}{\varphi_i^V\left(T,P,\mathbf{y}\right)} \qquad (2.128)$$

In cases where the pure component i has no vapour pressure, or when its pure component properties are difficult to find, the **asymmetric convention** is used for the solutes (equation 2.119) , to yield:

$$K_i = \frac{H_{i,s}^\sigma(T)\wp_i^\infty(T,P)}{P}\frac{\gamma_i^H(T,\mathbf{x})}{\varphi_i^V(T,P,\mathbf{y})} \qquad (2.129)$$

It is worth noting that, depending on the mixture and pressure conditions, the heterogeneous approach allows a number of simplifications that are of great use for understanding phase behaviour and identifying trends:

- When the process **pressure is low** (below 0.5 MPa), all fugacity coefficients may be considered equal to one. In addition, the Poynting corrections may then also be neglected. This considerably simplifies equations (2.128) and (2.129). They become:

$$K_i = \frac{P_i^\sigma(T)}{P}\gamma_i(T,\mathbf{x}) \qquad (2.130)$$

and

$$K_i = \frac{H_{i,s}^\sigma(T)}{P}\gamma_i^H(T,\mathbf{x}) \qquad (2.131)$$

Equation (2.130) is very often used for low pressure non-ideal mixture, and its effect on the phase diagram will be discussed in section 3.4.1 (p. 160).

Equation (2.131) is generally further simplified assuming that the solute is very diluted, and as such that $\gamma_i^H(T,\mathbf{x})$ is very close to unity:

$$K_i = \frac{H_{i,s}^\sigma(T)}{P} \qquad (2.132)$$

This last equation is sometimes known as the **simplified Henry's law** [32].

- When the liquid phase forms **an ideal mixture**, then the activity coefficient in equation (2.130) becomes unity, and Raoult's law is found:

$$K_i = \frac{P_i^\sigma(T)}{P} \qquad (2.133)$$

Using (2.126), this law is also written as:

$$x_i P_i^\sigma(T) = y_i P \qquad (2.134)$$

where the right hand side of the equation is sometimes called the partial pressure of component i. Since the sum of the partial pressures equals the total pressure, (2.134) leads to:

$$P = \sum_i x_i P_i^\sigma(T) \qquad (2.135)$$

Equation (2.135) clearly shows the consequence of Raoult's law on the vapour-liquid equilibrium on an isothermal *Pxy* plot (as shown in figure 2.5 in section 2.1.3.2, p. 30): the bubble pressure is a straight line. This is in fact what is observed for mixtures of like components (e.g. alkanes of similar size). It is important to note that Raoult's law implies that the composition be given in mole fraction, which is the second reason why molar fractions, rather than weight fractions are used in thermodynamic calculations.

For *liquid-liquid equilibria* the distribution coefficient required is defined by:

$$K_i^{\alpha\beta} \triangleq \frac{x_i^{\alpha}}{x_i^{\beta}} \qquad (2.136)$$

If the excess approach is used, we obtain from equation (2.115) the expression:

$$K_i^{\alpha\beta} = \frac{f_i^{L*} \gamma_i^{\beta}}{f_i^{L*} \gamma_i^{\alpha}} = \frac{\gamma_i^{\beta}\left(\mathbf{x}^{\beta}\right)}{\gamma_i^{\alpha}\left(\mathbf{x}^{\alpha}\right)} \qquad (2.137)$$

The pure component liquid fugacity (the vapour pressure) has no effect on liquid-liquid phase split. The activity coefficient model is often identical in both phases. The only difference between nominator and denominator in equation (2.137) is due to the composition difference between the two phases.

It may be of interest to note that the **heterogeneous approach** (different model in different phases) can also be used for liquid-liquid phase split (in particular when the phases are very different, as for example an aqueous and an organic phase). The asymmetric convention can then be employed, using a Henry constant for calculating the fugacity of dilute components (e.g. hydrocarbons in the aqueous phase). In this case, equation (2.137) becomes, in the same way as (2.132) (where a specific activity coefficient model is used in each phase):

$$K_i^{\alpha\beta} = \frac{H_{i,s}^{\sigma}\left(T\right)}{P_i^{\sigma}\left(T\right)} \frac{\gamma_i^{H,\beta}\left(T,\mathbf{x}^{\beta}\right)}{\gamma_i^{\alpha}\left(T,\mathbf{x}^{\alpha}\right)} \qquad (2.138)$$

The ratio of activity coefficients is in this case often neglected, as the concentration in the dilute phase (β) is very small, and the other phase (α) is considered almost ideal.

C. Flash calculation (the set of equations and unknowns)

The discussion proposed below focuses on the case of two-phase vapour-liquid equilibria. The same principles are also valid, however, for liquid-liquid, liquid-solid or vapour-solid equilibrium calculations. The equilibrium coefficient must then be defined as the ratio of molar compositions of the two phases present. When more than two phases are present, the number of equations and number of unknowns increases but the basic principles remain the same. More details concerning the algorithmic implementation of the calculations are available elsewhere (Rachford and Rice [35], Michelsen [36, 37]).

In any of the equations (2.127), (2.128) or (2.129), clearly the liquid and vapour compositions ($\mathbf{x}$ and $\mathbf{y}$) must be known before the distribution coefficient can be calculated. But since the objective of the phase equilibrium calculation is precisely to calculate these compositions, it is clear that an iterative algorithm should be used.

Duhem's phase rule (section 2.1.4, p. 41) indicates that two state variables are sufficient to calculate the composition and properties of all phases present, provided that the feed composition is known. In the remainder of this document, the feed compositional vector is written as $\mathbf{z}$ in order to differentiate it from the liquid ($\mathbf{x}$) and the vapour ($\mathbf{y}$) compositions. Consequently, we will call the phase equilibrium calculation (flash) depending on the type of the two state variables given (see also Table 2.7).

a. PT, Tθ or Pθ flash

Any phase property can be calculated knowing its composition, pressure and temperature. The vapour fraction (ratio of mole number in the vapour phase with respect to the total mole number, ($\theta = N^V/N$) must also be known in order to evaluate the material balance. Hence, the basic equations are identical for *PT, Tθ* or *Pθ* flash calculations:

- The iso-fugacity condition (2.111):

$$f_i^L = f_i^V \tag{2.139}$$

which, using one of equations (2.127), (2.128) or (2.129) for calculating K_i, can be written as:

$$x_i K_i = y_i \tag{2.140}$$

Note that there are two phases in equilibrium, resulting in as many equations as components in the mixture ($\mathcal{N}$). If there had been three phases, the iso-fugacity condition would have given rise to 2 $\mathcal{N}$ equations, and so on ($\mathcal{N}$ more equations for each additional phase).

- The mass balance equations:

$$\dot{F}z_i = \dot{L}x_i + \dot{V}y_i \Rightarrow z_i = (1-\theta)x_i + \theta y_i \tag{2.141}$$

where $\dot{F}$, $\dot{L}$ and $\dot{V}$ are respectively the molar feed flow, the liquid flow and the vapour flow.

If more phases had been present, there would have been more terms on the right hand side of the equation, but the number of equations would be the same.

As unknowns, we have the composition vector of the phases (liquid *x*, and vapour *y*), in addition to the vapour fraction θ (in the case of a *PT* flash). This makes $2\mathcal{N}+1$ unknowns in the case of a two-phase flash. An additional equation is required to solve the problem. It is found by the simple consideration that the sum of all molar fractions must be one. This sum can be applied to the liquid ($\sum_{i=1}^{\mathcal{N}} x_i = 1$), or to the vapour ($\sum_{i=1}^{\mathcal{N}} y_i = 1$). In order to keep the equation general, it is replaced by:

$$\sum_{i=1}^{\mathcal{N}} x_i - \sum_{i=1}^{\mathcal{N}} y_i = 0 \tag{2.142}$$

Rachford and Rice (1952) [35] propose to substitute (2.140) into (2.141) resulting in:

$$\begin{cases} x_i = \dfrac{z_i}{1+\theta(K_i-1)} \\[4mm] y_i = \dfrac{K_i z_i}{1+\theta(K_i-1)} \end{cases} \tag{2.143}$$

These values of x_i and y_i can now be substituted in (2.142), which yields:

$$\sum_{i=1}^{\mathcal{N}} \frac{(K_i-1)z_i}{1+\theta(K_i-1)} = 0 \tag{2.144}$$

which is known as the Rachford-Rice equation very often used inside computer algorithms.

Summing up, the unknowns in a **PT flash** are K_i and θ (*i.e.* $\mathcal{N}+1$ unknowns), and the equations to be solved are:

$$\begin{cases} K_i = \dfrac{\varphi_i^L}{\varphi_i^V} \\ \displaystyle\sum_{i=1}^{\mathcal{N}} \dfrac{(K_i - 1)z_i}{1 + \theta(K_i - 1)} = 0 \end{cases} \tag{2.145}$$

One possible procedure to solve the equations (known as successive substitution) is as follows:

1. Estimate the missing piece of information, θ
2. Estimate the distribution coefficients K_i
3. Use (2.143) to calculate x_i and y_i
4. Improve the evaluation of K_i using equations (2.127), (2.128) or (2.129)
5. Evaluate a better θ from (2.144) [a]
6. If θ is different from its previous value, return to 3, otherwise the answer is reached.

[a] A Newton-Raphson method is suitable for this equation [38], [39].

Similarly, a **Pθ flash** (bubble or dew temperature calculation for example), the unknowns are K_i and T (*i.e.* $\mathcal{N}+1$ unknowns), and the equations to be solved are again:

$$\begin{cases} K_i = \dfrac{\varphi_i^L}{\varphi_i^V} \\ \displaystyle\sum_{i=1}^{\mathcal{N}} \dfrac{(K_i - 1)z_i}{1 + \theta(K_i - 1)} = 0 \end{cases} \tag{2.146}$$

Note that, for bubble point calculations, $\theta = 0$, and $z_i = x_i$; so (2.144) becomes:

$$\sum_{i=1}^{\mathcal{N}} K_i x_i = 1 \tag{2.147}$$

Similarly, for dew point calculations, $\theta = 1$, and $z_i = y_i$; so (2.144) becomes:

$$\sum_{i=1}^{\mathcal{N}} (y_i / K_i) = 1 \tag{2.148}$$

One possible procedure to solve the equations is as follows:

1. Estimate the missing piece of information, T
2. Estimate the distribution coefficients K_i
3. Use (2.143) to calculate the unknown phase composition
4. Solve for the temperature using simultaneously equations for K_i calculations (for example (2.127), (2.128) or (2.129)) and (2.147) or (2.148).

A $T\theta$ flash is conceptually identical to the $P\theta$ flash, except that the unknown is the pressure P instead of the temperature T.

Example 2.7 Distillation column

A mixture of light hydrocarbons is processed in a distillation column (see figure 2.15). Compositions of distillate obtained from the total condenser and of the residue at the bottom of the column are as follows (see table 2.18):

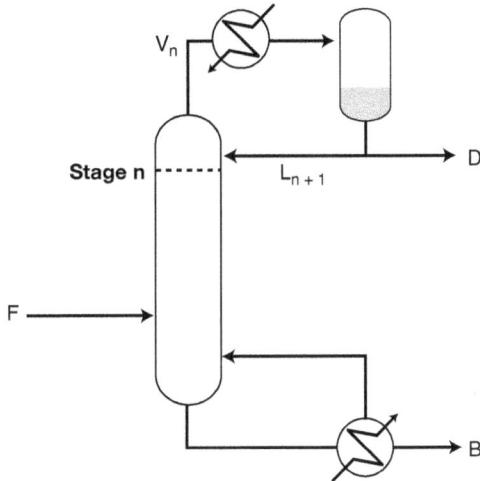

Figure 2.15

Sketch of a distillation column.

Table 2.18 Distillate and residue composition for example 2.7

Component	Distillate	Residue
Propane	0.23	
iso-Butane	0.67	0.02
n-Butane	0.10	0.46
iso-Pentane		0.15
n-Pentane		0.37

a. What is the column pressure to obtain the specified distillate if temperature in the condenser drum is 120 °F? (The pressure drop in column, exchanger and drum will be neglected).

b. What is the temperature in the upper stage and the composition of the liquid pouring off this plate (stage n on the figure)?

c. What is the temperature in the reboiler?

Use the Scheibel and Jenny diagram (figure 2.16) to evaluate the distribution coefficients.

Analysis:

a. In the condenser, the liquid composition is known. Since a liquid in equilibrium with a vapour is at its bubble point, the pressure is the bubble pressure. The pressure is a constant in the column.

b. The overall mass balance around the condenser indicates that all flows have the same composition. Hence, the vapour leaving the top plate of the column has the distillate composition. In addition, this vapour is saturated and is therefore at its dew point. As pressure is known (as calculated in a.), temperature has to be computed (dew temperature) and liquid composition will be part of the answer.

c. The residue from the reboiler is also a liquid phase. Pressure is unchanged, so bubble temperature must be calculated.

   Components are all light hydrocarbons and we are told to use the Scheibel and Jenny procedure. This procedure assumes an ideal mixture (equilibrium coefficients are independent of composition) and is more suitable than Raoult's law at expected working pressure (10 bar).

   All calculations are relative to vapour-liquid equilibrium and none of the hydrocarbons in the mixture is at supercritical conditions.

Solution:

a. A bubble pressure is calculated using (2.147), $\sum_{i=1}^{\mathcal{N}} K_i x_i = 1$. In order to be able to read the distribution coefficients K_i in the Scheibel & Jenny diagram, both pressure and temperature are needed. A first pressure must be guessed, so distribution coefficients can be read on the nomograph. Raoult's law can be used as a first approximation:

$$P = \sum_{i=1}^{\mathcal{N}} P_i^\sigma x_i$$. The estimated pressure is 8.5 atm.

Now a line must be drawn between the points T = 120 °F and P = 8.5 atm. The three K_i values are read on the figure. On the first attempt, the sum is not equal to 1 (see table 2.19). A different guess must be made. A basic trick consists in multiplying the old pressure by the sum of the vapour phase compositions ($P^{(n)} = P^{(n-1)} \sum_{i=1}^{\mathcal{N}} K_i \, x_i$). Due to the poor accuracy of the graphical method, the solution can be considered to be reached in the second iteration. The bubble point of the mixture is 8.4 atm.

This procedure is summarised in table 2.19.

Table 2.19   Bubble pressure calculation procedure at 120 °F

Component	x_i	P_i^σ	$K_i(8.5\,\text{atm})$	$y_i = K_i x_i$	$K_i(8.4\,\text{atm})$	$y_i = K_i x_i$
Propane	0.23	16.5	1.68	0.387	1.7	0.39
iso-Butane	0.67	6.3	0.8	0.543	0.82	0.55
n-Butane	0.10	4.6	0.6	0.062	0.62	0.06
Sum/Result		8.5		0.991		1.00

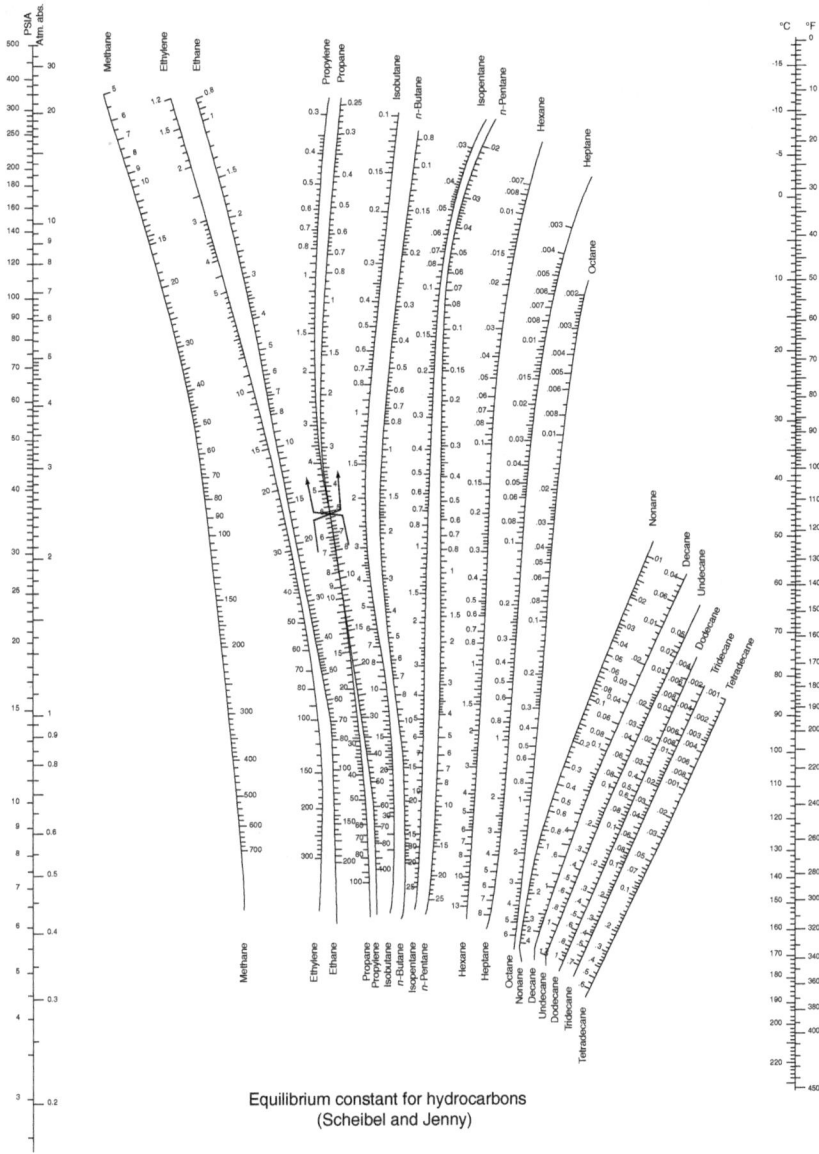

Equilibrium constant for hydrocarbons
(Scheibel and Jenny)

Figure 2.16

Scheibel and Jenny [40] nomograph for light hydrocarbons.

b. Since the pressure drop is neglected, the entire column is assumed to be at 8.4 atm. We will start the calculation with the distribution coefficients estimates from a. A dew point is calculated and the criteria $\sum_{i=1}^{n} y_i/K_i = 1$ (equation 2.148) must therefore be satisfied.

At the first iteration (table 2.20), as the composition of the liquid phase is not correct (sum equals 1.113), the composition is normalised, allowing us to calculate a hypothetical iso-butane distribution coefficient ($K_{iso-butane}$ = 0.67/0.734 = 0.91). A line is drawn at the same pressure through this point and the other values of K_i are read from the diagram and introduced in the table. Convergence is reached in almost two iterations, with a final temperature of 132 °F, as summarised in table 2.20.

Table 2.20   Dew temperature calculation procedure at 8.4 atm

Component	y_i	$K_i(120\,°F)$	$x_i = y_i/K_i$	$x_i' = x_i / \sum_{j=1}^{\mathcal{N}} x_j$	$K_i(132\,°F)$	$x_i = y_i/K_i$
Propane	0.23	1.7	0.135	0.121	1.85	0.124
iso-Butane	0.67	0.82	0.817	0.734	0.91*	0.736
n-Butane	0.10	0.26	0.161	0.145	0.72	0.139
Sum/Result			1.113			0.999

* calculated using $K_i = y_i/x_i' = 0.67/0.734$

c. The reboiler is the equipment where liquid at the bottom of the column exits the system while vapour in equilibrium is reinjected. This is a new bubble point problem where the pressure is known, and the temperature is to be calculated. In this case again, convergence is reached when $\sum_{i=1}^{\mathcal{N}} K_i x_i = 1$.

After the first iteration (table 2.21), the vapour composition is normalised and a K_i (in this case n-butane is used) is estimated. The line is drawn through this value and the pressure of the column. In this case, two iterations are necessary. The distribution coefficients are given in table 2.21 for the two intermediate calculations and the final result.

Table 2.21   Bubble temperature calculation procedure at 8.4 atm, for the residue composition

Component	x_i	$K_i(132\,°F)$	$y_i = K_i y_i$	$y_i' = y_i / \sum_{j=1}^{\mathcal{N}} y_j$	$K_i(207\,°F)$	$K_i(197\,°F)$	$y_i = K_i y_i$
iso-Butane	0.02	0.91	0.018	0.037	1.78	1.70	0.034
n-Butane	0.46	0.72	0.331	0.671	1.46*	1.37	0.629
iso-Pentane	0.15	0.32	0.048	0.097	0.77	0.74	0.111
n-Pentane	0.37	0.26	0.096	0.195	0.66	0.63	0.233
Sum/Result			0.493	1.000			1.007

* calculated using $K_i = y_i'/x_i = 0.671/0.46$

The temperature of the reboiler is found to be 197 °F.

This example is discussed on the website:
http://books.ifpenergiesnouvelles.fr/ebooks/thermodynamics

b. Flash where either P, T or θ is provided plus another property

In all cases, if the basic equations (2.144) along with the K_i calculation methods (2.127), (2.128) or (2.129) are to be used, pressure, temperature and vapour fraction need to be known. In this type of problem, however, only one of the three variables is given. An additional unknown must therefore be taken into account. In order to solve the problem, an additional equation must also be available. This extra equation can be written as a balance equation using the additional property that is given:

If volume is given (isochoric flash):

$$(1-\theta)v^L + \theta v^V = v \qquad (2.149)$$

If enthalpy is given (isenthalpic flash):

$$(1-\theta)h^L + \theta h^V = h \qquad (2.150)$$

If entropy is given (isentropic flash):

$$(1-\theta)s^L + \theta s^V = s \qquad (2.151)$$

If internal energy is given (closed system flash):

$$(1-\theta)u^L + \theta u^V = u \qquad (2.152)$$

Note that this means that a method must be provided for the calculation of volume, enthalpy or entropy for each phase as a function of pressure, temperature and phase composition (see section 4.1, p. 252).

As an example, if we can illustrate a **PH flash** (isenthalpic flash as in a distillation column), the unknowns are K_i, T and F (i.e. $\mathcal{N}+2$ unknowns), and the equations to be solved are:

$$\begin{cases} K_i = \dfrac{\varphi_i^L}{\varphi_i^V} \\[2mm] \displaystyle\sum_{i=1}^{N} \dfrac{\left(K_i-1\right)z_i}{1+\theta\left(K_i-1\right)} = 0 \\[2mm] (1-\theta)h^L + \theta h^V = h \end{cases} \qquad (2.153)$$

One possible procedure to solve the equations is as follows:
1. Estimate the missing piece of information, T and θ
2. Estimate the distribution coefficients K_i
3. Use (2.143) to calculate x_i and y_i
4. Solve for the temperature using the full system (2.153): $\mathcal{N}+2$ equations with $\mathcal{N}+2$ unknowns.

   The K_i values are calculated as shown with equation (2.127) or with (2.128) or (2.129) depending on the chosen method.

Example 2.8 Enthalpy balance in a column

An adiabatic distillation column is used to separate a mixture of n-butane (1) and n-heptane (2). The liquid feed is introduced directly on the third stage as shown on figure 2.17. The column operates at an isobaric pressure of 2.26 atm. Some additional pieces of information concerning the characteristics of the feed and the surrounding stages are also given in table 2.22:

Table 2.22   Data of example 2.8

Position	Feed	Stage 2	Stage 4
Temperature (°C)	47.5		
Vapour molar flow (mol s^{-1})	0	43.5	
Liquid molar flow (mol s^{-1})	100	0.46	118.2
Vapour molar composition y_1	0.5	0.565	0.990
Liquid molar composition x_1		0.969	

For the simplicity of the model, some basic expressions have been selected for enthalpy calculation and for distribution coefficient predictions: the expressions are $h = AT + B$ (with T in °C, and h in cal mol^{-1}) for both phases and $Ln(K) = a/T + b$ (with T in K). These equations apply to each component and the values of the coefficients are found in table 2.23:

Table 2.23   Parameters for example 2.8

Component	Vapour enthalpy		Liquid enthalpy		Distribution coefficient	
	A_i^V	B_i^V	A_i^L	B_i^L	a_i	b_i
n-Butane	23.3	5470	34	0	-2530.4	8.5426
n-Heptane	39.7	9128	54	0	-124.6	10.412

a. What is the temperature of stage number 2?

b. What is the temperature of stage number 4?

c. What are the temperature and compositions of stage number 3?

Analysis:

The properties to be evaluated are the temperature of the various equilibria on the three stages around the feed. Pressure is fixed and part of the compositions is given.

a. For the 2nd stage, the liquid and vapour composition are known, so the distribution coefficients are directly found. The temperature can be calculated using the relationship between K_i and T.

b. For the 4th stage, composition of the liquid is known, meaning that a bubble point has to be determined.

c. On the 3rd stage, all incoming flows are known, so the overall composition and the total enthalpy may be calculated. A PH flash must be solved using the set of equations (2.153).

The fluids are regular light hydrocarbons. All properties are well known. Particular expressions are recommended and no binary interactions are considered (ideal mixture, i.e. $h = h_1^* x_1 + h_2^* x_2$).

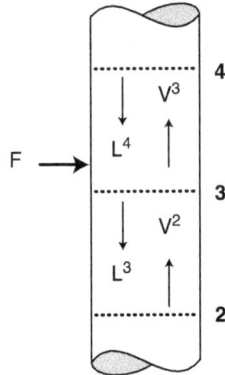

Figure 2.17

Section of the column with stream and plate nomenclature

Solution:

a. Stage number 2 is defined in both phases. So the distribution coefficients of each component can be calculated (it is a binary mixture!).

$$K_1 = \frac{y_1}{x_1} = \frac{0.969}{0.565} = 1.715 \text{ and } K_2 = \frac{y_2}{x_2} = \frac{0.031}{0.435} = 0.071$$

Normally, knowledge of T on the 2nd stage will give both K_1 and K_2, so there are two equations for one unknown. They lead to:

$$T_{2,1} = \frac{a_1}{\ln(K_1) - b_1} = 316.98 \text{ K and } T_{2,2} = \frac{a_2}{Ln(K_2) - b_2} = 315.98 \text{ K. The average is } 316.48 \text{ K.}$$

b. For the 4th stage, a bubble point calculation has to be undertaken. The iterative procedure must start with an initial value (perhaps the temperature of stage number 2, or any other chosen by the user). If the temperature of stage 2 is used, equation (2.147) yields:

$$\sum_i K_i x_i = \exp\left(\frac{-2530.4}{316.08} + 8.5426\right) 0.93 + \exp\left(\frac{-4124.6}{316.08} + 10.412\right) 0.07 = 1.596$$

The temperature must be lowered to reduce the sum. After a few iterations, a value of 298.66 K is found for stage number 4.

c. Material and energy balances around the 3rd stage must be written.

$$\begin{cases} \dot{F}x_1^F + \dot{L}^4 x_1^{L^4} + \dot{V}^2 y_1^{V^2} = \dot{L}^3 x_1^{L^3} + \dot{V}^3 y_1^{V^3} \\ \dot{F}x_2^F + \dot{L}^4 x_2^{L^4} + \dot{V}^2 y_2^{V^2} = \dot{L}^3 x_2^{L^3} + \dot{V}^3 y_2^{V^3} \\ \dot{F}h^{L,F} + \dot{L}^4 h^{L^4} + \dot{V}^2 h^{V,V^2} = \dot{L}^3 h^{L^3} + \dot{V}^3 h^{V,V^3} \end{cases}$$

This implies that enthalpies of each flow must be calculated. For example, the enthalpy of the vapour leaving stage 2 is expressed as:

$$h^{V,V^2} = y_1^{V^2} h_1^{xV,V^2} + y_2^{V^2} h_2^{xV,V^2} = 0.969 \times 6470 + 0.031 \times 10832 = 6605 \text{ cal mol}^{-1}$$

Using this technique, the material and enthalpy content of each of the streams entering the 3rd plate ($\dot{F}$, $\dot{L}^4$ and $\dot{V}^2$) can be calculated and summed, as shown in table 2.24.

Table 2.24   Material and enthalpy content of the stream entering plate 3

Component	$\dot{F}$	$\dot{V}^2$	$\dot{L}^4$	$\dot{F}+\dot{L}^4+\dot{V}^2$
n-Butane (mol s^{-1})	50	42.15	109.93	202.08
n-Heptane (mol s^{-1})	50	1.35	8.27	59.62
Total nC$_4$ + nC$_7$ (mol s^{-1})	100	43.5	118.2	261.7
h (cal mol^{-1})	2090	6605	903	**2304**
H (cal s^{-1})	209000	287336	106740	**603077**

Introducing the equilibrium condition in the above material balances (2 first equations), they can be treated as equation (2.141) leading to the Rachford-Rice equation. For a binary mixture, the vapour fraction can be calculated directly:

$$\frac{(K_1-1)z_1}{1+\theta(K_1-1)}+\frac{(K_2-1)z_2}{1+\theta(K_2-1)}=0 \Rightarrow \theta=-\frac{(K_1-1)z_1+(K_2-1)z_2}{(K_1-1)(K_2-1)}$$

where z_1 and z_2 are the global mole fractions on plate 3. They can be found from the global flow rates shown in table 2.24: z_1 = 202.08/261.7 = 0.772 and z_2 = 59.62/261.7 = 0.228.

We are now left with two equations (the enthalpy balance and the above Rachford-Rice equation) with two unknowns (θ and T). The complete algorithm becomes very simple:

• assume a temperature,
• calculate K$_1$ and K$_2$,
• solve the Rachford-Rice equation to find θ,
• calculate the enthalpies of streams V^3 and L^3,
• compare the sum of these two outlet enthalpies with the inlet enthalpy calculated in table 2.24.

A complete iteration is shown in table 2.25:

Table 2.25   Results of one iteration, using an assumed temperature of 40 °C, yielding a vapour fraction θ = 0.4366, and a total enthalphy of 989.444 kcal s^{-1}

Component	$\dot{F}+\dot{L}^2+\dot{V}^4$	z_i^3	K_i^3	x_i^3	y_i^3	$\dot{L}^3$	$\dot{V}^3$	$\dot{L}^3+\dot{V}^3$
n-Butane (mol s^{-1})	202.08	0.772	1.587	0.615	0.976	90.62	111.46	202.08
n-Heptane (mol s^{-1})	59.62	0.228	0.063	0.385	0.024	56.83	2.79	59.62
Total nC$_4$ + nC$_7$ (mol s^{-1})	261.7	1		1	1	147.45	114.25	261.7
h (cal mol^{-1})	2304					1668	6507	**3781**
H (cal s^{-1})	603077					103761	602748	**989444**

As observed in the results above, calculated at a temperature of 40 °C, the vapour fraction of the flash is 0.4366 and the total enthalpy of $\dot{L}^3+\dot{V}^3$ is 989444 cal s^{-1}, greater than the value of 602478 cal s^{-1} for $\dot{F}+\dot{L}^4+\dot{V}^2$. The temperature must therefore be less than 40 °C. A few additional iterations lead to a final value of 34.29 °C with the corresponding compositions and vapour fraction as seen in table 2.26:

Table 2.26 Final iteration, at 34.29 °C and vapour fraction $\theta = 0.1904$

Component	$\dot{F} + \dot{L}^4 + \dot{V}^2$	z_i^3	K_i^3	x_i^3	y_i^3	$\dot{L}^3$	$\dot{V}^3$	$\dot{L}^3 + \dot{V}^3$
n-Butane (mol s^{-1})	202.08	0.772	1.367	0.722	0.986	152.93	49.15	202.08
n-Heptane (mol s^{-1})	59.62	0.228	0.050	0.278	0.014	58.94	0.69	59.62
Total nC_4 + nC_7 (mol s^{-1})	261.7	1		1	1	211.87	49.84	261.7
h (cal mol^{-1})	2304					1357	6327	**2304**
H (cal s^{-1})	603077					103761	602748	**603077**

This example is discussed on the website:
http://books.ifpenergiesnouvelles.fr/ebooks/thermodynamics

c. Flash where P, T and θ are unknown

When none of the basic variables needed in the basic equations (2.144) along with the K_i calculation methods (2.127), (2.128) or (2.129) is available, they all must be calculated, resulting in a total of $\mathcal{N}+3$ unknowns (K_i, P, T et θ). Hence, $\mathcal{N}+3$ equations are required, or one more compared with (2.153). Two additional balance equations (from (2.149) through (2.152)) are therefore required: one for each of the known state variables.

As an example, for solving a flash at fixed volume and internal energy (*UV* **flash** – this case is found when solving pipe transport problems), the unknowns are K_i, T, P and θ (i.e. $\mathcal{N}+3$ unknowns), and the equations to be solved are (provided that (2.127) is used for calculating K_i):

$$
\begin{cases}
K_i = \dfrac{\varphi_i^L}{\varphi_i^V} \\[2mm]
\displaystyle\sum_{i=1}^{n} \dfrac{\left(K_i - 1\right) z_i}{1 + \theta\left(K_i - 1\right)} = 0 \\[2mm]
(1-\theta)u^L + \theta u^V = u \\[1mm]
(1-\theta)v^L + \theta v^V = v
\end{cases}
\qquad (2.154)
$$

One possible procedure to solve the equations is as follows:

1. Estimate the missing pieces of information, T, P and θ
2. Estimate the distribution coefficients K_i
3. Use (2.143) to calculate x_i and y_i
4. Solve for the temperature using the full system (2.154): $\mathcal{N}+3$ equations with $\mathcal{N}+3$ unknowns

2.2.3.2 Practical applications of phase equilibrium

In process engineering, phase equilibrium must be calculated in different types of problems. In this section, we show how the practical applications will lead to the use the concepts developed above.

A. Distillation or stripping: separation according to volatility (VLE)

When the problem requires separation according to the relative volatility of the components, the most important property to be calculated is the equilibrium coefficient (see (2.127), (2.128) or (2.129)) of each component. The relative volatility is then defined as:

$$\alpha_{ij} = \frac{K_i(T,P,\mathbf{x})}{K_j(T,P,\mathbf{x})} \tag{2.155}$$

The concept of relative volatility is essential in distillation processes. When pressure is low, but non-ideal mixtures are considered, equation (2.130) yields:

$$\alpha_{ij} = \frac{P_i^\sigma \gamma_i(T,\mathbf{x})}{P_j^\sigma \gamma_j(T,\mathbf{x})} \tag{2.156}$$

The activity coefficients can become significant and modify the relative volatility between two components considerably. An example is discussed in example 3.12 (p. 164). The relative volatility between *n*-butane and 1-3 butadiene changes significantly upon adding acetonitrile, which is therefore used as an extraction solvent. This solvent has a low volatility, and therefore remains in the liquid phase throughout the distillation column. While butane and butadiene are almost inseparable by classical distillation (1,3 butadiene is very slightly more volatile than *n*-butane), they can be separated due to their difference in volatility using a technique known as **extractive distillation**. Butane is then significantly more volatile than 1,3-butadiene.

For low pressures and ideal mixtures, the relative volatility may be approximated as (using (2.133)):

$$\alpha_{ij} = \frac{P_i^\sigma}{P_j^\sigma} \tag{2.157}$$

B. Extraction: separation according to chemical affinity (LLE)

Liquid-liquid extraction methods are also based on phase separation and as such equations (2.110) and (2.111) remain valid. Nevertheless, two liquid phases are present simultaneously, and the distribution coefficient now is defined by (2.136). The fugacities are calculated either using the residual approach (2.113), or using the excess approach (2.116). In principle, both methods can be used for calculating a liquid-liquid phase split. Considering the complexity of liquid-liquid phase behaviour (results are very sensitive to small changes in parameters), the excess property models are often more accurate for this type of calculation.

The reason why a liquid phase split appears is not related to the vapour pressure of the pure components, but rather to the chemical affinity among the components. This has been discussed above: the liquid-liquid distribution coefficient defined in equation (2.137). On the condition of having good mixing rules, equations of state can also be used.

Note that, as a consequence of this, the corresponding states principle section 2.2.2.1.C (p. 57) (which provides a very convenient method to calculate hydrocarbon properties and uses the critical point to describe its volatility) is not adapted to predict liquid-liquid phase splits. Consequently, the traditional pseudo-component description is not adapted to the calculation of these types of equilibrium. The liquid phase separation is in fact sensitive to chemical affinity, which is more accurately modelled using the activity coefficient models, further discussed in section 3.4.2 (p. 171).

C. Separation using crystallisation (LSE or VSE)

Crystallisation or solid phase formation is based on another physical phenomenon. When the temperature is sufficiently low, molecular vibrations can no longer keep the molecules apart in the liquid phase. Instead, they start "piling up" in a configuration that depends on the molecular structure. This is why crystallisation temperatures are totally unrelated to the volatilities of the components (except when comparing similar families where the structure remains identical). The pure component crystallisation properties are further discussed in section 3.1.1 (p. 102).

Crystallisation is a very convenient separation technique as its product is generally very pure: in the solid phase, components do not mix as in a liquid. In section 4.2.1.2 (p. 270), the types of solids that may be encountered are discussed.

Phase equilibrium in the presence of a solid phase uses the same equations as for all phase equilibria, *i.e.* the iso-chemical potential (2.110) or iso-fugacity (2.111) condition along with mass balances. The fluid phase fugacity is calculated using either the residual (equation of state) or the excess (activity coefficient) approach. However, for the solid phase fugacity, the most general approach is the excess approach (2.123) with (2.125), as discussed above. The residual approach can not be used here.

D. Risk of appearance of a new phase

On some occasions, it is essential to know the risk of appearance of an additional, unwanted phase. As an example, the appearance of a liquid phase in a vapour flow can strongly perturb the operation of compressors, turbines, heat exchangers, or, in reservoir conditions, decrease the permeability in porous material. In addition, if this liquid is an aqueous phase, it tends to concentrate corrosive components (acids, etc.) and as a result creates important risks of corrosion. If the incipient phase is solid, the risk of plugging equipment is evident.

From a thermodynamic point of view, this question is investigated looking at the phase diagram boundaries. Phase boundaries or solubilities are essentially two identical ways to envision the very same phenomenon. This can be understood when looking at a *Txy* diagram (see figure 2.4 for example). The bubble curve is the boundary between the pure liquid phase and the two phase vapour-liquid region.

The question can be investigated in four different ways:

1. It is useful to draw the full phase diagram to visualise the pressure-temperature (and perhaps composition) path that the system encounters in the process. Specific algorithms for phase diagram calculation exist. An example of such curves is given in figure 2.6 and figure 2.11. If presented in the pressure-temperature plane, they often include a critical point, and as such require the use of a homogeneous method (equation of state). If presented as *Txy* or *Pxy* plots, in the absence of a critical point, they can be obtained quite easily for vapour-liquid equilibria. For liquid-liquid or liquid-solid equilibrium, the calculations are more complex and automatic tools are not readily available.

2. The simplest approach consists in searching the boundary of the phase envelope by performing either $T\theta$ (search for the bubble or dew pressure) or $P\theta$ (search for the bubble or dew temperature) calculations. The actual system pressure and temperatures are then compared with those resulting from the calculation. If the incipient phase is a second liquid or a solid, bubble or dew calculations cannot be used. If it is a solid phase, the temperature when the first crystal appears is calculated. In the case of paraffinic crudes, the terminology used is the "Wax Appearance Temperature" (WAT).

3. When the incipient phase is assumed to be pure, it is also possible to compare the fugacity of the component in the incipient phase with its fugacity in the bulk system. Whenever the former fugacity drops below the latter, phase separation should occur (in particular for solid phases, but also water condensation).

4. The most elegant approach, when the system composition and two properties corresponding to those listed in table 2.7 are provided (for example pressure and temperature), is to perform a so-called stability calculation. Its purpose is to evaluate whether the single phase is thermodynamically stable. This algorithm (Michelsen [41]) evaluates whether, considering the constraints of the system, a second incipient phase can be formed, which would result in a lower Gibbs energy of the system. It is generally mathematically rather intricate, since being close to a phase boundary, the set of equations to be solved presents several minima that may be close to each other. The true solution is the global minimum, but the solver may stop with a local solution.

Obviously, in all cases, a suitable model is required both for the bulk phase and the incipient phase. As will be discussed in section 3.5.1 (p. 226), this type of calculation is rather sensitive to the correct representation of the components at both ends of the phase behaviour considered:

- For vapour-liquid calculations, these are the lowest volatility and highest volatility components.
- For liquid-liquid calculations, these are the components that most readily separate (with the lowest affinity for each other). The presence of a third component that acts as a co-solvent may influence the phase behaviour dramatically however.
- For fluid-solid calculations, these are the components with the highest crystallisation temperature on the one hand and the majority component on the other hand.

Example 2.9 Risk of water condensation in a gas stream

A light hydrocarbon mixture (CH_4: 80%; C_2H_6: 15%; C_3H_8: 5% in molar percent) is contaminated with water. The mixture is available at 200 kPa but due to severe weather conditions there is a risk of low temperature. Is there a real risk of condensation of water in the line?

Analysis:

Ambient temperature is in all possible cases greater than the critical temperature of methane. Ethane and propane could condense but the vapour pressure, even at 0 °C is greater than 200 kPa, so no hydrocarbons are expected to condense.

The pressure is given and a condensation temperature must be found. It is a dew point calculation.

Components are light hydrocarbons and water. Water is known not to mix with hydrocarbons in the liquid phase.

Phases are vapour and liquid. The liquid phase will contain only water.

Solution:

If water condenses the aqueous phase can be considered as pure, so with a composition equal to unity. Pressure is low so the Raoult approximation is valid. Hence, the fugacity of the liquid, incipient phase may be approximated with the vapour pressure of pure water. The fugacity of water in the bulk (vapour) phase is equal to its partial pressure. Hence, if the partial pressure of water reaches the vapour pressure at a given temperature, water will condense:

$$P_{H_2O}^{partial} = Py_{H_2O} = P_{H_2O}^{\sigma}(T)$$

The form of the DIPPR equation for vapour pressure cannot be solved analytically in T. A trial and error procedure has to be implemented.

$$\ln\left(P_{H_2O}^{sat}\right) = 73.649 - \frac{7258.2}{T} - 7.3037\ln(T) + 4.1653.10^{-6}T^2$$

The maximum water content before liquid drops out is then found as $y_{H_2O} = \dfrac{P_{H_2O}^{\sigma}}{P}$. Table 2.27 shows some results. If the temperature is below 0 °C then the sublimation pressure of water should be used and ice would appear.

Table 2.27   Maximum water content of a gas at the total pressure of 200 kPa

Temperature (°C)	Temperature (K)	Vapour pressure of water (Pa)	Maximum water content (%)
0	273.15	610	0.31%
5	278.15	872	0.44%
10	283.15	1227	0.61%
15	288.15	1705	0.85%
20	293.15	2339	1.17%
25	298.15	3170	1.59%
30	303.15	4248	2.12%
35	308.15	5630	2.82%
40	313.15	7386	3.69%
45	318.15	9596	4.80%
50	323.15	12352	6.18%
55	328.15	15760	7.88%
60	333.15	19940	9.97%
65	338.15	25030	12.51%
70	343.15	31181	15.59%
75	348.15	38564	19.28%

This example is discussed on the website:
http://books.ifpenergiesnouvelles.fr/ebooks/thermodynamics

E. Relative amounts of the phases

For a number of industrial applications, the composition of the phases is of lesser importance, but the true issue is the relative amount of each phase. An example where this is the true property to be looked for is found in multiple phase flow calculations, whether in pipelines or fluidised bed.

The procedure used to calculate the relative amounts of the phases is identical to that explained above (2.142) for vapour-liquid equilibrium. For liquid-liquid equilibrium the same kind of equation would be used

$$\sum_{i=1}^{n} \frac{\left(K_i^{\alpha\beta} - 1\right) z_i}{1 + \theta^{\alpha}\left(K_i^{\alpha\beta} - 1\right)} = 0 \tag{2.158}$$

where $\theta^{\alpha} = N^{\alpha}/\left(N^{\alpha} + N^{\beta}\right)$ is the ratio of matter in phase α.

The choice of the correct model will essentially depend on the components and their interactions, as discussed in chapter 4.

- In the case of vapour-liquid equilibrium, the vapour fraction will be more sensitive to the correct solubility calculation of the light components in the liquid than to the description of the volatilisation of the heavy end. Indeed, the vapour phase generally contains only a very small quantity of heavy components, while gases can make up a large fraction of the liquid phase. As a result, the engineer's attention should focus on the interactions between the gases and the bulk of the liquid.
- In the case of liquid-liquid equilibrium, the same issue should be considered as for the phase boundary calculation (hereabove section D, p. 83): the relative concentrations of the components that show the least affinity should be examined first. The presence of a co-solvent will be the next important issue.
- The calculation of the relative amounts of the solid and fluid phases requires, in addition to the crystallisation properties of the pure components, a good description of the solid phase. Considering that many types of solid phase may exist (see section 4.2.2.3, p. 281), this issue may become complex, in particular when several solid phases coexist. For this issue, we refer to more detailed textbooks [14]).

2.2.4 Chemical equilibrium

In some processes, thermodynamic calculations are needed in order to determine how the equilibrium composition changes with pressure and temperature as a result of one or several chemical reaction(s). This assumes that equilibrium is quickly reached, or that no kinetic limitations exist.

Yet even when the reaction rate is not large, and must therefore be evaluated, it may be useful to describe it as the sum of positive and negative contributions that cancel out at the equilibrium conditions. These conditions must therefore be known.

2.2.4.1 Basic principles for chemical equilibrium calculations

The starting point for chemical equilibrium calculations is to identify all components that exist or may be formed in the process. It can be written as a chemical reaction in the following way:

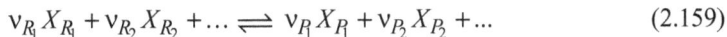

$$v_{R_1} X_{R_1} + v_{R_2} X_{R_2} + ... \rightleftharpoons v_{P_1} X_{P_1} + v_{P_2} X_{P_2} + ... \quad (2.159)$$

where the indices R stand for reactants and P for products. This equation can also be written in a more general way as:

$$\sum_i v_i X_i = 0 \quad (2.160)$$

in which products have a positive stoichiometric coefficient ($v_{i,product} = v_P$) and the reactants a negative one ($v_{i,reactant} = -v_R$).

Equilibrium is reached when the Gibbs energy is minimal at fixed pressure and temperature:

$$dG|_{T,P} = \sum_i \mu_i dN_i \leq 0 \quad (2.161)$$

In addition, the material balance must be respected. These can be written as stoichiometric equations:

$$\sum_i v_i N_i = 0 \quad (2.162)$$

where N_i is the number of moles of component i, and v_i its stoichiometric coefficient.

For any given reaction, the extent of reaction is defined as ξ, such that, when reaction is proceeding, the change in number of moles of any component i can be written as:

$$dN_i = v_i d\xi \quad (2.163)$$

Using (2.163), equation (2.161) can be written as:

$$dG|_{T,P} = \sum_i \mu_i v_i d\xi \leq 0 \quad (2.164)$$

which must be zero at thermodynamic equilibrium. This means that at equilibrium:

$$\sum_i \mu_i v_i = 0 \quad (2.165)$$

must be valid. This approach is called *stoichiometric*.

An alternative, *non stoichiometric* approach, considers a global minimisation of the Gibbs energy written as $G = \sum n_i \mu_i$.

No stoichiometric coefficients are then used. This method, illustrated in example 2.12, is of great use when the chemicals can react without restriction and when phase equilibrium is simultaneous with reaction equilibrium. For catalysed reactions, where some reaction paths are favored compared to others, this global minimisation cannot be used: a stoichiometric approach is needed.

If the chemical potentials are expressed in terms of fugacities, according to definition (2.54), and using a reference state (superscript [0]) that can be chosen arbitrarily for each component (provided it is at system temperature):

$$\mu_i = \mu_i^0 + RT \ln \frac{f_i}{f_i^0} \tag{2.166}$$

and (2.165) becomes:

$$RT \sum_i v_i \ln \frac{f_i}{f_i^0} = -\sum_i v_i \mu_i^0 \tag{2.167}$$

which can also be written as:

$$\prod_i \left(\frac{f_i}{f_i^0}\right)^{v_i} = \exp\left(\frac{-\sum_i v_i \mu_i^0}{RT}\right) = K \tag{2.168}$$

This is the definition of the well-known chemical equilibrium constant. It is thus clear that this constant depends on the reference state chosen and on the temperature of the system. It is however independent of the pressure and the system composition.

It is important to insist that the choice of reference state is absolutely arbitrary for each component. Table 2.28 gives some usual reference states.

The most common choice for the pure component reference state is the pure component, in the ideal gas state, at 0.1 MPa pressure (1 bar). In that case, equation (2.166) is written as:

$$\mu_i = \mu_i^0\left(T, P^0 = 0.1 \text{ MPa}\right) + RT \ln \frac{f_i}{P^0} \tag{2.169}$$

This is why P° is sometimes omitted, and the fugacity expressed in bar.

Table 2.28   Usual reference states in chemical reactions (definition of superscript ^{0})

Type	Temperature	Pressure	Composition	Phase
Non-ionic	T_{System}	101.325 kPa (1 atm) or 100 kPa	Pure	Same phase as that in which the reaction takes place (Vapour → Ideal gas) (Liquid → Pure liquid) (Solid → Pure solid)
Ionic	T_{System}	101.325 kPa (1 atm) or 100 kPa	Infinite dilution in solvent	Usually liquid

In order to calculate the chemical equilibrium constant, the chemical potentials of the individual components in their reference states must be known. This is where the formation properties are used: we can state that the chemical potential of a molecule i in its reference state can be written from the chemical potentials of the component elements in their own reference states, as follows:

$$\mu_i^0 = \sum_e n_{e,i} \mu_e^0 + \Delta G_{f,i}^0 \tag{2.170}$$

where the sum is taken over all elements e, and $\Delta G_{f,i}{}^0$ is the Gibbs energy of formation of component i. This property takes into account both the reference state of the component (ideal gas, generally at 298.15 K) and the reference states of the elements (which depend on the element: H_2 as an ideal gas for hydrogen; O_2 as an ideal gas for oxygen, graphite for carbon.). The Gibbs energy of formation is tabulated and further discussed along with the other formation properties in section 3.1.1 (p. 102).

Considering that the mass balance over all elements must be respected: $\sum_i v_i n_{e,i} = 0$, combining equations (2.170) and (2.168) yields:

$$K = \exp\left(\frac{-\sum_i v_i \Delta G_{f,i}{}^0}{RT}\right) = \exp\left(\frac{-\Delta G_r^0}{RT}\right) \tag{2.171}$$

where ΔG_r^0 is the Gibbs energy of reaction, that depends on temperature in the same way as $\Delta G_{f,i}{}^0$. The temperature dependence of K is a direct consequence of the Gibbs-Helmholtz equation (2.36):

$$\frac{\partial\left(\dfrac{\Delta G_r^0}{T}\right)}{\partial\left(\dfrac{1}{T}\right)}\Bigg|_P = \Delta H_r^0(T) \tag{2.172}$$

where $\Delta H_r^0 = \sum_i v_i \Delta H_{f,i}{}^0$ is the enthalpy of reaction that is a related to the enthalpy of formation of the individual species. The temperature dependence of the chemical equilibrium constant is therefore:

$$\frac{\partial \ln K}{\partial\left(\dfrac{1}{T}\right)}\Bigg|_P = -\frac{\Delta H_r^0}{R} \tag{2.173}$$

According to Van't Hoff's theorem, we state that if the enthalpy of reaction is positive (the products have a higher enthalpy than the reactants; this is an endothermic reaction), the equilibrium constant increases with temperature. In other words, if we increase the temperature the reaction responds with higher conversion, to consume heat and partially counteract the temperature increase.

In an exothermic reaction (negative heat of reaction) the opposite is true: if temperature is increased, the reaction will respond by favouring the reverse reaction, resulting in lower conversion. This is also known as Le Chatelier's principle, which is a generalisation of Van't Hoff's theorem [42].

2.2.4.2 Model requirements for chemical equilibrium

A. Fugacities or activities

We can see from equation (2.168) that to calculate the chemical equilibrium the fugacity of the components involved must be known. The calculation method is identical to that described above for the phase equilibrium calculation. Note that in principle only a single phase is involved here. In cases where the phase and chemical equilibria are both calculated simultaneously the approach used can be either homogeneous or heterogeneous.

In the **low pressure vapour phase**, fugacities may be approximated with partial pressures, in which case the chemical equilibrium (2.168) is calculated using:

$$K = \prod_i \left(y_i \frac{P}{P^0} \right)^{v_i} = \left(\frac{P}{P^0} \right)^{\sum_i v_i} \prod_i \left(y_i \right)^{v_i} \tag{2.174}$$

This equation illustrates that if the sum of the stoichiometric coefficient is zero, pressure will have no effect on the conversion. However, if the reaction produces more species than it consumes ($\sum_i v_i > 0$), then conversion decreases with pressure (K is independent of pressure, hence $\prod_i \left(y_i \right)^{v_i} \sim \left(\frac{P}{P^0} \right)^{-\sum_i v_i}$). On the other hand, if there are fewer products than

---

Example 2.10 Effect of the pressure on the methane reforming reaction

In the methane reforming reaction, water and methane react to yield hydrogen gas and carbon monoxide. The reaction is catalysed so that it can be considered in thermodynamic equilibrium. It occurs in the vapour phase. The equilibrium constant at 800 K is 0.032 if the reference state is taken at 1 bar. What is the effect of pressure on this reaction?

Analysis:

Pressure and temperature are assumed to be known.

The reaction occurs in the vapour phase, and at rather high temperature (this is an endothermic reaction, so its advancement is increased at high temperature). To a first approximation, we can therefore state that the reaction occurs in an ideal gas. The equation that relates pressure with the equilibrium condition is then (2.174).

Solution:

The exact stoichiometry of the reaction must first be determined:

$$H_2O + CH_4 = 3\,H_2 + CO$$

The sum of the stoichiometric coefficients, $\sum_i v_i = 3+1-1-1 = 2$ is positive, indicating that

the more the reaction proceeeds, the larger the number of moles. When pressure increases, Le Chatelier's principle then tells us that the extent of reaction decreases.

This example is discussed on the website:
http://books.ifpenergiesnouvelles.fr/ebooks/thermodynamics

---

reactants, then the opposite is true, and increasing the pressure will increase the conversion. This is also in accordance with Le Chatelier's principle.

In **a liquid phase**, the fugacities may be expressed using (2.85), so that the equilibrium equation becomes:

$$K = \prod_i \left(\gamma_i x_i f_i^{L*}\right)^{\nu_i} \approx \prod_i (a_i)^{\nu_i} \prod_i \left(P_i^\sigma\right)^{\nu_i} \tag{2.175}$$

Where $a_i = \gamma_i x_i$ is the activity of component i. Note that if the reaction occurs at high pressure, Poynting corrections should be used to correct the vapour pressures P_i^σ, according to (2.116). In (2.117), the vapour pressures depend only on temperature. Hence, if a new equilibrium constant is chosen as $K' = K \prod_i \left(P_i^\sigma\right)^{-\nu_i}$, equation (2.175) can be written as

$$K' = \prod_i (a_i)^{\nu_i} \tag{2.176}$$

If the mixture is ideal, the activity can be replaced by mole fractions. In some cases, for ideal mixtures, a molar concentration is used ($\left[i\right]$, expressed in mol L^{-1}). In this case, the equilibrium constant should be further corrected using the molar volume of the liquid:

$$K' = \prod_i \left(\left[i\right]v_m\right)^{\nu_i} = v_m^{\sum \nu_i} \prod_i \left[i\right]^{\nu_i} \tag{2.177}$$

B. Formation properties

As discussed earlier [equation (2.171)], the equilibrium constant is computed from the Gibbs energy of reaction at the system temperature. The Gibbs energy of reaction is itself computed from the Gibbs energies of formation of each of the components involved in the reaction, according to:

$$\Delta G_r^0 = \sum_i \nu_i \Delta G_{f,i}^0 \tag{2.178}$$

The Gibbs energies of reaction are provided in many databases at 298.15 K (25 °C). In order to calculate it at another temperature using (2.173), the enthalpy of reaction must be known:

$$\ln K(T) = \ln K(T_0) + \int_{1/T_0}^{1/T} -\frac{\Delta H_r^0(T)}{R} d\left(\frac{1}{T}\right) \tag{2.179}$$

The enthalpy of reaction is computed from the enthalpies of formation of the components:

$$\Delta H_r^0 = \sum_i \nu_i \Delta H_{f,i}^0 \tag{2.180}$$

As shown in (2.179), the enthalpy of reaction may depend on temperature. Its temperature dependence is expressed using the heat capacity of formation of each component:

$$\frac{\partial \Delta H_r^0}{\partial T} = \sum_i \nu_i \frac{\partial \Delta H_{f,i}^0}{\partial T} = \sum_i \nu_i \Delta C_{P_{f,i}}^0(T) \tag{2.181}$$

where the heat capacity of formation is, like all formation properties, the difference between the heat capacity of the component i in its reference state (usually ideal gas), and the elements it is composed of:

$$\Delta C^0_{P_{f,i}}(T) = C^0_{P_i}(T) - \sum_e n_{e,i} C^0_{P_e}(T) \qquad (2.182)$$

Hence, using the material balance over all elements, $\sum_i v_i n_{e,i} = 0$, (2.181) is equivalent to:

$$\frac{\partial \Delta H_r^0}{\partial T} = \sum_i v_i C^0_{P_i}(T) = \Delta C^0_{P,r} \qquad (2.183)$$

This is why the heat capacity of the elements is not used when calculating the effect of temperature in the heat of reaction. However, the heat capacity of each component in its reference state (ideal gas) must be known.

Example 2.11 Effect of temperature on the chemical equilibrium constant

Calculate the equilibrium of the dodecane dehydrogenation reaction as a function of temperature, at a total pressure of 3 bar and the ratio of the partial pressure of H_2 and hydrocarbons is equal to 6 ($\alpha = P_{H_2}/(P_{HC})$):

$$C_{12}H_{24} \rightleftharpoons C_{12}H_{22} + H_2$$

Analysis:

Required properties:

Pressure is known. The system composition must be calculated as a function of temperature (which is therefore known). In order to apply the chemical equilibrium condition (2.168), two types of properties must be computed: fugacities and Gibbs energy of reaction at the reference state.

Mixture:

It is considered here that only hydrogen and hydrocarbons are of interest.

The hydrogen pure component properties are readily available in any database.

This may not always be the case for the long chain alkanes and alkenes. The *n*-dodecane and *n*-dodecene molecules are available in the DIPPR database that is used here. Use of a group contribution model will also be illustrated since it may be of interest for long-chain molecules.

Although the molar mass of the components are very different, all are non-polar. We will consider that the mixture is ideal for this example. This is justified below.

Phase conditions:

In this example, the main argument for the choice of the fugacity model is based on the observation that the pressure is low and that in the reaction condition the mixture is in its vapour phase. Under these conditions, the ideal gas approximation can be used and the fugacities written as partial pressures (hence ideal mixture behaviour). Thus (2.168) becomes (2.174).

Solution:

In order to calculate the equilibrium constant, the component formation properties are required. Table 2.29 provides these values as found in the DIPPR database.

Table 2.29 Data of example 2.11:
Formation properties at 298.15 K, according to DIPPR

Component	$\Delta H_{f,i}^{0}$ (kJ mol^{-1})	$\Delta G_{f,i}^{0}$ (kJ mol^{-1})	$\Delta S_{f,i}^{0}$ (kJ mol^{-1} K^{-1})	S_{i}^{0} (kJ mol^{-1} K^{-1})
n-dodecane	− 290.7	49.81	− 1.1443	0.62415
1-dodecene	− 165.4	136.1	− 0.6006	0.6185
Hydrogen	0	0	0	0.13057

Note that the entropy of formation is not given by the database, but that it is here calculated using its dependence on enthalpy and Gibbs energy:

$$\Delta S_{f,i}^{0} = -\frac{\Delta G_{f,i}^{0} - \Delta H_{f,i}^{0}}{T^0}$$

This value is different from the absolute entropy that is also provided in the database (here given as S_i^0 but not used).

Since gaseous hydrogen is considered as the component that represents the element *H*, its properties of formation are zero by definition.

Using (2.178), we can now calculate the Gibbs free energy of reaction at 298.15 K: ΔG_r^0 = 86.74 kJ mol^{-1}, and therefore the equilibrium constant (2.171): in K = − 34.99. This value is very small, indicating that at room temperature, the reaction equilibrium is such that the alkane is much more stable than the alkene. To observe dehydrogenation, the temperature must be raised, as demonstrated by the positive heat of reaction ΔH_r^0 = 125.26 kJ mol^{-1} (indicating an endothermic reaction: the higher the temperature, the larger the conversion).

The temperature-dependence of the equilibrium constant is written according to (2.173), where ΔH_r^0 itself depends on temperature according to (2.183). This is why the ideal gas heat capacity of the components must be taken into account. Using the model of Passut and Danner (see section 3.1.1.2, p. 109):

$$C_p(T) = A + BT + CT^2 + DT^3 + ET^4$$

the parameters are found in the database, and provided in table 2.10:

Table 2.30 Heat capacity parameters for the Passut and Danner expression (C_p units are J mol^{-1} K^{-1})

Component	A	B	C	D	E
n-dodecane	− 8.02	1.14384	− 0.0000629	1.316 × 10^{-7}	0
1-dodecene	− 26.011	1.22282	− 0,00008206	2.334 × 10^{-7}	0
Hydrogen	2.54	0.0194	− 0.000365	3.00 × 10^{-8}	− 8.18 × 10^{-12}

We can now integrate equation (2.179) to find the value of the equilibrium constant as a function of temperature. The result is shown in figure 2.18. We see that a high temperature is required in order to have a visible conversion.

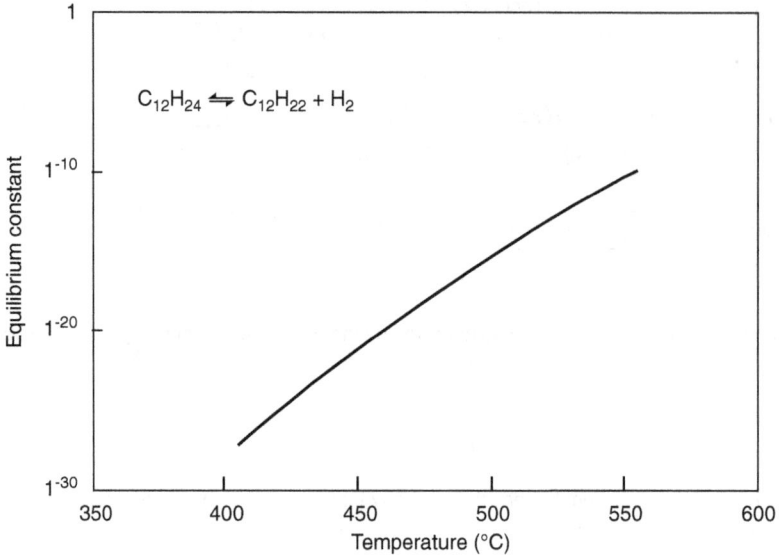

Figure 2.18

Equilibrium constant of the dehydrogenation of *n*-dodecane as a function of temperature.

The mixture equilibrium composition as a function of temperature can be calculated using (2.174) and the information provided in the problem statement (the pressure is 3 bar and the molar ration $H_2/HC = 6$). This will yield three equations with three unknowns (the partial pressures):

$$\begin{cases} K = \dfrac{P_{C_{12}^=} P_{H_2}}{P_{C_{12}}} \\[2ex] \alpha = \dfrac{P_{H_2}}{P_{C_{12}^=} + P_{C_{12}}} \\[2ex] P = P_{C_{12}^=} + P_{C_{12}} + P_{H_2} \end{cases}$$

It is for example possible to calculate the conversion

$$\frac{P_{C_{12}^=}}{P_{C_{12}} + P_{H_2}} = \frac{(1+\alpha)K}{P\alpha + (1+\alpha)K}$$

This example is discussed on the website:
http://books.ifpenergiesnouvelles.fr/ebooks/thermodynamics

C. Comment on the simultaneous phase and chemical equilibrium problem

Note that in some cases, simultaneous phase and chemical equilibrium must be calculated.

This occurs for example when an acid gas (H_2S or CO_2) is dissolved in an aqueous phase. In the aqueous environment, the chemical equilibria are given by:

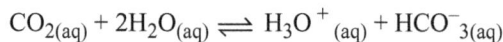

$$CO_{2(aq)} + 2H_2O_{(aq)} \rightleftharpoons H_3O^+_{(aq)} + HCO^-_{3(aq)}$$

and

$$H_2S_{(aq)} + H_2O_{(aq)} \rightleftharpoons H_3O^+{}_{(aq)} + HS^-{}_{(aq)}$$

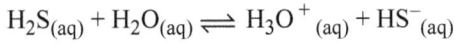

These reactions explain why these gases dissolve well in water. Their solubility increases further if a base is present in the aqueous phase (an alkanolamine, for example) as it is employed for acid gas treatment.

Calculation of equilibrium under these circumstances requires no properties other than those already discussed (fugacities or chemical potentials, and formation properties). The mathematical algorithm for solving such a system becomes rather complex however. It is not the purpose of this book to discuss these difficulties in detail (see for more details Michelsen [41]).

Example 2.12 Chemical looping

Carbon dioxide is one of the main contributors to global warming. CO_2 emissions must be reduced to minimise the rise in the earth's temperature. During classical combustion, CO_2 is mixed with nitrogen. This dilution makes subsequent treatment, such as reinjection, very difficult. One alternative is to carry out selective oxidation of a fuel gas like methane on a metal oxide (e.g. nickel) to produce heat for energy production. The carbon dioxide resulting from this reaction is almost pure and can therefore be captured more easily. The reduced metal can be reoxidised with air for subsequent chemical looping.

Possible reactions are:

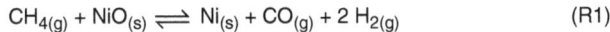

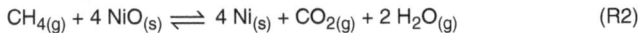

$$CH_{4(g)} + NiO_{(s)} \rightleftharpoons Ni_{(s)} + CO_{(g)} + 2\ H_{2(g)} \tag{R1}$$

$$CH_{4(g)} + 4\ NiO_{(s)} \rightleftharpoons 4\ Ni_{(s)} + CO_{2(g)} + 2\ H_2O_{(g)} \tag{R2}$$

Operating conditions are atmospheric pressure and temperatures from 800 °C to 1200 °C. The objective is to determine the position of chemical equilibrium. Nickel and nickel oxide are solids while all other components are in the gaseous phase.

As the reaction scheme is not defined, we use the non stoichiometric approach. Equilibrium is reached when the Gibbs energy is minimised.

$$G\left(T,P,n_i^\varphi\right) = \sum_\varphi^\phi \sum_i^{\mathcal{N}} n_i^\varphi \mu_i^\varphi$$

where ϕ is the number of phases and $\mathcal{N}$ the number of components.

For solids, the chemical potentials are independent of pressure:

$$\mu_i^S(T,P) = \mu_i^S(T)$$

and for gases (with ideal gas approximation):

$$\mu_i^G(T,P) = \mu_i^\circ(T) + RT\ln\left(\frac{f_i}{P^\circ}\right) = \mu_i^\circ(T) + RT\ln\left(\frac{y_iP}{P^\circ}\right)$$

The Gibbs energy of formation is approximated by a simple linear expression (T in K and μ_i° in kcal mol^{-1}).

$$\Delta G_{fi}^\circ(T) = A + BT = \mu_i^\circ(T)$$

for which the parameters are shown in table 2.31.

Table 2.31   Parameters for the approximation of the chemical potential of formation in example 2.12 [43]

Component	A	B
CH_4	− 19.662	0.024
CO	− 26.388	− 0.0216
CO_2	− 94.133	− 0.0005
H_2	0	0
H_2O	− 58.442	0.0123
Ni	0	0
NiO	− 56.714	0.021

The problem is now to find the minimum of the Gibbs energy that changes with the composition. This minimum depends on pressure and temperature. Following restrictions apply to satisfy the conservation of each of the atoms.

$$n_C^T = n_{CH_4} + n_{CO_2} + n_{CO}$$

$$n_H^T = 4n_{CH_4} + 2n_{H_2O} + 2n_{H_2}$$

$$n_O^T = n_{H_2O} + 2n_{CO_2} + n_{CO} + n_{NiO}$$

$$n_{Ni}^T = n_{NiO} + n_{Ni}$$

Once the overall quantities of each atom present in the system have been specified, the Gibbs energy can be minimised for each temperature and pressure. Various commercial libraries can be used for this purpose, including the Excel "Solver" add-in. A measure of the degree of conversion of methane can be used to monitor the reaction effectiveness:

$$\eta = \frac{n_{CO} + n_{CO_2}}{n_{CH_4} + n_{CO} + n_{CO_2}}$$

Results are as shown in tables 2.32 and 2.33 for either of the above reaction R1 and R2:

Table 2.32   Results of the Gibbs energy minimisation for one molecule of methane reacting with one molecule of nickel oxide

T (°C)	800	900	1000	1100	1200
CH_4	0.0825	0.0311	0.0130	0.0061	0.0032
CO	0.8912	0.9613	0.9845	0.9930	0.9964
CO_2	0.0263	0.0076	0.0025	0.0009	0.0004
H_2	1.7788	1.9144	1.9635	1.9826	1.9909
H_2O	0.0562	0.0235	0.0105	0.0052	0.0028
Ni	1.0000	1.0000	1.0000	1.0000	1.0000
NiO	0.0000	0.0000	0.0000	0.0000	0.0000
Conversion η (%)	91.75	96.89	98.70	99.39	99.68

As we can appreciate from table 2.32, with a CH_4/NiO ratio of 1 to 1, all the oxide is reduced. Conversion rises with temperature. If a CH_4/NiO ratio of 1 to 4 is used (table 2.33), the CH_4 is completely converted and a very small proportion of nickel remains in NiO form. Other analyses can be carried out changing the CH_4/NiO ratio to see the effect on CO and H_2 production.

Table 2.33 Results of the Gibbs energy minimisation for one molecule of methane reacting with four molecules of nickel oxide

T (°C)	800	900	1000	1100	1200
CH_4	0.0000	0.0000	0.0000	0.0000	0.0000
CO	0.0058	0.0090	0.0131	0.0179	0.0234
CO_2	0.9942	0.9910	0.9869	0.9821	0.9766
H_2	0.0109	0.0116	0.0124	0.0130	0.0136
H_2O	1.9891	1.9884	1.9876	1.9870	1.9864
Ni	3.9833	3.9794	3.9746	3.9691	3.9630
NiO	0.0167	0.0206	0.0254	0.0309	0.0370
Conversion η (%)	100.00	100.00	100.00	100.00	100.00

These results are in agreement with the proposed equations. For the N_iO/CH_4 ratio equal to one, CO is produced, and for a ratio equal to four, CO_2 is observed.

This example is discussed on the website:
http://books.ifpenergiesnouvelles.fr/ebooks/thermodynamics

REFERENCE LIST

[1] Smith, J. M., Van Ness, H. C. and Abbott, M. M. "Introduction to Chemical Engineering Thermodynamics", Sixth Edition; Mc Graw Hill, Inc **2001**.

[2] Elliott, J. R. and Lira, C. T. "Introductory Chemical Engineering Thermodynamics"; Prentice Hall PTR: Upper Saddle River, NJ, **1999**.

[3] Prausnitz, J. M., Lichtenthaler, R. N. and Gomes de Azevedo, E. "Molecular Thermodynamics of Fluid Phase Equilibria", 3rd Ed.; Prentice Hall Int., **1999**.

[4] O'Connell, J. P. and Haile, J. M. "Thermodynamics: Fundamentals for Applications" 1st Ed.; Cambridge University Press, **2005**.

[5] Ledanois, J. M., Lopez de Ramos, A. L. and Bronner, C. "Quantities, Dimensions and Units Conversions" *Revue de l'Institut Français du Pétrole* **1995**, **50**, 5, 685-716.

[6] Benedict, R. P. "Fundamentals of Temperature, Pressure and Flow Measurements" 3rd Ed.; Wiley-Interscience, **1984**.

[7] Ben-Naim, A. "Entropy Demystified: The Second Law Reduced to Plain Common Sense"; World Scientific Publishing Co.; Singapore, **2008**.

[8] Howard, I. K. "H is for Enthalpy, Thanks to Heike Kamerlingh Onnes and Alfred W. Porter" *Journal of Chemical Education* **2002**, **79**, 6, 697-698.

[9] Howard, I. K. "H is for Enthalpy – The Author Replies" *Journal of Chemical Education* **2003**, **80**, 5, 486-486.

[10] Wright, J. D., Johnson, A. N. and Moldover, M. R. "Design and Uncertainty Analysis for a PVTt Gas Flow Standard" *Journal of Reseach of the Institute of Standards and Technology* **2003**, **108**, 1, 21-47.

[11] NBS "Table of Chemical Properties" *Journal of Physics and Chemical Reference Data* **1982**, **11** (2-Supplement).

[12] Rowley, R.L., Wilding, W.V., Oscarson J.L., Yang Y., Zundel, N.A., Daubert, T.E. and Danner, R.P. "DIPPR® Data Compilation of Pure Compound Properties", Design Institute for Physical Properties; AIChE, New York, **2003**.

[13] Antoine, Ch. "Tension des vapeurs: nouvelle relation entre les tensions et les températures" *Comptes rendus hebdomadaires des séances de l'Académie des Sciences,* **1888**, 107, 681-684.

[14] de Swaan Arons, J. and de Loos, T. "Phase Behavior: Phenomena, Significance and Models"; Ed. Sandler, S. I.; Marcel Dekker, Inc., **1994**.

[15] Modell, M. and Reid, R. C. *Thermodynamics and its applications,* 2nd Ed.; Prentice Hall: Englewoods Cliffs, N.J., **1983**.

[16] Boyle, R. "Spring of the air" in *Boyle papers Online*; Royal Society: London, **1660**.

[17] Mariotte, E. "De la nature de l'air"; Paris, **1679**.

[18] Gay-Lussac, L. J. "Recherches sur la dilatation des gaz et des vapeurs" *Annales de Chimie,* **1802**, 43, 137.

[19] Clapeyron, E. "Mémoire sur la puissance motrice de la chaleur" *Journal de l'Ecole Royale Polytechnique* **1834**, **14**, 23, 153-190.

[20] van der Waals, J. D. "Over de continuiteit van den gas en vloeistoftestand (On the Continuity of the Gas and Liquid State)", Hoogeschool te Leiden, **1873**.

[21] van der Waals, J. D. "The Equation of State for Gases and Liquids" *Nobel Lectures in Physics* **1910**, 1, 254-265.

[22] Clausius, R. "Sur une détermination générale de la tension et du volume des vapeurs saturées" *Comptes rendus hebdomadaires des séances de l'Académie des sciences* **1881**, 97, 619-625.

[23] Berthelot, D. "Sur les thermomètres à gaz, et sur la réduction de leurs indications à l'échelle absolue des températures"; Gauthier-Villars: Paris, **1903**.

[24] Pitzer, K. S., Lippman, D. Z., Curl, R. F., Huggins, C. M. and Petersen, D. E. "The Volumetric and Thermodynamic Properties of Fluids. II. Compressibility Factor, Vapor Pressure and Entropy of Vapourisation" *Journal of the American Chemical Society* **1955**, **77**, 13, 3433-3440.

[25] Lee, B. I. and Kesler, M. G. "A Generalized Thermodynamic Correlation Based on Three-Parameter Corresponding States" *AIChE Journal* **1975**, **21**, 3, 510.

[26] Kay, W. B. "Density of Hydrocarbon Gases and Vapors at High Temperatures and Pressures" *Industrial & Engineering Chemistry* **1936**, **28**, 9, 1014-1019.

[27] O'Connell, J. P. "Thermodynamics of Gas Solubilitiy in Mixed Solvents" *Industrial & Engineering Chemistry Fundamentals* **1964**, **3**, 4, 347-351.

[28] Fogg, P. G. T. and Gerrard, W. "Solubility of Gases in Liquids"; J. Wiley & Sons, Chichester, **1991**.

[29] Tsonopoulos, C. "Thermodynamic Analysis of the Mutual Solubilities of Normal Alkanes and Water" *Fluid Phase Equilibria* **1999**, 156, 21-33.

[30] Dhima, A., de Hemptinne, J. C. and Jose, J. "Solubility of Hydrocarbons and CO_2 Mixtures in Water under High Pressure" *Industrial & Engineering Chemistry Research* **1999**, **38**, 8, 3144-3161.

[31] Carroll, J. J. "Use Henry's Law for Multicomponent Mixtures" *Chemical Engineering Progress* **1992**, **88**, 8, 53-58.

[32] Carroll, J. J. "What is Henry's Law?" *Chemical Engineering Progress* **1991**, **87**, 9, 48-52.

[33] Coutinho, J. A. P. "A Thermodynamic Model for Predicting Wax Formation in Jet and Diesel Fuels" *Energy & Fuels* **2000**, **14**, 3, 625-631.

[34] Vidal, J. "Thermodynamics: Applications in Chemical Engineering and the Petroleum Industry"; Editions Technip: Paris, **2003**.

[35] Rachford, H. H. and Rice, J. D "Procedure for Use of Electrical Digital Computers in Calculating Flash Vapourisation Hydrocarbon Equilibrium" *Journal of Petroleum Technology*, **1952**, **1**, 4, 19-20.

[36] Michelsen, M. L. "The Isothermal Flash Problem.1. Stability" *Fluid Phase Equilibria* **1982**, **9**, 1, 1-19.

[37] Michelsen, M. L. "The Isothermal Flash Problem.2. Phase-Split Calculation" *Fluid Phase Equilibria* **1982**, **9**, 1, 21-40.

[38] Walas, S. M. "Phase Equilibria in Chemical Engineering", Butterworth Publishers, **1985**.

[39] Ledanois, J. M., Lopez A.L., Pimentel J.A. and Pironti F. "Capitulo 3: Ecuaciones Implícitas" in *Métodos Numéricos Aplicados en Ingeniería*, McGraw-Hill Interamericana, **2000**.

[40] Scheibel, E. G. and Jenny, F. J. "Representation of Equilibrium Constant Data" *Industrial & Engineering Chemistry* **1945**, **37**, 1, 80-82.

[41] Michelsen, M. L. and Mollerup, J. *Thermodynamic Models: Fundamental and Computational Aspects,* 1st Ed.; Tie-Line Publications, **2004**.

[42] Olivera-Fuentes, C. and Colina, C. "Stability, Displacement and Moderation of Chemical Equilibrium: Rediscovering Le Chatelier's Principle", International Conference on Engineering Education, Coimbra, Portugal, Sept. 3-7, **2007**.

[43] Stull, D., Westrum, E. and Sinke, G. "The Chemical Thermodynamics of Organic Compounds", Robert E. Krieger Publishing Company, Malabar, Florida, **1987**.

3

From Components to Models

Chapter 3 will discuss how the fluid composition affects the thermodynamic calculations. As already pointed out in the first chapter, discussion brings up both the issue of models as that of their parameters *i.e.* the fluid description. These two issues cannot be dissociated.

In table 1.1, it has been proposed to subdivide this analysis criterion into three sub-questions:

- What is the nature of the components? This question will lead us to investigate the pure component data, how they are presented, how they can be validated.
- What type of molecular interactions may occur between the components? This question will bring us to investigate how a component behaves when it is surrounded by other components. Many thermodynamic models attempt to describe these interactions.
- Finally, the concept of key component(s) will be introduced. In some cases, this is not crucial, but in the case where the fate of impurities is important, it will be shown how different the analysis becomes.

In order to help answering these questions, we shall stress that:

In one way or another, the model-parameter combination must always be compared (either for regression or for validation) to experimental data.

The first two sections are therefore devoted to the description of the type of data that may exist, the way they are presented and some discussions on how to evaluate their quality (pure components and mixtures). The third section is entirely focused on the determination of model parameters by fitting on experimental data. The fourth section will discuss the thermodynamic models, focusing on their molecular construction.

Finally, in the fifth section, the concept of "key component" will be discussed and the mixture data to be evaluated first will be identified (it is often almost impossible to consider all binary data in a multicomponent mixture).

3.1   PURE COMPONENTS: PROPERTIES AND PARAMETERS

Model parameters may or may not have a physical meaning, yet, in all cases, they originate from a comparison of the physical behaviour with the model calculations.

The objective of this section is to help the reader understand the behaviour of pure components, the relationships that may exist among their properties and thus provide some criteria for assessing the coherence of the available data. First, the pure component properties and parameters will be discussed. Secondly, the types of components that can be encountered are presented, and the resulting choices that can be made for calculating the properties.

3.1.1   Pure components properties and parameters

Each component comes with its set of properties and parameters. The parameter values can either directly be measured (physical properties), or may be fitted on pure component properties (as liquid molar volume and saturated vapour pressure for example). In this section, the most frequently used properties will be listed and some rules will be given in order to help the engineer evaluate whether or not the data are realistic.

Obviously, numerous criteria can be used to evaluate how a molecule should behave. We provide the following guidelines however:

- Look at the molar mass, since it is a direct measure of how "heavy" a molecule is.
- Look at its ability to form hydrogen bonds. Bonds will form when hydrogen has lost its electrons due to the proximity of a heteroatom, generally an oxygen but it may be any heteroatom (nitrogen, sulfur). The most trivial example of this type of bonds can be found in alcohols or water, where the oxygen has two lone electron pairs while the hydrogen has a positive charge.
- Even though the molecule may not form hydrogen bonds, it may be polar due to a non-homogeneous distribution of its electrons. Examples include, once again, alcohols (dipoles), as well as esters, ethers (dipoles), aromatic rings and carbon dioxide (quadrupoles).
- Next, the branching of the molecule is important. The more branched, the more it tends to behave as if it were shorter in length. Conversely, the more branched a molecule is, the less likely it is to crystallise easily.

The well-known book entitled "The Properties of Gases and Liquids" [1] provides predictive calculation methods for many of the properties discussed below.

3.1.1.1   Physical properties as model parameters

The parameters and their significance obviously depend on the model used. Many of them have no true physical meaning. Yet, a number of physical properties of the pure components are used as model parameters. Some of general interest are listed here:

A. Molar mass

Although this may not be obvious for a process engineer, thermodynamic methods are all based on molar dimensions. Hence, as the process data are generally expressed on a mass basis, it is essential to have access to the molar mass of the components in order to transform values to the correct units. This property was previously called molecular weight or molecular mass (M_w).

The molar mass is easy to calculate from the molecular structure and increases with the number of atoms in the molecule.

B. Saturated properties at the normal boiling point

The normal boiling point (NBP) is defined as the thermodynamic state at which the pure component boils at atmospheric pressure (101.325 kPa). Obviously, the normal boiling temperature is one the most important properties used to describe the volatility of a component. Other important properties defined at atmospheric pressure are the enthalpy of vapourisation and the liquid molar volume at this point.

C. Saturated properties at room/ambient temperature

Room or ambient temperature is generally defined as 298.15 K (25 °C), although it is useful to check the source of each publication (288.15 K and 293.15 K are used as well). This term is also associated with standard conditions (see section 2.1.2, p. 26). The boiling temperature, the molar volume and the enthalpy of vapourisation at room temperature are different from those measured at the normal boiling point. Other parameters are the vapour pressure at ambient temperature and the solubility parameter. The latter was introduced by Hildebrand, and defined as the square root of the ratio of the internal energy of vapourisation and the liquid molar volume:

$$\delta_i = \sqrt{\frac{\Delta u_i^{\sigma}}{v_i^*}} \qquad (3.1)$$

This parameter is used in the regular solution theory (section 3.4.2.2.B, p. 175), but it is often considered as an interesting quantity that expresses the cohesion energy among molecules in the pure component. Note that the SI unit is $(J\,m^{-3})^{1/2}$ or $Pa^{1/2}$ and special care is required for conversion if the data are expressed in other units.

This property is sometimes provided for components that are supercritical at room temperature ($T_c < 298.15$ K). DIPPR then provides the value at the component normal boiling point. For components that are solid at 298.15 K, they provide it at the triple point temperature.

Sometimes, the enthalpy of vapourisation is known rather than the internal energy. Using the definition of enthalpy, and assuming that the liquid volume is negligible compared to the vapour volume $\left(\dfrac{RT}{P}\right)$, the solubility parameter (3.1) can be re-written: $\delta_i = \sqrt{\dfrac{\Delta h_i^{\sigma} - RT}{v_i^*}}$.

D. Critical parameters and the corresponding states principle

Critical pressure P_c, critical temperature T_c and critical volume v_c (or compressibility factor $Z_c = \dfrac{P_c v_c}{R T_c}$) are the physical properties that define the critical point of the pure component. The first two are used in all models that are based on the corresponding states principle.

According to this principle (discussed in section 2.2.2.1.C, p. 57), residual thermodynamic properties of the components of a given family behave in the same way if expressed as a function of the reduced temperature $T_r = T/T_c$ and reduced pressure. $P_r = P/P_c$. This principle is extremely helpful for hydrocarbons (non-polar components).

Figure 3.1 illustrates how the critical coordinates vary for hydrocarbon molecules. Note that molecules of similar size, but exhibiting polar behaviour (CO_2) or forming hydrogen bonds (CH_3OH) have noticeably higher values of their critical point (both larger pressure and temperature). A good classification proposed by Ewell *et al.* (1944 [2]) helps understand the kinds of different hydrogen bonds between these molecules.

Within the same family, the critical temperature increases with molar mass, while the critical pressure decreases with molar mass. It may be useful to state that the value of the critical compressibility factor is generally located in a limited range, between 0.25 and 0.32.

Note that the critical point can only be measured for small-sized molecules (typically, for hydrocarbons below C_{16}). For heavier molecules, the critical temperature becomes so high that the molecules become unstable and decompose (crack) before reaching their critical temperature. Nevertheless, these parameters remain important if the corresponding states principle is to be used. Hence, extrapolating correlations have been developed (see for example using group contributions (in section 3.1.2.2, p. 124).

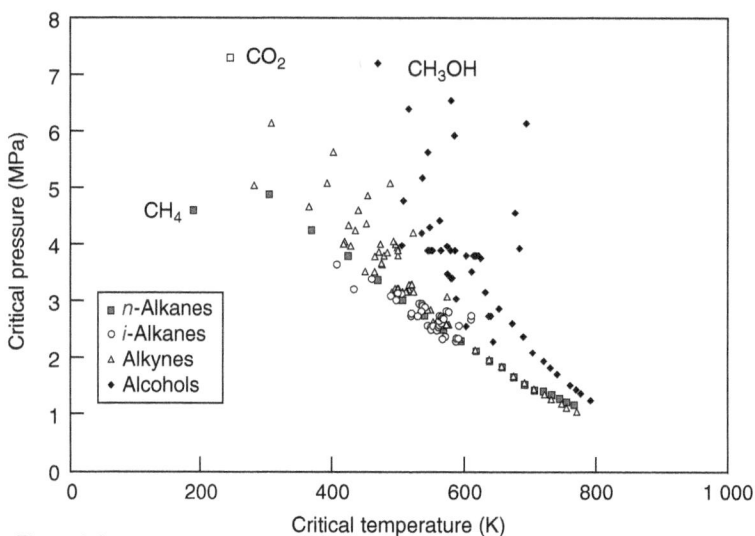

Figure 3.1

Relation between critical pressure and critical temperature for different families of molecules.

E. The acentric factor

This factor is defined by Pitzer [3], based on the reduced vapour pressure ($P_r^\sigma = P^\sigma / P_c$) at the reduced temperature of $T_r = 0.7$:

$$\omega \overset{\Delta}{=} -\log_{10}\left(P_r^\sigma\right)_{T_r=0.7} - 1 \qquad (3.2)$$

This quantity was introduced in order to improve the corresponding states principle. As an example, the acentric factor is used in the vapour pressure equation of pure components proposed by the same Pitzer, or in the Lee-Kesler method (section 3.4.3.3.C.d, p. 202), or in the definition of parameters of cubic equation of state.

The acentric factor is almost zero for methane, and increases with normal boiling point, as shown in figure 3.2. It can reach values above one for heavy molecules. Although a general trend can be observed with boiling point for alkanes, it breaks down when dealing with hydrogen-bonding molecules (alcohols) or polar molecules.

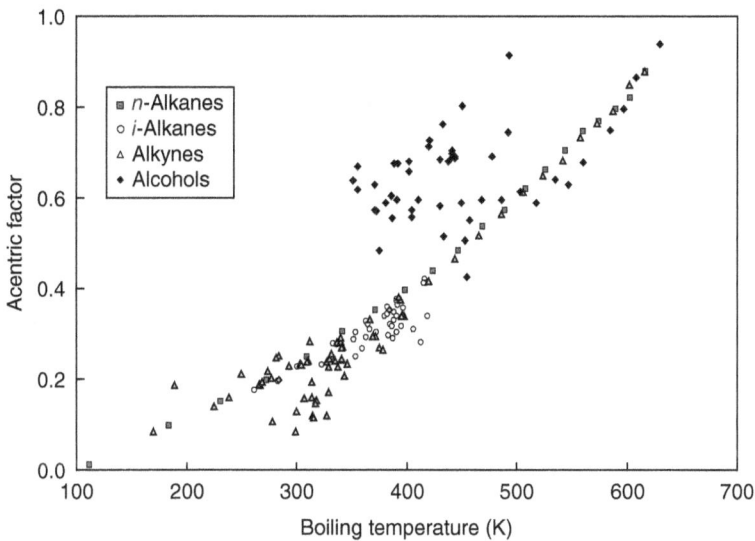

Figure 3.2

Acentric factor as a function of normal boiling temperature for some families of molecules.

F. Crystallisation properties at atmospheric pressure

Crystallisation is the process of formation of a solid from a fluid phase. The inverse process is called melting/fusion (from solid to liquid) or sublimation (from solid to vapour). The word deposition is also sometimes used to describe the change from vapour directly to solid. The crystallisation temperature or melting temperature, enthalpy of crystallisation and solid state molar volume are required to calculate equilibrium with a solid phase. Note that some components (carbon dioxide is the best example), sublimate at atmospheric pressure.

The crystallisation (or fusion) conditions, as shown in table 3.1, depend much more on steric properties than on molar mass. This explains why the ordering of components by normal crystallisation temperature differs from that by normal boiling point: benzene crystallises at a much higher temperature than the other aromatics, because of its regular structure. This is why *p*-Xylene can be separated from its isomers using crystallisation.

Table 3.1   Melting temperature, enthalpy and volume of some hydrocarbons [4]

Component	Melting temperature T_F (K)	Melting enthalpy Δh_F (J mol^{-1})	Melting volume change Δv_F (m^3 mol^{-1})
p-Xylene	286.4	1.71×10^4	1.86×10^{-5}
Cyclohexane	279.7	2.74×10^3	8.30×10^{-6}
Benzene	278.7	9.87×10^3	1.04×10^{-5}
o-Xylene	248	1.36×10^4	1.17×10^{-5}
nC_{10}	243.5	2.87×10^4	2.71×10^{-5}
m-Xylene	225.3	1.16×10^4	1.56×10^{-5}
nC_9	219.7	1.55×10^4	2.17×10^{-5}
CO_2	216.6	9.02×10^3	8.21×10^{-6}
nC_8	216.4	2.07×10^4	1.97×10^{-5}
nC_7	182.6	1.41×10^4	1.39×10^{-5}
Cyclopentane	179.3	6.09×10^2	1.20×10^{-5}
Toluene	178.2	6.64×10^3	5.90×10^{-6}
nC_6	177.8	1.31×10^4	1.26×10^{-5}
Methylcyclohexane	146.6	6.75×10^3	1.66×10^{-5}
nC_5	143.4	8.40×10^3	1.65×10^{-5}
nC_4	134.9	4.66×10^3	9.17×10^{-6}
Methylcyclopentane	130.7	6.93×10^3	1.40×10^{-5}

nC_x stands for normal-alkane with x carbon atoms

The larger the molar mass, the greater likelihood that the component may crystallise at process temperatures. It is therefore important to be able to extrapolate these properties to heavier components.

Laux *et al.* [5] propose correlations that are related to the refractive index. More recently, Briard *et al.* [6] have presented an interesting review of the state of the art concerning the crystallisation properties (figure 3.3). They correlate the melting temperature (expressed in K) of *n*-alkanes to the number of carbon atoms n_C with the expression:

$$T_F = 412.3 \frac{n_C - 0.624}{n_C + 5.880} \tag{3.3}$$

The inset in figure 3 shows that the even-numbered and odd-numbered alkanes do not have exactly the same behaviour. Nevertheless, equation (3.3) provides a correct approximation of the experimental values.

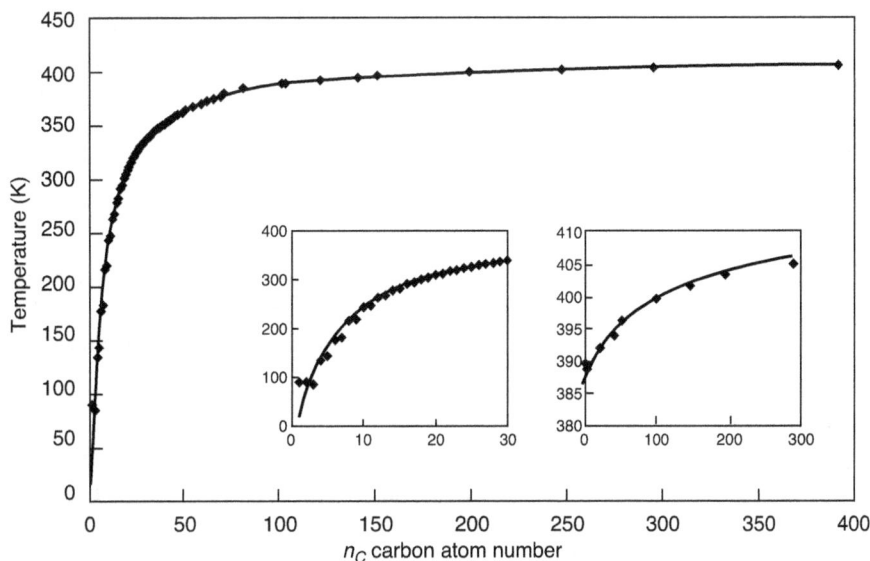

Figure 3.3

Melting points of pure *n*-alkanes as a function of carbon number, from Briard [6]. Reprinted with permission from [6]. © 2003, American Chemical Society.

Although the same effect (difference between odd- and even-numbered alkanes) is observed for the enthalpy, Briard *et al.* [6] propose a single correlation (enthalpy expressed in J mol^{-1}):

$$\Delta h_F = 4099 n_C - 10977 \tag{3.4}$$

Note that solid-solid phase transitions may also occur, in particular for long chains (above C_{20}) [6]. This type of information may be very helpful for modelling the appearance of wax at temperatures well below the normal crystallisation temperatures. Although data do exist, they are not always easy to find [7].

G. Triple point data

The triple point (T_t, P_t) is that point on the phase diagram where three phases coexist for a single component (usually associated with vapour, liquid and solid as shown in section 2.1.3.1, p. 28). The triple point pressure is often much lower than atmospheric pressure. Nevertheless, as crystallisation temperature varies very little with pressure, the triple point temperature and the atmospheric crystallisation temperature is very often identical, as well as the other fusion or crystallisation properties.

H. Formation properties

The Gibbs energy of formation and the enthalpy of formation of the ideal gas at standard conditions (see section 2.2.4.2.B, p. 91) are required for chemical equilibrium calculations. They are related through:

$$\Delta G_{f,i}^0(T) = \Delta H_{f,i}^0(T) - T\Delta S_{f,i}^0(T) \tag{3.5}$$

They are defined as the property change (Gibbs energy, Enthalpy or Entropy) upon formation of the component from its elements in their standard states. Note that the temperature dependence of this property is not linear as could be derived from relation (3.5). The use of these properties and their temperature dependence is further discussed in section 2.2.4.2.B (p. 91).

I. Absolute entropy

The entropy at standard temperature and atmospheric pressure may also be found. Note that this is different from the formation entropy! It provides the value of the entropy, as defined by the third principle of thermodynamics, which states that at absolute temperature of zero, the entropy of any component is zero.

J. Refractive index

Although the refractive index is not truly a thermodynamic property, some authors [8, 9] have found that it may prove to be extremely usefull in the prediction of other properties. It is defined as the ratio of the velocity of light in vacuum to that in the given substance. The values used in petroleum engineering generally concern the frequency of light of the sodium D line, and the substance is taken at 293.15 K and atmospheric pressure. Its main advantage is that it can be measured in a fairly straightforward manner using a refractometer. The trend of the refractive index as a function of hydrocarbon molar mass is given in figure 3.4.

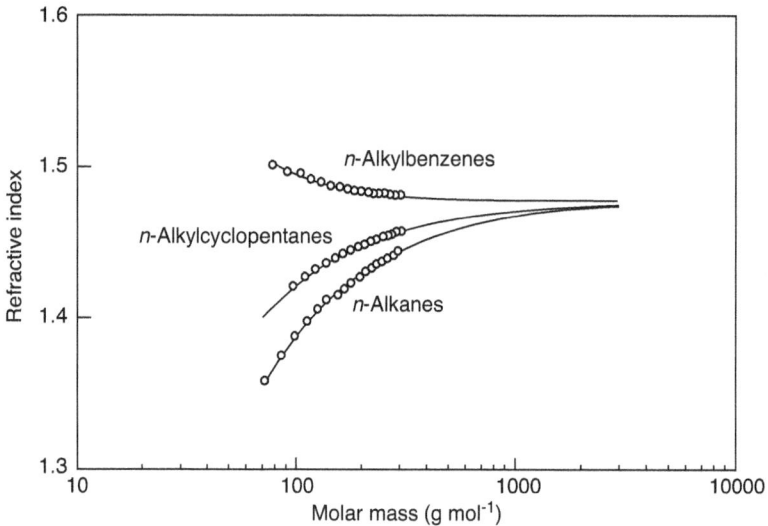

Figure 3.4

Refractive index as a function of molar mass for a number of hydrocarbons, taken from Riazi 2005 [10].

K. Molecular descriptors

Molecular dipole or quadrupole moments, radius of gyration and the like are generally not considered as model parameters. With the advent of molecular simulation tools and of models based on statistical mechanical considerations, however, such parameters that relate directly to the molecular structure may become increasingly used (e.g. SAFT models use segment diameters and well-depth energies [11, 12] sections 3.4.3.2, p. 192 and 3.4.3.5, p. 216).

3.1.1.2 Temperature dependent physical properties

Different types of temperature dependent properties may be considered. Three of them are saturation curves, *i.e.* curves that provide the boundary, in the *PT* phase diagram, between different phases:

The **vapour pressure** shows the boundary between vapour and liquid.

The **sublimation** curve shows the boundary between vapour and solid.

The **crystallisation** curve shows the boundary between liquid and solid.

For all of these curves, Clapeyron's law must always be valid:

$$\left.\frac{dP}{dT}\right|_{\sigma} = \frac{\Delta h^{\sigma}}{T\Delta v^{\sigma}} \tag{3.6}$$

It relates the slope of the curve with the enthalpy and volume difference upon phase change. This relationship is of key importance for evaluating whether the data are of good quality.

Many other properties have important variations with respect to the temperature. This is the case with density, of interest only for liquids and solids. Of considerable importance for energy calculation is the heat capacity, for ideal gas, liquid and solid phases. Another energy parameter is the heat of vapourisation (related to vapour pressure as mentioned above). The second virial coefficient is also a function of temperature.

A. Vapour pressure curve

Along the vapour pressure curve, Clapeyron's law is often transformed according to Clausius, who assumed that the vapour phase can be considered as an ideal gas, that the liquid volume can be neglected and that the enthalpy of vapourisation is constant. In this case, equation (3.6) is written as:

$$\frac{d\ln P^{\sigma}}{d\left(1/T\right)} \approx -\frac{\Delta h^{\sigma}}{R} \tag{3.7}$$

Although this expression is of great help for providing trends and orders of magnitude, it should not be used for exact calculations. This equation is nevertheless important in determining whether a series of data makes sense. In addition, it illustrates the fact that on a plot showing the logarithm of pressure as a function of the inverse of absolute temperature, the vapour pressure line should be close to a straight line, whose slope is a measure of the enthalpy of vapourisation.

The vapour pressure is one of the most important properties required for phase equilib-rium calculations. Many correlations exist, but only a few of them will be presented. Remember that if no parameters exist, they must be determined from experimental data. It is also important to mention some predictive equations constructed using only critical coordi-nates and the acentric factor. These equations may be used where there is no other alterna-tive, particularly in the case of petroleum fluids. They may also be useful for evaluating the quality of the data.

a. Two-point equation

This is the straightforward application of Clausius's equation (3.7):

$$\ln P^\sigma = \ln P_1^\sigma + \left(\frac{1}{T}-\frac{1}{T_1}\right)\frac{\ln P_2^\sigma - \ln P_1^\sigma}{\frac{1}{T_2}-\frac{1}{T_1}} \tag{3.8}$$

If the critical point and the acentric factor are known, using (3.2), this equation may also be written as:

$$\log_{10} P_r^\sigma = -\frac{7}{3}(1+\omega)\left(\frac{1-T_r}{T_r}\right) \tag{3.9}$$

where subscript r stands for reduced coordinates: $T_r = T/T_c$ and $P_r = P/P_c$. Also known as the Wilson equation [13], it is useful to estimate distribution coefficient for hydrocarbons (see section 3.5.1.1, p. 226).

It may be of interest to note that using (3.9) and (3.7) it is possible to relate acentric factor with an average value of the heat of vapourisation:

$$\Delta h^\sigma \approx 2.3025 \times \frac{7}{3}(1+\omega)\,RT_c \tag{3.10}$$

b. Three-point equation

Note that the preceding concept can be generalised to a curve passing through any number of points. A quadratic version passing through the following special points (critical point $\{x_1, y_1\}$, normal boiling point $\{x_2, y_2\}$ and triple point $\{x_3, y_3\}$) is very easy to implement as shown by Ledanois *et al.* [14]. The corresponding equation can be obtained by Newton's polynomial development:

$$\ln P^\sigma = y_1 + (x-x_1)\frac{y_2-y_1}{x_2-x_1} + (x-x_1)(x-x_2)\frac{\frac{y_3-y_2}{x_3-x_2}-\frac{y_2-y_1}{x_2-x_1}}{x_3-x_1} \tag{3.11}$$

where $x_i = 1/T_i$ and $y_i = \ln P_i^\sigma$.

c. Antoine equation

The Antoine equation [15] is probably the most well known vapour pressure equation:

$$\ln P^\sigma = A + \frac{B}{C+T} \tag{3.12}$$

It is simple and there are numerous sets of parameters available in databases. The logarithm may be decimal or Neperian/natural. Pay attention to the units corresponding to the parameters selected. This equation has poor extrapolation behaviour, however, and must only be used within the temperature range provided with the parameters. The B parameter is always negative and some versions of this equation are found with a minus sign and a table with all B positive.

It is sometimes convenient to use equation (3.12) in dimensionless form, in which case P^σ is replaced by P_r^σ and T by T_r.

d. DIPPR equation

The DIPPR correlation used for the vapour pressure is as follows:

$$\ln P^\sigma = A + \frac{B}{T} + C \ln T + DT^E \tag{3.13}$$

We see that the first terms correspond to the linear expression that was developed in a semi-theoretical way by Clausius [equation (3.7)], but it has been corrected with two additional terms and three additional parameters. This equation provides a relatively good fit over the entire saturation domain, from the triple point to the critical point.

e. Wagner equation

The Wagner equation [16] is based on the critical coordinates:

$$\ln \frac{P^\sigma}{P_c} = \frac{T_c}{T}\left(A\tau + B\tau^{1.5} + C\tau^3 + D\tau^6\right) \tag{3.14}$$

with

$$\tau = 1 - \frac{T}{T_c} \tag{3.15}$$

It has one parameter less than the DIPPR equation, but due to its design, it always passes through the critical point, which guarantees a good accuracy for reduced temperatures close to 1. A generalised variant of the Wagner equation, with change in the exponents, has been proposed by Ambrose and Walton [17] and is recommended by the API-Technical Data Book of 1997 [18]:

$$\ln P_r^\sigma = -5.97616\tau + 1.29874\tau^{1.5} - 0.60394\tau^{2.5} - 1.06841\tau^5$$
$$+ \omega\left(-5.03365\tau + 1.11505\tau^{1.5} - 5.41217\tau^{2.5} - 7.46628\tau^5\right) \tag{3.16}$$
$$+ \omega^2\left(-0.64771\tau + 2.41539\tau^{1.5} - 4.26979\tau^{2.5} + 3.25259\tau^5\right)$$

f. Lee and Kesler equation [19]

A more ancient but simpler predictive equation was proposed by Lee and Kesler [19].

$$\ln P_r^\sigma = 5.92714 - \frac{6,09648}{T_r} - 1.28862\ln T_r + 0.168347T_r^6$$
$$+ \omega\left(15.2518 - \frac{15.6875}{T_r} - 13.4721\ln T_r + 0.43577T_r^6\right) \tag{3.17}$$

Example 3.1 Vapour pressure predictions of methane using different formulas

Compare the results of the various correlations with the DIPPR correlation.

In the following tables, the characteristic properties necessary for the evaluation of each expression is given. Remember that the normal boiling point is measured at 101.325 kPa. The various parameters of the equations are shown in tables 3.2 and 3.3:

Table 3.2   Properties of methane used as parameters in the vapour pressure equations

Parameters	T_c (K)	P_c (kPa)	ω (–)	T_t (K)	P_t (kPa)	T_b (K)
Value	190.56	4599	0.0115	90.69	11.7	111.66

Table 3.3   Parameters for various vapour pressure equations for Methane

Parameter	A	B	C	D	E
Antoine, eq. (3.12) – API	8.6775	– 911.234	–6.34		
DIPPR, eq. (3.13)	38.664	– 1314.7	– 3.3373	3.0155×10^{-5}	2

(Temperature in K or dimensionless-Pressure in Pa or dimensionless)

Analysis:

Pressure and temperature are the basic properties of the system.

Methane is a well-known light hydrocarbon.

Phases present are vapour and liquid all along the saturation curve.

Solution:

Tables for each equation are constructed and compared with the DIPPR set of calculated values. Note that the DIPPR equation does not predict exactly the critical point, the normal boiling point and the triple point; there are relative deviations around 0.2 to 0.6 %. Additionally, the acentric factor ('omega point' shown in figure 3.5) predicted by this equation is not exactly the same as that given in the database.

Relative differences are plotted in figure 3.5 over the full temperature range, from the triple point to the critical point for six different equations:

• the 3 points equation (3.11) that uses the critical point, the normal boiling point and the triple point;
• the 2 points equation (3.8) that uses the critical point and the triple point. Note that the Wilson equation is identical but uses the critical point and the 'omega point' instead. It is not shown in figure 3.5.
• The corresponding states equations of Ambrose (equation 3.16) and Lee-Kesler (equation 3.17) are also shown: they provide almost identical results above the 'omega point', but the more complex Ambrose equation extrapolates better to low reduced temperatures.
• The Antoine equation (3.12) is used either with a regression over the entire temperature domain, or using the parameters that are recommended by API Riazi (2005) [10], but no validity range is associated with it. It is clear that the difference becomes very high beyond 150 K, clearly indicating that this equation must be used only over the range 90-140 K.

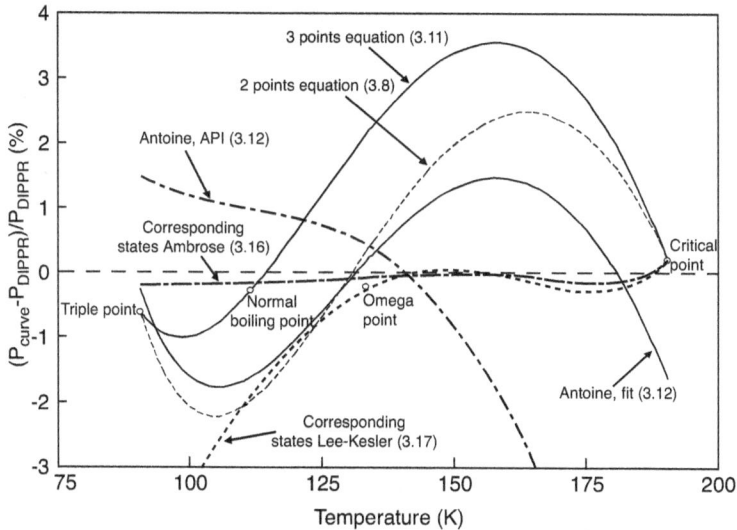

Figure 3.5

Precision of various equations for the vapour pressure of methane against the DIPPR equation.

This example is discussed on the website:
http://books.ifpenergiesnouvelles.fr/ebooks/thermodynamics

B. Sublimation curve

Theoretically, the Clausius approximation of the Clapeyron equation (3.7) is also applicable to the sublimation curve. All other comments can also be extrapolated, except that it does not end at the critical point, but at the triple point where the three physical states coexist (vapour, liquid and solid, see section 2.1.1.1, p. 24). Very often, the triple point pressure is very low resulting in the fact that this point is very difficult to observe experimentally. Consequently, sublimation data are rarely used.

C. Crystallisation (melting) fusion curve

The liquid and the solid volume are often very close to each other. Hence, using Clapeyron's equation (3.6) and data of table 3.1, it is easy to show that the crystallisation curve is almost vertical. Table 3.4 shows some examples of how much the temperature can vary with pressure. The temperature difference generally is close to 1 K for a pressure difference of 5 MPa. However, for some components (in particular small naphthenic molecules), the difference may become significant. This behaviour is associated with the change in volume upon crystallisation, probably because the molecular orientations vary in the liquid and solid phases.

Table 3.4   Effect of pressure on the crystallisation temperature of some hydrocarbons [4]

	T_F (K) (0.1 MPa)	T_F (K) (5 MPa)	T_F (K) (10 MPa)	ΔT_F (K) (5-10 MPa)
p-Xylene	286.4	288.0	289.5	1.5
Cyclohexane	279.8	283.9	288.2	4.3
Benzene	278.7	280.1	281.6	1.5
o-Xylene	248.0	249.0	250.1	1.1
nC_{10}	243.5	244.7	245.8	1.1
m-Xylene	225.3	226.8	228.3	1.5
nC_9	219.7	221.2	222.7	1.5
CO_2	216.5	217.5	218.4	0.9
nC_8	216.4	217.4	218.4	1.0
Cyclopentane	179.6	197.1	214.9	17.8
nC_7	182.6	183.5	184.4	0.9
Toluene	178.2	179.0	179.8	0.8
nC_6	177.8	178.7	179.5	0.8
Methylcyclohexane	146.6	148.4	150.2	1.8
nC_5	143.4	144.8	146.2	1.4
nC_4	134.9	136.2	137.5	1.3
Methylcyclopentane	130.8	132.1	133.4	1.3

nC_x stands for normal-alkane with x carbon atoms

D. Liquid molar volume

The liquid molar volume and specifically the saturation molar volume, changes from triple to critical point. At the critical point it must match the critical volume.

a. Rackett

The Rackett [20] equation is both one of the simplest and most accurate equations for liquid saturated molar volumes of hydrocarbons. Its accuracy decreases for hydrocarbons beyond C_{10} as shown in example 3.2. The equation is originally written as follows:

$$v^{L,\sigma} = \frac{RT_c}{P_c} Z_{RA}^{\left(1+\left(1-T_r\right)^{2/7}\right)}$$

(3.18)

where Z_{RA} is the Rackett compressibility factor. It is interesting to observe that Z_{RA} can be evaluated from a single experimental point of liquid density. This parameter is not always available, and may be calculated by the following expression, based on the acentric factor:

$$Z_{RA} = 0.29056 - 0.08775\omega$$

(3.19)

Note that correlation (3.18) requires no numerical parameters other than the critical coordinates. It is entirely based on the corresponding states principle and is used for petroleum fluids, as recommended in the API handbook [18].

b. DIPPR correlation

In its database, DIPPR uses the following expression for the saturated liquid molar volume:

$$v^{L,\sigma} = \frac{B^{\left(1+\left(1-\frac{T}{C}\right)^D\right)}}{A} \qquad (3.20)$$

Its range of validity spans the entire range from the triple point to the critical point provided that the parameters have been determined adequately.

c. Liquid volume pressure effect (Tait)

When high pressure applications are considered, it may be useful to have a correlation that allows calculation of liquid density as a function of pressure. The Tait [21] equation is typically used for that purpose. Knowing the liquid volume at P_0, the presure effect is given as:

$$\frac{v^L(P_0)-v^L(P)}{v^L(P_0)} = C\ln\left(\frac{B+P}{B+P_0}\right) \qquad (3.21)$$

The constants B and C should be adapted for each component. Nevertheless, for light hydrocarbons or petroleum fractions, the following COSTALD kind of equations are recommended (up to 65 MPa, according to [22]) for both constants:

$$\frac{B}{P_c} = -1 - 9.0702\left(1-T_r\right)^{1/3} + 62.45326\left(1-T_r\right)^{2/3} - 135.1102\left(1-T_r\right) + e\left(1-T_r\right)^{4/3} \qquad (3.22)$$

$$e = \exp\left(4.79594 + 0.250047\omega + 1.14188\omega^2\right)$$

and

$$C = 0.0861488 + 0.0344483\omega \qquad (3.23)$$

The API Technical Data Book [18] offers various expressions for B and C (Riazi, 2005 [10]). Other models suitable for compressed liquids can be found in Aalto *et al.* (1996, 2002) [23-25].

Example 3.2 Short quality evaluation of liquid molar volume correlations

A comparison of liquid molar volume for various hydrocarbons is proposed. Differences between the DIPPR correlation (eq. 3.20) and the Racket prediction (eq 3.18) will be analysed. Z_{RA} is calculated from equation (3.19).

Parameter values for the different components are found in table 3.5:

Table 3.5   Critical constant and parameters for liquid molar volume of some hydrocarbons (DIPPR [4])

Component	T_c (K)	P_c (kPa)	ω	A	B	C	D
Ethane	305.32	4872	0.0995	1.9122	0.27937	305.32	029187
n-Hexane	507.6	3025	0.3013	0.70824	0.26411	507.6	0.27537
Cyclohexane	553.8	4080	0.2081	0.88998	0.27376	553.8	0.28571
Benzene	562.05	4895	0.2103	1.0259	0.26666	562.05	0.28394
n-Decane	617.7	2110	0.4923	0.41084	0.25175	617.7	0.28571

Analysis:

The property given is temperature, or preferably a range of temperatures on the saturation curve.

Components are all hydrocarbons of different molar mass and different families.

The phase of interest is liquid along the saturation curve.

Model requirement:

Two different correlations are compared: the DIPPR correlation (fitted on experimental data) and the Rackett prediction. The relative deviation between the Rackett equation and the DIPPR saturated liquid volume correlation has been evaluated for a number of hydrocarbons.

Solution:

The typical uncertainty on volume measurements is close to 3% (figure 3.6). We can therefore conclude that, especially for light hydrocarbons, the Rackett equation is very useful. The deviation is larger for heavier hydrocarbons. The equation is not suitable for non-hydrocarbons.

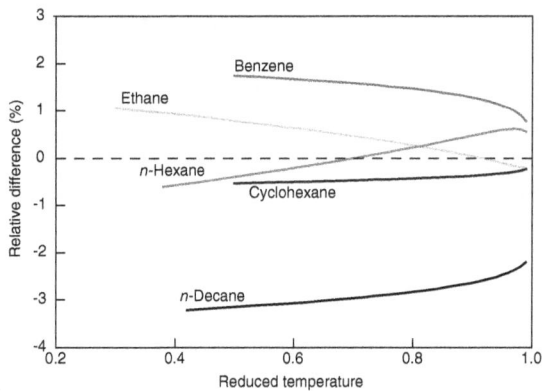

Figure 3.6

Evaluation of the Rackett equation with respect to the DIPPR correlations.

This example is discussed on the website:
http://books.ifpenergiesnouvelles.fr/ebooks/thermodynamics

E. Ideal gas heat capacity curve

In order to calculate thermal properties, the heat capacity of the pure components must be known as a function of pressure and temperature. The ideal gas heat capacity is essential when using the residual approach (section 2.2.2.1, p. 52).

Many equations have been proposed (Younglove [26, 27]); the Wilhoit equation is proposed by NIST and the TRC [28]; PPDS-Physical property data service [29]) but in practice, only two will be mentioned:

a. Polynomial expansion

The use of polynomial expansion has been proposed by Shomate (1946 [30]) and later by Passut and Danner in 1972 [31]. This simple polynomial function gives a good description of the ideal gas heat capacity:

$$c_p^\# = A + BT + CT^2 + DT^3 + ET^4 \tag{3.24}$$

This equation is unable to capture the ideal gas heat capacity behaviour over a large range of temperature, but is very simple to integrate.

b. DIPPR (Aly and Lee)

Using statistical mechanical arguments, Aly and Lee (1981, [32]) provide following equation:

$$c_p^\# = A + B\left[\frac{C/T}{\sinh(C/T)}\right]^2 + D\left[\frac{E/T}{\cosh(E/T)}\right]^2 \qquad (3.25)$$

This equation has the major advantage of providing very accurate results over a very wide range of temperatures (typically from 150 K up to 1500 K). It is however slightly more complex to use, especially considering that it must be integrated to calculate enthalpy and entropy. This equation is widely used in the DIPPR database.

Equations for enthalpy and entropy are then:

$$h^\#(T) - h^\#(T_0) = \left[AT + BT(C/T)\coth(C/T) + DT(E/T)\tanh(E/T)\right]_{T_0}^T \qquad (3.26)$$

$$s^\#(T, P_0) - s^\#(T_0, P_0) = \begin{bmatrix} A\ln(T) + B\{(C/T)\coth(C/T) - \ln(\sinh(C/T))\} \\ -D\{(E/T)\tanh(E/T) - \ln(\cosh(E/T))\} \end{bmatrix}_{T_0}^T \qquad (3.27)$$

c. Petroleum cuts

In order to calculate enthalpic properties with pseudo-components, we saw in section 2.2.2.1, p. 52 that the ideal gas heat capacity of each component is required.

Riazi (2005 [10]) recommends the use of the method by Lee-Kesler (section 3.4.3.3.C.d, p. 202) for this purpose:

$$c_P^\# = M_w\left[A_0 + A_1 T + A_2 T^2 - C\left(B_0 + B_1 T + B_2 T^2\right)\right] \qquad (3.28)$$

$$A_0 = -1.4779 + 0.11828 K_W$$

$$A_1 = -\left(1.4779 - 8.69326 K_W + 0.27715 K_W^2\right) \times 10^{-4}$$

$$A_2 = -2.2582 \times 10^{-6}$$

where

$$B_0 = 1.09223 - 2.48245\omega$$

$$B_1 = -\left(3.434 - 7.14\omega\right) \times 10^{-3}$$

$$B_2 = -\left(7.2661 - 9.2561\omega\right) \times 10^{-7}$$

$$C = \left[\frac{\left(12.8 - K_W\right)\left(10 - K_W\right)}{10\omega}\right]^2$$

where K_W is the Watson factor defined in (3.44) hereafter (section 3.1.2.3.C, p. 135) M_w is the molar mass and ω the acentric factor.

Example 3.3 Comparison of the ideal gas heat capacity prediction of pentane over the range 200-1500 K with the DIPPR equation

Two different polynomial forms (equations (3.24)) are compared with the DIPPR expression over the range 200-1500 K using different parameters. Both are obtained by regression of the DIPPR values, one over the complete range and the other over the range 400-1500 K. The resulting $C_p^\#$ are expressed in J kmol^{-1} K^{-1}. Parameters for the different equations are shown in table 3.6:

Table 3.6   Parameters for various ideal heat capacity equations for pentane

Equation	A	B	C	E
DIPPR (Aly and Lee)	88050	301100	1650.2	189200
Polynomial 400-1500 K	− 9352.073	516.5751	− 0.3096357	9.78224×10^{-5}
Polynomial 200-1500 K	18576.58	366.4169	− 0.03453294	$− 1.09886 \times 10^{-4}$

Analysis:

The properties are temperature and ideal gas heat capacity.

Component is pentane.

The phase is the ideal gas.

Solution:

A graph is constructed (figure 3.7) showing the relative difference between the two polynomials and the DIPPR equation (expression of Aly and Lee). The figure shows that it is impossible to reproduce the low and high temperatures simultaneously with the polynomial equation. Nevertheless, the fitting is adequate over a smaller temperature range.

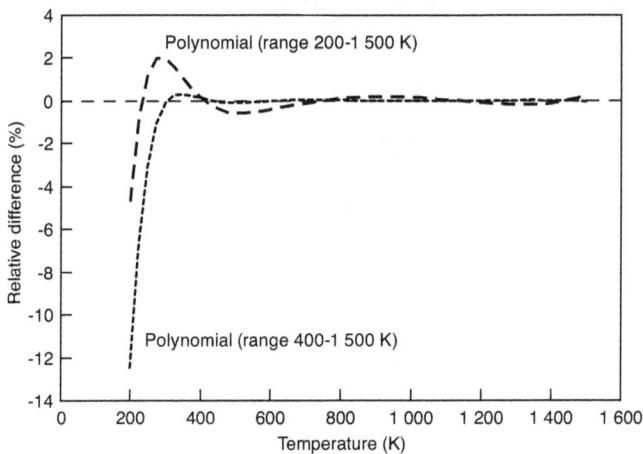

Figure 3.7
Typical deviation behaviour between the polynomial equations and the Aly and Lee equation for ideal gas heat capacity of pentane.

This example is discussed on the website:
http://books.ifpenergiesnouvelles.fr/ebooks/thermodynamics

F. Heat of vapourisation

The heat of vapourisation is the opposite of the heat of condensation. Since the heat content of a vapour is always larger than that of the liquid, hence, we will use the positive value defined below:

$$\Delta h^\sigma (T) = h^V (T) - h^L (T) \qquad (3.29)$$

The heat of vapourisation used in the Clausius-Clapeyron equation (3.7) is an average value: in reality, this property is temperature dependent and reaches zero at the critical point.

a. Watson equation

The Watson equation uses the relationship that has been shown to exist [33] between vapourisation properties in the neighbourhood of the critical point. It uses a critical exponent. The value of the exponent has been improved over time:

$$\frac{\Delta h^\sigma (T)}{\Delta h^\sigma (T_0)} = \left(\frac{1 - T_r}{1 - T_{r,0}} \right)^{0.38} \qquad (3.30)$$

Where T_0 is any temperature lower than the critical temperature, and $\Delta h^\sigma (T_0)$ is the corresponding enthalpy of vapourisation and $T_r = T/T_c$ is the reduced temperature. This expression is accurate over a relatively large temperature range.

b. DIPPR equation

The DIPPR equation uses essentially the same limiting behaviour, but allows the exponent to vary as a function of the reduced temperature:

$$\Delta h^\sigma (T) = A \left(1 - T_r \right)^{\left(B + CT_r + DT_r^2 + ET_r^3 \right)} \qquad (3.31)$$

c. Clapeyron equation

The Clapeyron equation (3.6) provides a relationship between three saturation properties: vapour pressure derivative, vapourisation volume and vapourisation enthalpy. This expression is always correct. Hence, if two of the properties are known, the third is calculated. Equation (3.6) can also be written as:

$$\Delta h^\sigma = -\frac{d \ln P^\sigma}{d \left(1/T \right)} R \Delta Z^\sigma \qquad (3.32)$$

This expression simplifies further for low vapour pressures since it can then be stated that $\Delta Z^\sigma = 1$, thus yielding the Clausius relationship (3.7).

Example 3.4 Short quality evaluation of enthalpy of vapourisation correlations

A comparison of vapourisation enthalpy correlations for two hydrocarbons is proposed. Differences between correlations are analysed.

This example is discussed on the website:
http://books.ifpenergiesnouvelles.fr/ebooks/thermodynamics

G. Second virial coefficient

Abundant literature exists concerning the temperature dependence of the second virial coefficient (see for example [34-37]). The reason for this is that its value is well-known for a large number of components and that it is an interesting source of independent information concerning the repulsive and attractive molecular potentials. It can be shown from statistical thermodynamics, that:

$$B(T) = -2\pi N_{Av} \int_0^\infty \left[\exp\left(-\frac{u(r)}{RT} \right) - 1 \right] r^2 dr \tag{3.33}$$

where $u(r)$ is the pair interaction energy as defined in section 3.4.3.2.A, p. 193 and r is the distance between the centers of two particules. N_{Av} is Avogadro's number.

The temperature dependence of the second virial coefficient is shown in figure 3.8. At low temperature, the value is negative, indicating that the attractive potential dominates. When the temperature rises, it crosses zero at the component Boyle temperature, and reaches an asymptotic positive value for very high temperatures. The latter value corresponds to the hard sphere volume of the molecular fluid.

a. DIPPR relationship

The DIPPR correlation for the second virial coefficient is a polynomial in inverse temperature. Not all terms are used, however:

$$B(T) = A + \frac{B}{T} + \frac{C}{T^3} + \frac{D}{T^8} + \frac{E}{T^9} \tag{3.34}$$

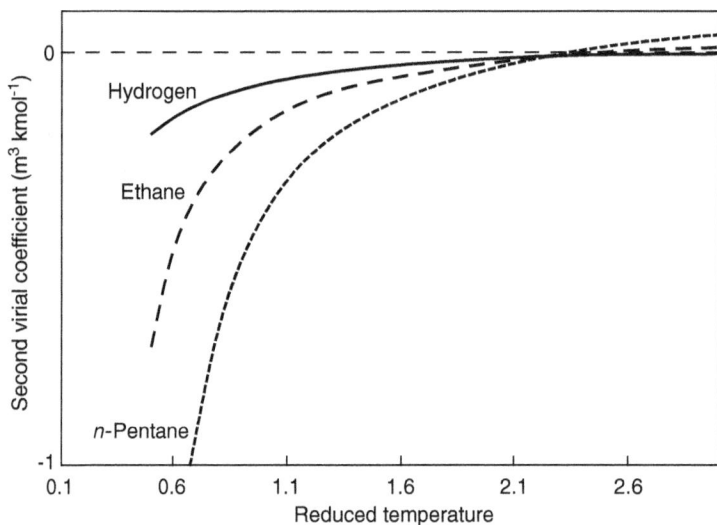

Figure 3.8

Temperature dependence of the second virial coefficient (from Vidal [38]) of a number of representative components.

Reprinted from [38], with permission from Elsevier.

Another classic formulation for the second virial coefficient is given by (Ambrose [44, 45]):

$$B(T) = A + B\exp\left(\frac{C}{T}\right) \tag{3.35}$$

The temperature validity range of this second equation is relatively small.

b. Corresponding states relationship

If no parameter is available for equation (3.34) or (3.35) and if hydrocarbon components are considered, *i.e.* for which the corresponding states principle is applicable, the correlation proposed by Tsonopoulos [39] can be used:

$$\frac{BP_c}{RT_c} = F^{(0)}(T_r) + \omega F^{(1)}(T_r) \tag{3.36}$$

with

$$F^{(0)}(T_r) = 0.1446 - \frac{0.330}{T_r} - \frac{0.1385}{T_r^2} - \frac{0.0121}{T_r^3} - \frac{0.000607}{T_r^8} \tag{3.37}$$

and

$$F^{(1)}(T_r) = 0.0637 + \frac{0.331}{T_r^2} - \frac{0.423}{T_r^3} - \frac{0.008}{T_r^8} \tag{3.38}$$

Note however that the accuracy of corresponding states methods is lower than that of equations whose parameters are directly adjusted on experimental data.

Example 3.5 Short quality evaluation of second virial coefficient

A comparison of various models for the second virial coefficient for hydrocarbons is proposed. Differences between correlations are analysed.

This example is discussed on the website:
http://books.ifpenergiesnouvelles.fr/ebooks/thermodynamics

3.1.2 Types of components

As stressed in the first chapter of this book, the quality of a thermodynamic model greatly depends on the numerical parameters used. The various chemical species are characterised through parameter values. Depending on the type of species at hand, the origin of these parameters may vary. In this work, we classify the components in three major families: database components, non-database components and petroleum fluid components.

3.1.2.1 Database components

All commercial simulators come with a default database that contains the most common parameters used in the models they provide. This is very convenient but may be dangerous if the results are extrapolated outside the range in which the proposed values have been generated. Sometimes, they provide a choice between several databases. Obviously, even using the same model, the numerical results will change depending on the database.

Databases are a collection of numerical values that relate to physical properties of pure compounds or mixtures. They can contain either directly measured data, model parameters, or both. They have been constructed using a large number of published or regressed data.

Commercial databases are available independent from simulators as well. It is important to mention the Thermodynamic Research Center (TRC: http://trc.nist.gov/) which is part of National Institute of Standards and Technology (NIST), a US government agency. It is an interesting source of data because it is free and can be found on the NIST website (http://webbook.nist.gov/chemistry/). Another important source of experimental data is the Dortmund Data Bank (DDBST: http://www.dortmunddatabank.com) distributed within the Detherm database by Dechema: http://i-systems.dechema.de/detherm. Its added value is essentially its very large number of mixture data. Other well-known databases are the Korean databank (http://infosys.korea.ac.kr/kdb/index.html), the Yaws databank (http://www.knovel.com), the International Union of Pure and Applied Chemistry (IUPAC: http://old.iupac.org/), the Danish Computer Aided Process-Product Engineering Centre database (CAPEC: http://www.capec.kt.dtu.dk/) [40], The Helvetica Physical Properties Sources Index (PPSI: http://www.ppsi.ethz.ch/en/) and the English Physical Property Data Service (PPDS: http://www.ppds.co.uk/). This list is not exhaustive.

Determining which is the best database is a complex task, which explains why the American Institute of Chemical Engineers (AIChE) came up with the Design Institute for Physical PRoperties (DIPPR: http://dippr.byu.edu/) project, whose purpose is to evaluate a large number of databases from different origins, and to recommend pure component data. The DIPPR database is nowadays recognised as one of the leading sources of pure component data. It contains (as of november 2008) 1944 compounds and values for 49 thermophysical properties (34 constant properties and 15 temperature dependent properties).

Obviously, because of the strategic importance of the physical property data for process simulations, many private companies have their own database which is not open to public access.

Example 3.6 Comparison of critical points and acentric factor from different databases

This example shows comparisons of critical data between different databanks. The suggested comparison is carried in two steps. The first one is a direct comparison between the values extracted from databases. The second step discusses the impact of these differences on one of the most important property for vapour - liquid equilibrium: the vapour pressure. Three hydrocarbons have been chosen in this example: vinylacetylene, 2-methyl-1-butene, trans 1,3-pentadiene. The critical temperatures and pressures and the acentric factors of two databases are given in table 3.7.

Table 3.7 Critical parameters and acentric factor of hydrocarbons

Database		Vinylacetylene	2-methyl-1-butene	Trans 1,3-Pentadiene
DIPPR	T_c (K)	454	465	500
	P_c (kPa)	4860	3447	3740
	ω (–)	0.1069	0.2341	0.1162
Process Simulation Package	T_c (K)	455.15	470.15	496.15
	P_c (kPa)	4894	3850	3992
	ω (–)	0.1205	0.234	0.1719

Analysis:

The properties are critical temperature, critical pressure and acentric factor.

The vapour pressure is estimated with the Lee and Kesler equation (section 3.1.1.2, p. 109) which is function of critical parameters and acentric factors.

Components are hydrocarbons, which are databank components.

Solution:

Table 3.8 gives the direct comparison between these values. We can observe that the critical temperature shows the smallest deviation between databases. In opposite, there is a large dispersion in acentric factor. This kind of behaviour is common when there is a direct comparison of different sources. Generally, light compounds have similar values in different Process Simulators, but heavy compounds, unstable compounds or branched isomers, which are less well characterised, can show larger deviations. It is important to bear in mind that some published values in databases are not experimental values. Some are predicted or extrapolated. For example, the critical point is not always measurable due to the cracking phenomena. The acentric factor is evaluated at reduced temperature equal to 0.7 but what is this temperature if the critical point is unknown?

Table 3.8   Deviation between the two databanks values

	Vinylacetylene	2-methyl-1-butene	Trans 1,3-Pentadiene
T_c (K)	1.15 (0.3%)	−5.15 (1.1%)	3.85 (0.8%)
P_c (kPa)	−34 (0.7%)	−403 (11.7%)	−252 (6.7%)
ω (−)	−0.013 (12.8%)	0.000 (0.03%)	−0.0557 (48%)

The direct comparison of these critical values is not the best criterion to check the validation of the databank. It is important to check the sensitivity of the complete model to evaluate how much the result is affected by a small deviation of the input. As an example, the critical point and the acentric factor are generally used in order to calculate the vapour pressure.

Using any corresponding states relationship (e.g. Ambrose, as discussed in exercice 3.1.) we can calculate the vapour pressure of the three compounds as a function of temperature in the reduced temperature range of 0.5 to 1.0. The relative deviations using these two databanks parameters are the following:

• 4.4 % for the vinylacetylene,
• 2.4 % for the 2-methyl-1-butene,
• 7.8 % for the trans 1,3-Pentadiene.

To validate a databank, it is important to compare the model with experimental data. Figure 3.9 shows such an example with the 2-methyl-1-butene, as the percentage deviation between the predicted vapour pressure and the experimental ones. From this type of plot, it is possible to identify the trends of the curves with respect to temperature: the DIPPR parameters will result in smaller average deviation in the temperature range surrounding 350K, but the extrapolation towards lower temperatures will not be garanteed. In opposition, the process package database provides overall an overestimation of the vapour pressure, but the tendency with temperature is probably more acceptable.

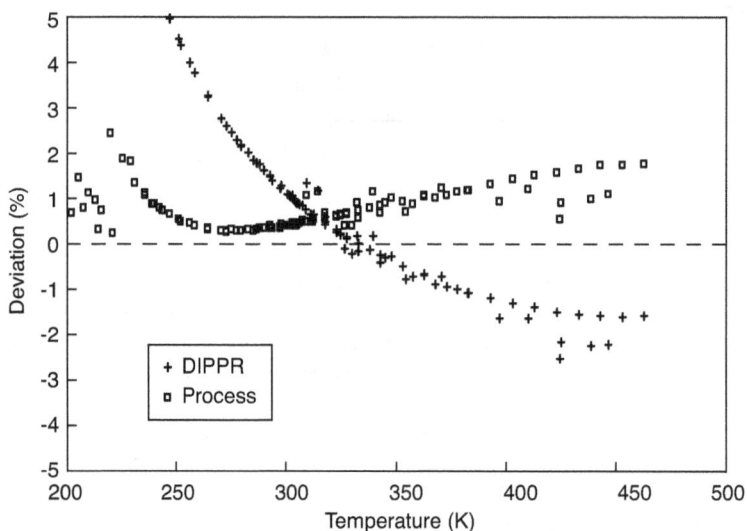

Figure 3.9

Deviation between the corresponding states predictions using the DIPPR and the process package database, using the experimental data for 2-methyl-1-butene.

The main conclusion of this example is that the user must not place blind trust in a value simply because it is published in a database. In addition, it should be kept in mind that process simulators have generally more than one databank, and for compatibility reasons, the older database is often the default one. It is important to compare different sources of information if they are available. If a predictive model is used to calculate a property, the physical meaning of the result must be compared with known trends or behaviour.

This example is discussed on the website:
http://books.ifpenergiesnouvelles.fr/ebooks/thermodynamics

3.1.2.2 Non-database components (group contributions)

Small molecules are almost all well-identified and available in databases. This is no longer the case beyond a certain molar mass. For hydrocarbons beyond C_{10}, the number of isomers becomes so large that it is impossible to evaluate all the parameters required to use a model. The process engineer has a number of options:

He may decide that the given component is not a key component (see section 3.5, p. 225, for further analysis of the concept of "key component") for his problem. In that case, he can consider that it behaves like another component whose properties are well-known. This is called "lumping" and will be discussed further in section 3.1.2.3.E, p. 137.

He may however decide that the given component is key to his problem, and that he needs to know its properties or parameters accurately. He will either need a predictive model or need to go to the laboratory to measure the required data. Generally, predictive models are not as accurate as fitted models. However, laboratory measurements are expensive and time-consuming. A final option is to use molecular simulation techniques which, through the definition of transferrable group potentials and a statistical analysis of a large number of molecular interactions, simulates the physical properties of simple systems (Ungerer *et al.* [41]). These tools no longer require excessive computing times, and can be used to calculate one or more physical properties of molecules that are not too large.

A group contribution method is based on the principle that all molecules are built from atoms linked together by bonds. Some particular group of atoms, together with their bonds produce behaviour characteristic to the molecules (e.g. the "OH" group results in hydrogen bonding). If a different molecule has the same groups, its behaviour will be similar to that of the first molecule.

The first group contribution methods were introduced by Lydersen in 1955 [42], followed by Benson [43]. Ambrose made contributions on this subject in 1978 and in 1979 [44, 45]. Joback proposed a dissertation in 1984 [46] followed by papers with Reid in 1987 [47]. Currently, the group of Gani (CAPEC) is among the most well-known for its contributions to this subject [48-52]. They have proposed a large number of methods to calculate pure component properties, such as critical properties, boiling properties, melting properties, and formation properties. For ideal gas properties (c_p), Nielsen *et al.* [40] proposed an extension of the Constantinou-Gani method [50]. The Marrero-Gani [49] method is further detailed below, as one of the most accurate example of this type of method.

The book by Poling *et al.* [1] summarises and evaluates in a very useful way the various methods that exist today. Their recommendations are extremely valuable to the process engineer. They were mainly concerned with pure component properties.

The recent paper of Gmehling (2009 [56]) provides a complete review of the entire history of Group Contributions.

A. Group Contribution for pure component properties

The group contribution methods are based on following steps:
1. Decompose the molecule into a number of constituting groups;
2. Add the group contributions for the selected properties;
3. Use a mathematical relationship, sometimes involving other properties, for calculating the final result.

a. Group decomposition

The group definition varies considerably from one method to another: the user must therefore be very cautious when counting the groups. Examples of molecular decomposition into groups are shown in figure 3.10.

In order to improve the accuracy, and distinguish among isomers, a number of authors use second order [50, 55] and sometimes third order groups (Marrero-Gani, 2001 [49]). This results in a correction on the first-order group contribution methods, based on the relative position of the groups with respect to one another. In the same spirit, Nannoolal [58] uses interaction parameters in order to correct the predictions for multifunctional components.

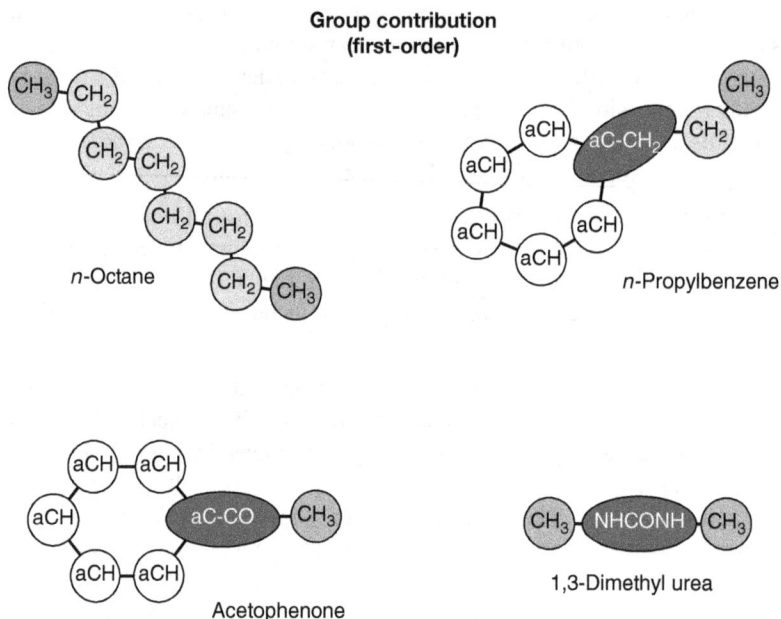

Figure 3.10

Group contribution by Marrero-Gani (first order molecules).

The main difficulty with these methods is how to locate the groups clearly. For example, the Marrero-Gani method includes 182 first-order groups, 122 second-order groups and 66 third-order groups. Some examples with their respective group construction and calculation are given for a few molecules in table 3.9.

Although very often, the group decomposition is based on practical considerations, some authors have proposed a theoretical foundation for group decomposition (Sandler[53]; Mavrovouniotis[54]).

b. Group contributions

Once the groups have been identified, their contribution to the required property X are added, using a simple sum:

$$F(X) = \sum_i N_i X_{1i} + \sum_j N_j X_{2j} + ...$$ (3.39)

where X_{1i} is the first order contribution for property X, of group i appearing N_i times in the molecule, X_{2j} is the second order contribution, etc. As examples, table 3.9 shows the group contribution calculation using the first order Marrero-Gani contributions for the molecules presented in figure 3.10.

Table 3.9 Sample calculation of first-order in the Marrero-Gani group contribution method for normal boiling temperature

Group	T_{bi}	N	N.T_{bi}	N	N.T_{bi}	N	N.T_{bi}	N	N.T_{bi}
		\multicolumn{2}{c}{**n-Octane**}	\multicolumn{2}{c}{**n-Propylbenzene**}	\multicolumn{2}{c}{**Acetophenone**}	\multicolumn{2}{c}{**1,3-dimethylurea**}
CH_3	0.8491	2	1.6982	1	0.8491	1	0.8491	2	1.6982
CH_2	0.7141	6	4.2846	1	0.7141				
aCH	0.8365			5	4.1825	5	4.1825		
aC-CH_2	1.4925			1	1.4925				
aC-CO	3.465					1	3.465		
NHCONH	8.9406							1	8.9406
$F(T_b)$			5.9828		7.2382		8.4966		10.6388

Calculated 1st-order			398.1		440.5		476.2		526.2
Database (DIPPR)			398.83		423.39		475.26		542.15

c. Final calculation

In a last step, the result of the additive contribution (equation 3.39) is used to calculate the final result, sometimes using another property. Table 3.10 shows, for example, the equations for the Marrero-Gani and the Joback methods.

> **Example 3.7 Evaluate critical temperatures using the group contribution methods of Joback and Gani**
>
> This example shows how to calculate a number of important properties using group contributions.
>
> Details are provided on the website:
> http://books.ifpenergiesnouvelles.fr/ebooks/thermodynamics

d. Choice among equations

A comparison of various group contribution methods for some well-known molecules is shown in tables 3.11 and 3.12.

It can be seen from the two tables that the quality of the various methods vary greatly. In general, it is recommended to evaluate the quality of the predictions within the same chemical family before using it for a given component. It is also good to keep in mind that all methods have been developed by regressing on available, *i.e.* low molecular weight components. When using any of these methods for property predictions of high molecular weight components, the deviation from the true behaviour may become important.

B. Group contribution for model parameters

It is worth noting that beyond pure component properties, group contribution methods have also been developed for equation parameters. The advantage of this second approach is that all properties that can be calculated by the equation are thus available.

Table 3.10   Functions used in the Marrero-Gani and Joback group contribution methods

Property	X (Marrero-Gani)	X (Joback)
Normal melting point, T_m (K)	$147.45 \ln\left(F\left(T_m\right)\right)$	$F\left(T_m\right)+122.5$
Normal boiling point, T_b (K)	$222.543 \ln\left(F\left(T_b\right)\right)$	$F\left(T_b\right)+198$
Critical temperature T_c (K)	$231.239 \ln\left(F\left(T_c\right)\right)$	$T_b\left[0.584+0.965 F\left(T_c\right)-\left(F\left(T_c\right)\right)^2\right]^{-1}$
Critical pressure, P_c (bar)	$5.9827+\left(\dfrac{1}{F\left(P_c\right)+0.108998}\right)^2$	$\left[0.113+0.0032 \times Na^{(1)} - F\left(P_c\right)\right]^{-2}$
Critical volume, v_c (cm^3 mol^{-1})	$F\left(v_c\right)+7.95$	$F\left(v_c\right)+17.5$
Standard Gibbs energy of formation at 298 K, Δg_f (kJ mol^{-1})	$F\left(\Delta g_f\right)-34.967$	$F\left(\Delta g_f\right)+53.88$
Standard enthalpy of formation at 298 K, Δh_f (kJ mol^{-1})	$F\left(\Delta h_f\right)+5.549$	$F\left(\Delta h_f\right)+68.29$
Standard enthalpy of vapourisation$^{(2)}$, Δh^σ (kJ mol^{-1})	$F\left(\Delta h^\sigma\right)+11.733$	$F\left(\Delta h^\sigma\right)+15.3$
Standard enthalpy of fusion at 298 K, Δh_F (kJ mol^{-1})	$F\left(\Delta h_F\right)-2.806$	$F\left(\Delta h_F\right)-0.88$

(1) *Na* is the number of atoms in the molecule.
(2) The enthalpy of vapourisation is calculated at 298 K by Marrero-Gani, and at the normal boiling temperature by Joback.

Table 3.11   Comparison of various prediction methods for critical temperature (K)

	n-Octane		*n*-Propylbenzene		Acetophenone		1,3-Dimethyl urea	
Ambrose-Original [44]	568.83	0.0%	636.93	-0.2%	706.19	-0.5%	793.84	0.9%
Ambrose-API [45]	568.83	0.0%	636.93	-0.2%	724.35	2.1%	883.55	12.3%
Joback [47]	569.27	0.1%	639.67	0.2%	702.94	-0.9%	786.70	-0.1%
Lydersen [57]	568.62	0.0%	638.57	0.0%	701.88	-1.1%	787.05	0.0%
Nannoolal-PC [58]	570.26	0.3%	639.55	0.2%	698.74	-1.5%	838.88	6.6%
Constantinou [50]	577.95	1.6%	639.59	0.2%	697.76	-1.7%	414.41	-47.4%
Forman-Thodos [60]	567.03	-0.3%	642.80	0.7%	693.23	-2.3%	392.17	-50.2%
Database	568.70		638.35		709.60		787.10	

Table 3.12 Comparison of various prediction methods for critical pressure (MPa)

	n-Octane		*n*-Propylbenzene		Acetophenone		1,3-Dimethyl urea	
Ambrose-Original [44]	2.48	– 0.5%	3.19	– 0.3%	3.83	– 4.5%	4.89	0.3%
Ambrose-API [45]	2.48	– 0.5%	3.19	– 0.3%	4.27	6.6%	9.20	88.9%
Joback [47]	2.54	1.8%	3.26	1.9%	3.95	– 1.6%	4.98	2.3%
Lydersen [57]	2.49	0.0%	3.22	0.6%	3.84	– 4.3%	4.87	0.0%
Nannoolal-PC [58]	2.56	2.7%	3.30	3.1%	4.09	2.0%	4.31	– 11.6%
Constantinou [50]	2.55	2.6%	3.20	0.1%	3.91	– 2.5%	8.00	64.2%
Forman-Thodos [60]	2.48	– 0.3%	3.20	– 0.1%	3.91	– 2.5%	5.90	21.1%
Marrero-Pardillo [61]	2.57	3.4%	3.20	0.0%	8.03	1901.3%	8.37	71.9%
Database	2.49		3.20		4.01		4.87	

The group contribution methods adapted to **equation of state** (EoS) parameters are extremely useful. Cubic equations of state are obviously the type most used. For the pure component parameters, Coniglio [62] has proposed an interesting method. For mixtures, the first to propose a group contribution method was Abdoul *et al.* [63]. More recently, the group of Jaubert has developed the so-called PPR78 EoS that is further discussed in section 3.4.3.4 (p. 204) for the Peng-Robinson or the Soave Redlich Kwong EoS (2004-2008 [64-71]).

Similarly, several research teams have proposed group contribution methods for the SAFT EoS [72-76]. The group contribution principle has also been applied, for polymer applications, to the lattice-fluid EoS [77, 78], and specific equations of state have been designed for group contributions, like the GC (Group-Contribution) EoS [79], which has been improved with an association term by Gros *et al.* [80] yielding the so-called GCA EoS.

The most well-known example of group contribution method for model parameters is the UNIFAC (Fredenslund, 1975 and 1977 [81-82]) for **activity coefficients** calculations. Other, less known, methods for predictive activity coefficient calculations are DISQUAC (DISpersive QUAsi Chemical) introduced by Kehiaian [83, 84] and ASOG (Analytical Solution of Groups) proposed originally by Derr and Deal (1969 [85]).

This approach can also be applied to other equations as, for example, the **Henry's constant or other infinite dilution properties** of hydrocarbons in water, as evaluated by Brennan *et al.* [86] or proposed and extended by Plyasunov *et al.* [87-89]. Similarly, Lin *et al.* [90] propose this type of group contribution approach to calculate octanol-water distribution coefficients.

3.1.2.3 Petroleum fluid components

Petroleum mixtures are composed of so many components that it is both impossible to identify them individually and out of the question to use several hundreds components in the modelling tools. Because of their industrial importance, a significant effort has been made to improve the molecular understanding of petroleum mixtures. The book by Riazi (2005) [10] contains a wealth of details on this subject. The one of Pedersen and Christiansen (2006) [91] can also be consulted.

For practical reasons, a number of so-called ***pseudo-components*** are defined and used as if they were pure components. These are in fact mixtures in their own right, but it is considered that the mixture property is unaffected by considering them as a single lump. The more pseudo-components are used in the model, the closer we can expect the calculated fluid properties to mimic the true properties. However, since many other uncertainties enter into the calculation, and due to the rapidly increasing computer time that may result, it is pointless increasing the number of components too much (depending on the application, a number between five and twenty is generally chosen).

As we will see, the characterisation procedure is generally based on a vapourisation curve, with the extensive use of the corresponding states principle. Consequently, it is only suitable for calculating vapour-liquid properties of hydrocarbon mixtures. If liquid-liquid or liquid-solid calculations must be performed, or if non-hydrocarbons are present in significant amounts in the mixture, this method cannot be used. An approach based on the chemical affinities must then be selected.

A. Existing types of vapourisation curves

As a starting point for defining a set of pseudo-components, a normalised vapourisation curve is generally provided. Different norms exist: the true boiling point (TBP), the ASTM-D86 and the simulated distillation (SD) curves are the most well-known. Correlations have been developed for transforming one curve into another. The curve used for pseudo-component definition is the TBP curve.

a. TBP (ASTM D2892)

The procedure used is a batch distillation, as illustrated in figure 3.11. The sample is located in a vessel whose temperature is continuously monitored. As it is progressively heated, the lightest components evaporate. They are separated in the column, equivalent to 15 theoretical plates, and the liquid distillate (expressed as a percentage of the feed) is collected, along with its temperature. The data are provided as condensation temperature as a function of volume percent distillate.

For heavy ends, in order to avoid cracking at high temperature, distillation at lower pressure is necessary. In this case the test method to be used is the ASTM D1160.

b. ASTM D86

This method is in fact a progressive evaporation, as opposed to the distillation procedure used in the True Boiling Point method. In this case (figure 3.12), the sample is progressively heated, and the vapour is condensed without further separation. The initial temperature is higher than that found in the TBP method, since it is close to the mixture bubble temperature. The condensed phase contains all the components that are in the vapour in equilibrium with the liquid phase, including the heavier ones. As a result, less separation is obtained using the D86 method due to the lack of reflux and this method can not be used to separate the components precisely according to their volatility.

c. Simulated distillation (ASTM D2887)

The ***Simulated Distillation*** (SD) method essentially uses a chromatographic technique to extract the various components according to their volatility. Because of its better reproducibility

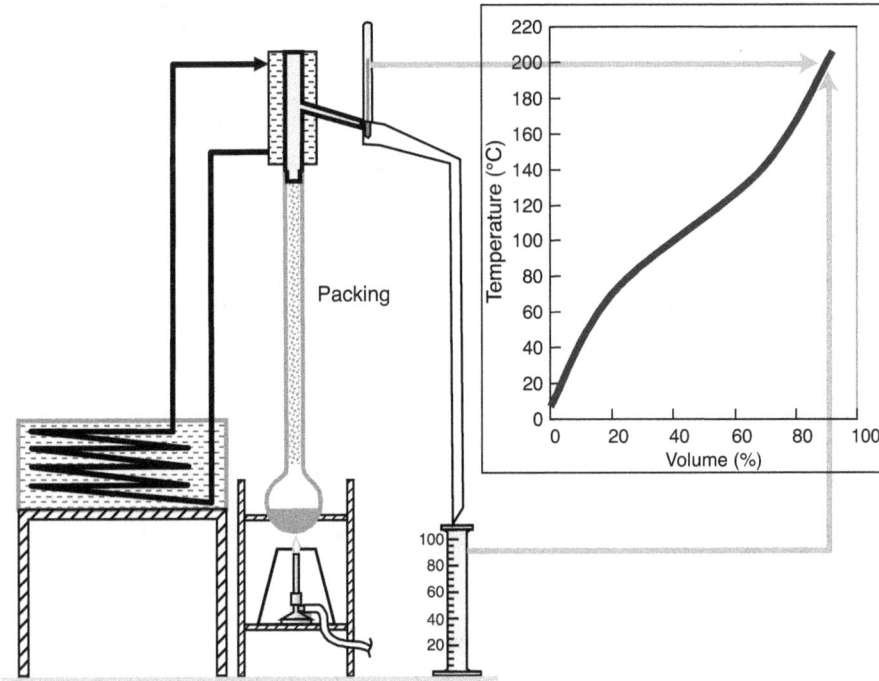

Figure 3.11

Sketch of the "True Boiling Point" (TBP-ASTM2892) procedure set-up.

than the TBP, it is nowadays the preferred method to characterise crude oils. The gas chromatography results are transformed in the so-called simulated distillation using the ASTM D2887 method. It is used for products with final boiling temperatures less than 1000 °F ($\approx$ 540 °C), but with a boiling range greater than 55 °C. Gas chromatography results are given as a function of weight fraction while the ASTM D86 and TBP are expressed in volume fractions.

d. Transformation between curves

For the definition of pseudo-components, only the TBP curve is used. If another vapourisation curve is provided, the data must be converted into TBP information as shown in figure 3.13.

In the sixth edition of the API technical data book (1997 [92]), Daubert proposes some mathematical transformation from ASTM D86 or simulated distillation to TBP. The technique is based on calculating the conversion of the central point (50% vol) from (for example) ASTM D86 to TBP. The consecutive points are then corrected by adding or subtracting a small difference (1994 [93]).

Riazi and Daubert have proposed a new simplified version [94]. Using their method, each temperature can be directly converted *via* a unique formula of the type:

$$T_{i,TBP} = a T_{i,ASTM}^b SG^c \tag{3.40}$$

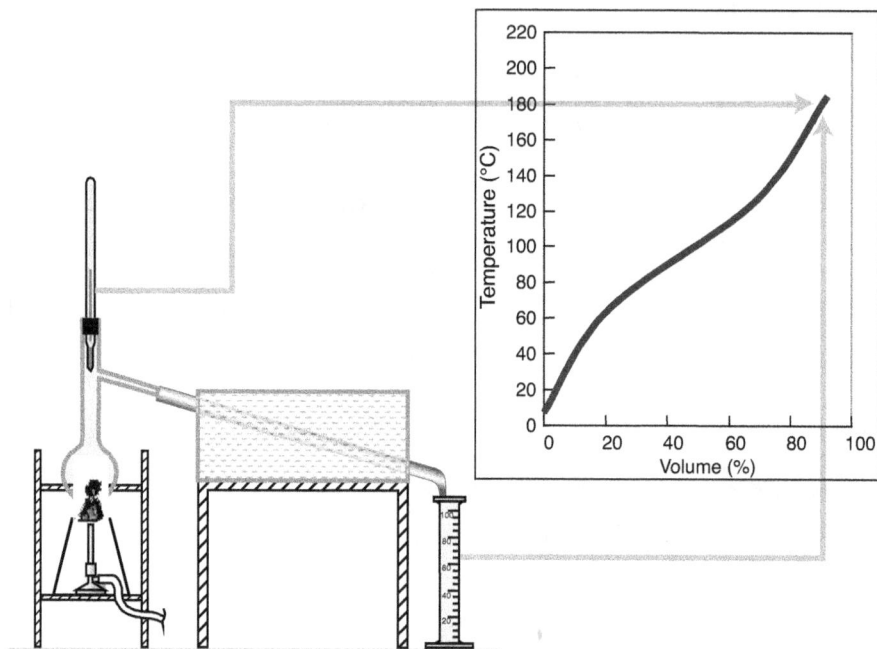

Figure 3.12

Progressive distillation procedure according to ASTM D86.

Coefficients a, b and c depend on the observed and desired curves and on the volumetric percentage of distillation. $T_{i,TBP}$ and $T_{i,ASTM}$ are temperatures (in K) of both curves, at a given volume %. Note that if c is zero, the older central conversion of Daubert's method is recovered. If the specific gravity SG (60 °F/60 °F) is not specified in the experimental data, it can also be predicted by an approximated formula.

When starting from a simulated distillation, the Daubert method is used, stating TBP (50% vol) = SD (50% wt). The difference between adjacent cut points (ΔT_{TBP}) is calculated from the equivalent difference on the simulated distillation curve (ΔT_{SD}):

$$\Delta T_{TBP,i} = c(\Delta T_{SD,i})^d \tag{3.41}$$

When pressure corrections are necessary, as it is the case for the ASTM D1160, a simple formula (from Wauquier, 1994 [95] based on information of Maxwell and Bonnel 1955) is used for the conversion of each measured temperature to the equivalent temperature at atmospheric pressure (the experimental pressure P_x is in mmHg).

$$T_{i,760} = \frac{748.1A}{1/T_{i,Px} + 0.3861A - 5.1606 \times 10^{-4}} \tag{3.42}$$

where

$$A = \frac{5.9991972 - 0.9774472\log_{10}\left(P_x\right)}{2663.129 - 95.76\log_{10}\left(P_x\right)} \tag{3.43}$$

A final correction should be taken into account if the result is higher than 366 K. Riazi [10] also offers alternative equations for the parameter A if the pressure P_x is less than 2 mmHg or greater than 760 mmHg.

> It is not recommended to use the interconversion formulae forth and back several times: at each conversion, some information is lost, and the resulting curve is therefore no longer useable.

In conclusion, a number of vapourisation curves can be used to characterise petroleum fluids and various empirical interconversion methods are available to obtain the equivalent TBP curve, the only curve suitable for the definition of pseudo-components (figure 3.13). These conversion methods are generally available in commercial process simulators.

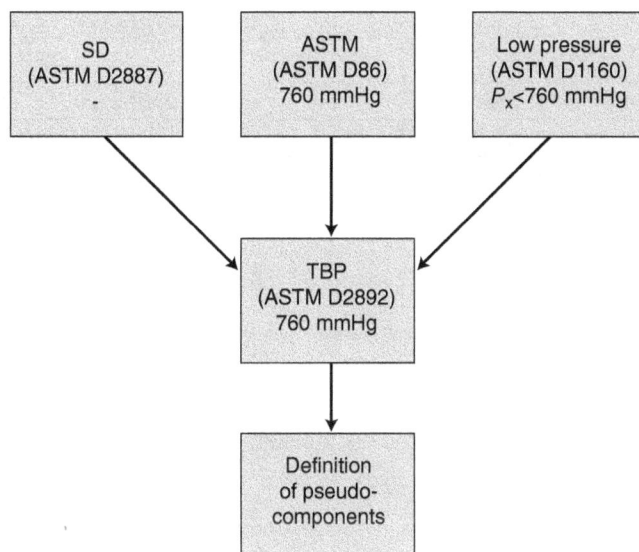

Figure 3.13
Order of use of interconversion equations.

B. Defining pseudo-components from the TBP curve

Once the TBP curve is available, it is used to define the pseudo-components as shown in example 3.8. The curve contains a number of points that may be scattered between 0% and 100% volume. The 0% temperature is not necessarily given. It is called the initial point (IP). The 100% point is even more rarely given, as it is generally impossible to distil the full sample. It is called the end point (EP).

Before any pseudo-component can be defined, this curve must be smoothed by a mathematical algorithm and cut into intervals. As every interval will represent a pseudo-component, it is recommended to cut the curve into evenly-spaced temperature intervals. The pseudo-components will therefore have evenly-spaced volatilities. Nevertheless, in some cases the curve is cut in equal volume intervals.

The smaller the temperature intervals, the more accurate the representation will be, but obviously the number of pseudo-components will increase and as a result the computing time. Suitable values are 10 K or 15 K intervals. The curve with the plotted intervals looks as shown in figure 3.14.

Example 3.8 Diesel fuel characterisation

A diesel fuel has been characterised in laboratory by a TBP curve, as shown in table 3.13 and figure 3.14. For simulation purposes, this diesel must be split in 10 °C cuts. Find the volume percentage corresponding to each cut.

Table 3.13   Sample data for the TBP of a Diesel

Fraction distillate (%)	0	3	7	17	29	40	50	59
Boiling temperature (°C)	123	162	185	206	226	244	262	279
Fraction distillate (%)	67	73	80	85	88	92	100	
Boiling temperature (°C)	295	309	324	338	350	362	375	

Analysis:

The properties given are the boiling temperature as a function of distilled fraction.

Components are a mixture of a large number of components and will be evaluated as pseudo-components.

Phases are vapour and liquid at atmospheric pressure.

Solution:

The curve is obtained by an interpolation method. One of the best choices is to use a cubic spline interpolation. Once the natural condition for extrapolation is chosen, the solution of the linear system gives all the curvatures at all points. This information is used to calculate the coefficients of each cubic polynomial.

The cuts are constructed on a 10 °C width basis which requires an inverse interpolation. An iterative procedure must be implemented. The volume percent corresponding to each cut is given in the following table 3.14 and figure 3.14.

Table 3.14   Percent volume of each cut obtained by the spline approximation of the TBP sample data

Cut (°C)	123-133	133-143	143-153	153-163	163-173	173-183	183-193	193-203
% volume	0.693	0.718	0.778	0.912	1.244	2.068	3.446	5.248
Cut (°C)	213-223	223-233	233-243	243-253	253-263	263-273	273-283	283-293
% volume	5.925	6.101	6.108	5.709	5.416	5.297	5.282	4.978
Cut (°C)	303-313	313-323	323-333	333-343	343-353	353-363	363-373	373-375
% volume	4.408	4.717	3.969	2.746	2.568	3.607	6.002	1.559

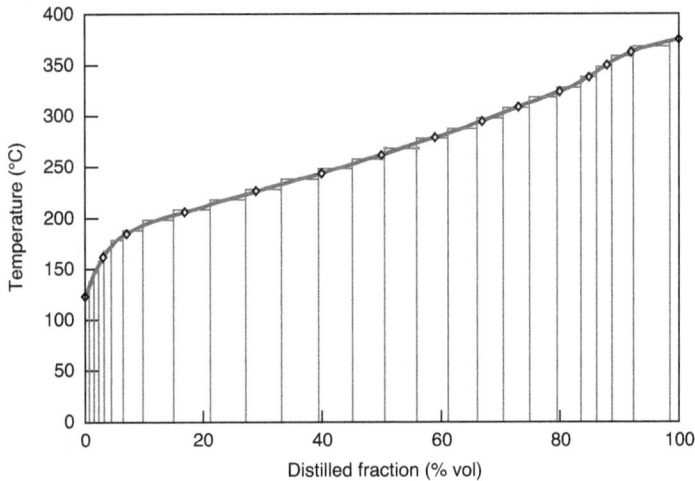

Figure 3.14

The TBP curve of example 3.8 smoothed and cut into evenly-spaced temperature intervals.

The example discussed on the website:
http://books.ifpenergiesnouvelles.fr/ebooks/thermodynamics, and extended this calculation to pseudo-component characterisation.

C. Characterising the pseudo-components

From figure 3.14, each pseudo-component can be given an average boiling temperature. This is not sufficient to perform thermodynamic calculations, however. At least one additional property must be known, which is generally the density. The density profile may be known if the samples collected in the TBP experiment are further analysed for their density. Very often, this is not the case. The most common method to obtain a density profile is to use a calculation procedure based on the "Watson factor" K_W method. This new parameter, K_W, is defined as follows [96]:

$$K_W = \frac{\left(1.8T_b\right)^{1/3}}{SG} \tag{3.44}$$

where T_b is the component boiling temperature (in K), and SG is its standard specific gravity (60 °F/60 °F). This property represents a way to "measure" the "average" chemical family of the mixture. Typically, paraffins have a K_W factor close to 13, naphthenes 12 and aromatics 10. Hence, if it is assumed that the mixture contains the same blend of chemical families, irrespective of the boiling temperature, it is reasonable to state that the K_W factor is a constant.

Its value is found using (3.44), where SG is the average measured specific gravity and T_b is an average value, that can be determined from the TBP curve using for example [10]:

$$T_{b(average)} = \frac{T_{b(10\%v)} + 2T_{b(50\%v)} + T_{b(90\%v)}}{4} \tag{3.45}$$

The density of each cut can now be calculated, by inversing equation (3.44):

$$SG_{pseudo} = \frac{\left(1.8 T_{b(pseudo)}\right)^{1/3}}{K_W} \tag{3.46}$$

Knowing the boiling temperature and the density (or specific gravity) of each cut, several correlations exist for calculating a number of pure component characteristic values that are required in thermodynamic models. Table 3.15 illustrates, for a number of them, what these correlations can calculate.

Table 3.15   Some correlations for calculating pseudo-component properties

	Winn (1957 [97])	Cavett (API) (1962 [98])	Kesler-Lee (1976 [99])	Twu (1984 [100])	Riazi(API) (1980/2005 [10])	Others
M			x	x	x	
T_c	x	x	x	x	x	
P_c	x	x	x	x	x	
$Z_c\,(v_c)$				x	x	Rackett
ω			$T_b/T_c > 0.8$		x	Edmister
$v^{L,\sigma}(T)$		x			15 °C	Rackett
$c_p^{\#}(T)$	x	x	x			
$P^{\sigma}(T)$		x				

The detailed expressions for the various correlations are not provided here, but can be found in Riazi (2005) [10]. They are also available in the supporting files of example 3.8. All use as input the normal boiling point and the specific gravity.

Of the different methods available, the Twu method (1984) [100] is most generally recommended: its construction, building the calculation as a perturbation on the *n*-alkane properties, provides a strong basis for safe extrapolation. It starts calculating the properties of the equivalent *n*-alkanes having the same boiling temperature. Next, a correction factor is introduced taking into account the difference between the *n*-alkane and the pseudo-component specific gravity.

D. Property calculations

The corresponding states principle, described in section 2.2.2.1.C, p. 57, is well-adapted to the calculation of the residual properties of hydrocarbon components. It is generally assumed that pseudo-components originate from petroleum fluids, which means that this principle can be used.

Cubic equations of state are more often used for phase equilibrium calculations (section 3.4.3.4, p. 204), while the Lee Kesler equation of state (section 3.4.3.3.C.d, p. 202) is recommended for single phase properties. The ideal gas heat capacity is calculated, for example, from equation (3.46) in section 3.1.1.2.E.c, p. 117.

Simple correlations based on the corresponding states principle have been discussed in section 3.1.1.2 of this chapter (p. 109). In particular, for vapour pressure, the Pitzer (3.17) or Ambrose (3.16) equations are applicable. For liquid molar volumes, the Rackett equation (3.18) is well adapted.

As an alternative method, we can mention the API method 7B4.7 which is recommended for calculating directly the liquid phase heat capacity [101] [18] as a third order polynomial, whose parameters depend on the Watson factor and the specific gravity.

E. Lumped pseudo-components

When many individual components are identified, it may happen that using all of them makes the calculation procedure much too long, while not improving the accuracy greatly. A lumping procedure is then used. The idea is to combine several known components (obtained for example by gas chromatography up to C_{11} or other equivalent techniques for heavier fractions) and to assume that they behave in a similar manner for the calculation required.

a. Lumping

The criterion used for "grouping" elements under "same" characteristics may lead to very different collections as can be seen from an example taken from Ruffier-Meray *et al.* [103]. In figure 3.15 a small part of a gas chromatographic analysis (from n-C_9 to n-C_{10}) is split in three different ways: a) according to boiling temperatures, b) according to number of carbon atoms and c) according to chemical nature (paraffins, naphthenes and aromatics).

This "membership" of a family must be defined by a mathematical criterion. Montel and Gouel (1984 [102]) have therefore proposed to following definition of the "distance"

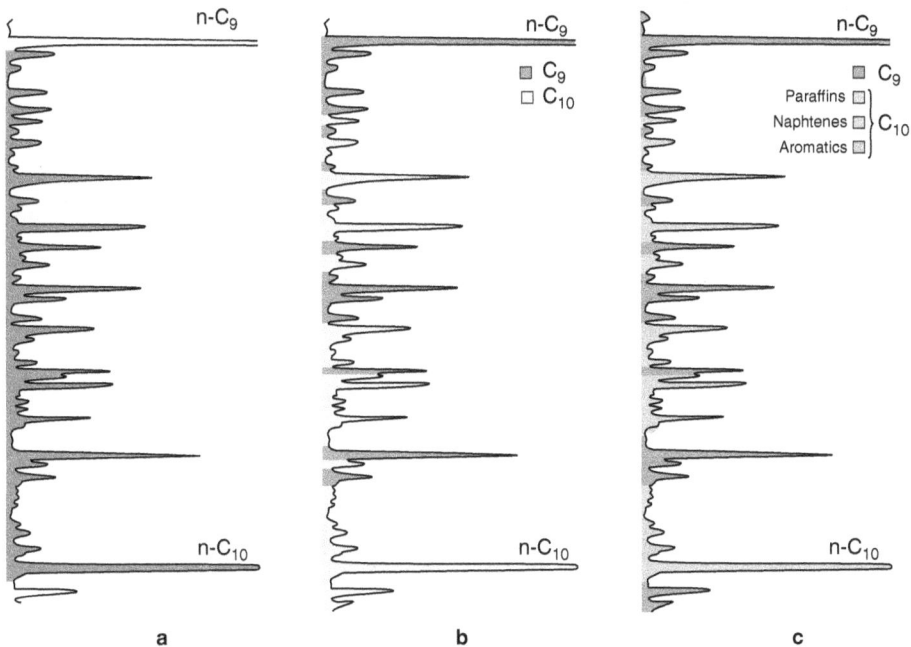

Figure 3.15

Different lumping results depending on the chosen criterion (from [103]).

between two components, from a combination of various scaled properties associated to the weighting factor w_k:

$$d_{ij} = \sum_{k=1}^{n_{properties}} w_k \left| r_{i,k}^2 - r_{j,k}^2 \right| \tag{3.47}$$

The variable $r_{i,k}$ are the scaled properties based on the minimum and maximum of each property k of point i ($P_{i,k}$) on the total of components.

$$r_{i,k} = \frac{P_{i,k} - P_{min,k}}{P_{max,k} - P_{min,k}} \tag{3.48}$$

The selected properties and the respective weight factors can be chosen at will (either pure component properties, or, better, equation of state parameters).

The lumping procedure is essentially based on the following steps (Montel and Gouel 1984 [102], Ruffier-Meray *et al.* 1990 [103]):

1. Identify a lumping criterion and the number of lumps that are required. The criterion can be a single property (e.g. number of carbon atoms), or a combination of several properties in which case a "distance" must be defined between the various components.
2. The components are distributed among the various lumps according to their nearest distance criterion.
3. The barycentres are calculated for each lump based on this first distribution.
4. With these new values, components are dispatched again as in point 2 until all barycentres are stable (no component switches to a different lump).

Once the lumps are defined, their characteristic parameters are computed on the basis of a mixing rule, using the parameters of the original components. The original formulae proposed by Montel and Gouel are given below:

$$T_c = \frac{\sum_i \sum_j x_i x_j T_{c,i} T_{c,j} / \sqrt{P_{c,i} P_{c,j}}}{\sum_i x_i T_{c,i} / P_{c,i}} \tag{3.49}$$

$$P_c = \frac{T_c}{\sum_i x_i T_{c,i} / P_{c,i}} \tag{3.50}$$

$$M_W = \sum_i x_i M_{W_i} \tag{3.51}$$

$$\omega = \sum_i x_i \omega_i \tag{3.52}$$

Leibovici (1993 [105]) observed that it is more significant to define the lump parameters in such a way as to obtain the same equation of state parameters (see section 3.4.3.4, p. 204, for a further description). Using a cubic equation of state, this yields:

$$\frac{T_c^2}{P_c} = \sum_i \sum_j x_i x_j \frac{T_{c,i} T_{c,j}}{\sqrt{P_{c,i} P_{c,j}}} \sqrt{\alpha_i (T_c) \alpha_j (T_c)} \left(1 - k_{ij}\right) \qquad (3.53)$$

$$\frac{T_c}{P_c} = \sum_i x_i \frac{T_{c,i}}{P_{c,i}} \qquad (3.54)$$

These two equations are very similar to those of Montel and Gouel. It is easy to observe that division of (3.53) by (3.54) gives an implicit equation with T_c as a unique unknown. If the product $\sqrt{\alpha_i (T_c) \alpha_j (T_c)} \left(1 - k_{ij}\right)$ is equal to 1, equation (3.49) is recovered. Equations (3.50) and (3.54) are then identical.

b. Delumping

In some cases, it may be important to recover the detailed composition after one or several phase equilibrium calculations have been performed. This is known as "delumping" and has been investigated by Leibovici (1996 [105]), Leibovici *et al.*, 1998-2000 [106] Nichita *et al.* (2006-2007 [107-109]).

If a cubic equation of state is used (presented in section 3.4.3.4, p. 204), without binary interaction parameters (BIP), the formula to calculate the distribution coefficient can be found to be:

$$\ln\left(K_i\right) = \Delta C_0 + \Delta C_1 \sqrt{a_i} + \Delta C_2 b_i \qquad (3.55)$$

where the quantities ΔC_0, ΔC_1 and ΔC_2 are obtained from phase properties only (*i.e.* they are independent of the individual components). Hence, it is sufficient to know the individual $\sqrt{a_i}$ and b_i to calculate the distribution coefficient, using (3.55).

If the BIPs are non zero, coefficients ΔC_0, ΔC_1, ΔC_2 should be obtained by regression.

3.1.3 Screening methods for pure component property data

Experimental data should always be used for validating the calculations or regressing parameters. Yet, it may happen that some data cannot be found or that their quality is questionable. In this case, rules for screening these values are most welcome. Various authors have discussed this question [110, 111].

Note that today, molecular simulation becomes a tool that may be used with great advantage to provide new "pseudo-experimental" data [112, 113]. The detailed implementation of these techniques, which are numerical, but based on a true physical understanding of the fluid phase behaviour [41] will not be described here. As such, it can be used to complete the experimental databases or help assess the physical soundness of the values.

3.1.3.1 Internal check

Any correlation between properties can be used to complete the picture. The correlations most useful in this respect are that of Clausius (3.7), Clapeyron (3.6) (remember that the Clapeyron relationship is valid for any two-phase equilibrium), the two- or three-point relationships for vapour pressure (3.9), or the Rackett equation (3.18) when the liquid volume is concerned.

As an example, the data discrimination method proposed by Wilsak and Thodos [114] can be used for vapour pressure data. The principle is to use a high resolution graphical representation of the deviation from a simple corresponding states presentation of the Clapeyron plot. An illustration of such analysis is proposed in example 3.6. Any two point relationship can be used, as also discussed in example 3.1. The experimental data that significantly deviate from the plot must be regarded with great caution.

3.1.3.2 External check

If the component under consideration belongs to a well-identified chemical family, trends can be analysed, based on the behaviour of the same property for another representative of the same family. An example of such a plot is shown in figure 3.16, where the liquid molar volume of n-alkanes is shown as a function of the carbon number. From this plot, it is clear that the liquid molar volume of ethane is not in line with that of the other components. It is therefore reasonable to assume that the value is incorrect.

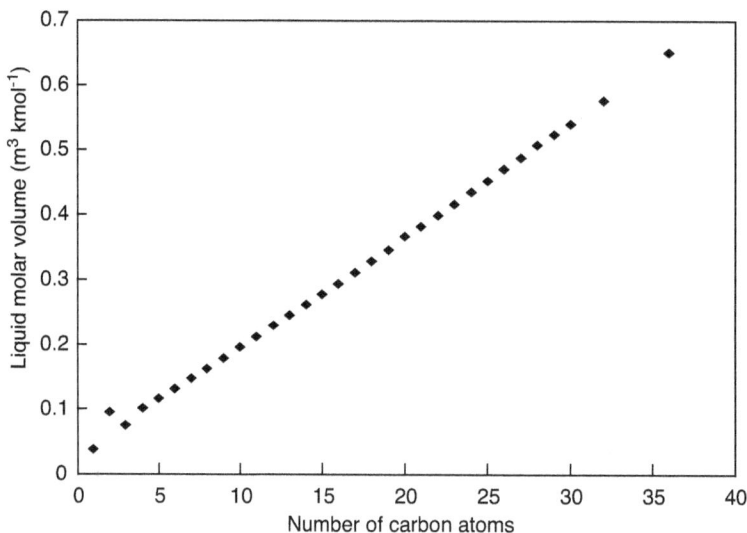

Figure 3.16

Liquid molar volumes of *n*-alkanes from a given database.

This is the basis of how group contribution methods are constructed. These methods allow some extrapolation of the properties of low molecular weight components to high molecular weight components. If taken too far however, this type of extrapolation may yield catastrophic results.

Example 3.9 Vapour pressures of di-alcohols

When comparing the vapour pressures of di-alcohols using DIPPR, figure 3.17 is found. It is expected that the lightest component has the highest vapour pressure, and the heaviest component the lowest. Yet, the graph shows curves that cross each other, clearly indicating something is wrong.

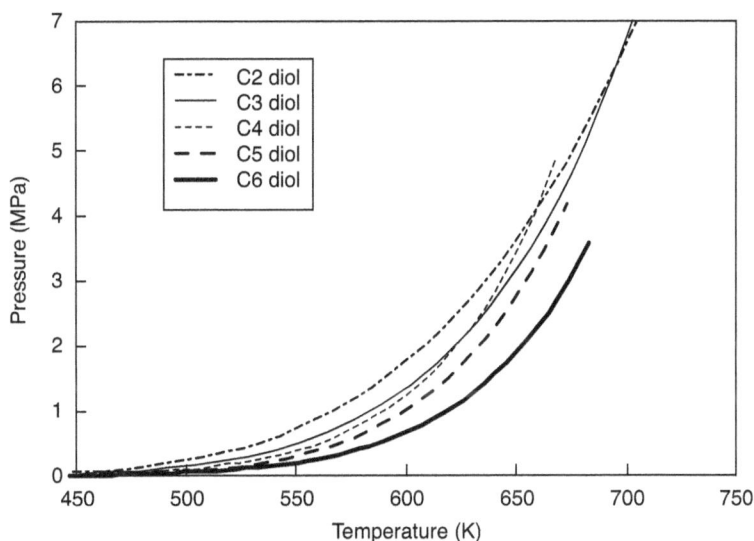

Figure 3.17

Vapour pressure behaviour of the series of diols where the alcohol groups are at the end of the chain, according to the DIPPR correlations.

In order to analyse the data, the Clapeyron equation is used: two pieces of information are required in order to fully define the Clapeyron straight line. For example, we can choose normal boiling temperature (one point) and enthalpy of vapourisation at this temperature (slope). Doing so, each component is represented by a point on figure 3.18. For each component, the vapourisation enthalpy can be obtained, either using the recommended value from DIPPR, or using the slope of the vapour pressure curve at the normal boiling temperature. These two values do not coincide. Whenever possible, an additional source of data (here CRC) was used.

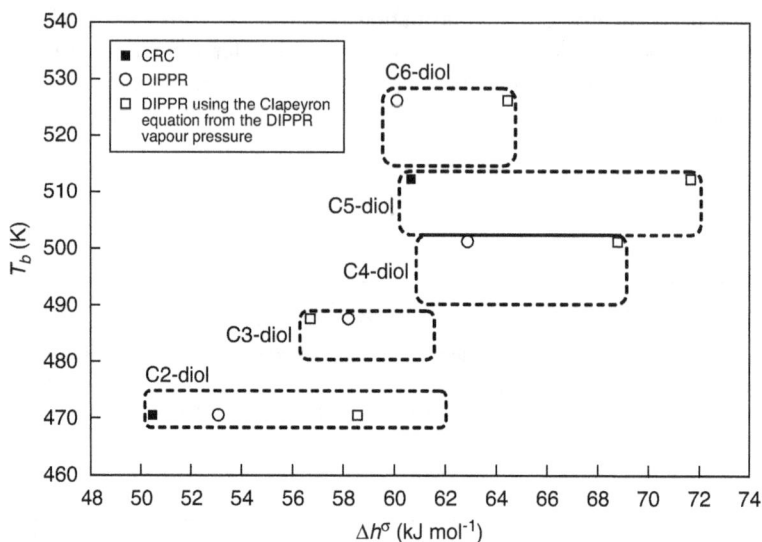

Figure 3.18

Plot of the diols (one –OH group at each end of the alcane chain) according to the Clapeyron representation (normal boiling point *vs* heat of vapourisation at the normal boiling point).

The plot clearly shows that the data do not follow a regular trend, which they should do. It is therefore highly recommended to collect new experimental information.

This example is discussed on the website:
http://books.ifpenergiesnouvelles.fr/ebooks/thermodynamics

3.2 MIXTURES: PROPERTIES AND PARAMETERS

We can make a distinction between the parameters related to individual components and those related to a pair of components (generally called "binary interaction parameters" BIP). Less often, ternary parameters may be used. Parameters are rarely measurable properties. They must be determined using a regression on experimental data. Obviously, the pure component parameters are obtained from pure component properties, binary interaction parameters from binary mixture properties and so on.

Data fitting will be further discussed in section 3.3 (p. 148). In this section, the type of data that exist will be presented briefly together with theoretical tools that can help assess their quality. The textbook by Weir and de Loos [115] may help understand how these data have been obtained. Several databases exist that contain a large number of such data, the most well-known probably being the Detherm database (http://i-systems.dechema.de/detherm, see also section 3.1.2.1, p. 121), as well as the NIST (http://trc.nist.gov/) or the Korean database (http://infosys.korea.ac.kr/kdb/index.html).

In order to evaluate the quality of the data, it is essential to try to collect data from different origins. For example, the ends of a binary phase equilibrium diagram are the pure component properties. Similarly, it may be useful to compare infinite dilution activity coefficient data with bubble pressure data.

The thermophysical Properties division at NIST has developed some rules and automatic tools, in particular the NIST Thermo Data Engine (TDE) which provides critically evaluated data. It is available at the web page http://trc.nist.gov/tde.html [116].

In a second stage, once a proper model has been selected and a first regression has been done, it is possible to identify "outliers", which most often arise from bad quality data.

3.2.1 Vapour-liquid equilibrium data

The data most often used are the vapour-liquid equilibrium data, and most databases that propose mixture data will provide following choices:

3.2.1.1 *TPx* (isobaric or isothermal)

These data correspond to basic vapour-liquid equilibrium. Pressure and temperature are reported. Sets of data give the bubble temperature as a function of liquid composition at a fixed pressure (isobaric data), or inversely, the bubble pressure as a function of liquid composition maintaining the temperature constant (isothermal data). Vapour composition is ignored in this kind of data.

3.2.1.2 Gas solubility

This type of data is essentially identical to the previous case (only the liquid phase composition is given). The way it is presented may be quite different, however. Many authors measure the Henry constant and use a variety of magnitudes and units to express it. Others use the Bunsen coefficient or the Ostwald coefficient [117].

A. Molar or mass fraction

When pressure and temperature are given, the most straightforward unit is the gas concentration in the liquid. It can be expressed in molar fraction x_g, weight fraction w_g. These are defined as (g is gas, s is solvent; n is the molar number and m is the mass):

$$x_g = \frac{n_g}{n_g + n_s} \text{ or } w_g = \frac{m_g}{m_g + m_s} \tag{3.56}$$

B. Molar or mass ratios

Pay attention to the distinction between fractions and ratios: the units are identical, but the basis for ratios is the amount of solvent, rather than the total amount of material for fractions:

$$L_g^x = \frac{n_g}{n_s} \text{ or } L_g^w = \frac{m_g}{m_s} \tag{3.57}$$

C. Henry Constants

Since the gas solubility is essentially proportional to the pressure, it is often convenient to use a quantity that considers the ratio of pressure and solubility. The Henry constant has been defined rigorously in section 2.2.3.1.A, p. 66, using the fugacity in the vapour phase. Yet, it is often assumed that this fugacity can be written as a partial pressure. In this case, the Henry constant becomes:

$$H_g = \frac{y_g P}{x_g} \tag{3.58}$$

When the solvent is assumed not to evaporate, equation (3.58) states effectively that the Henry constant is inversely proportional to the mole fraction solubility. It is clear that this constant therefore has units of pressure. However, if x_g is expressed in weight fraction or in volume concentration instead of molar fraction, the units must be adapted. Sometimes the pressure is implicitly assumed to be atmospheric, in which case the pressure disappears in the units: generalised use of the Henry constant may result in confusion.

D. Bunsen coefficients

The Bunsen coefficient is defined as the volume of gas reduced to standard conditions (273.15 K and 101 325 Pa) which is absorbed per unit volume of solvent (at the measurement temperature), under a partial pressure of 1 atmosphere.

$$\alpha = \frac{v_g(T_0, P_0)}{v_s(T)} \approx \frac{v_g(T, P_0)}{v_s(T)} \frac{T_0}{T} \tag{3.59}$$

The right hand side of this equation assumes that the ideal gas law and Henry's law are obeyed. If the latter is true, the mole fraction x_g can be calculated as (v_s^* is the molar volume of the pure solvent):

$$x_g = \frac{\alpha}{\alpha + \left(RT_0 / v_s^* P_0\right)} \tag{3.60}$$

E. Ostwald coefficients

This coefficient is defined as the volume of solvent needed to dissolve a given volume of gas at a given temperature and pressure:

$$L_g^v = \frac{V_g}{V_s} \tag{3.61}$$

Here, the definition of volume must be taken carefully since the volume of solvent depends on temperature and pressure. If the gas is considered ideal and if Henry's law is applicable, the mole fraction x_g can be calculated from the Ostwald coefficient using:

$$x_g = \left(1 + \frac{RT}{L_g^v P_g v_s^*}\right)^{-1} \tag{3.62}$$

where P_g is the gas partial pressure and v_s^* is the solvent molar volume at experimental temperature and pressure.

F. Kuenen coefficients

This is the volume of gas, reduced to 273.15 K and 1 atmosphere pressure, dissolved at a partial gas pressure of 1 atmosphere in 1 gram of solvent. Consequently, it is expressed in normal volume.

The book of Hefter and Tomkins on determination of solubilities and the various ways of expressing its values (2003 [118]) is recommended for further details as well as the page of NIST http://srdata.nist.gov/solubility/intro.aspx.

3.2.1.3 *TPxy* (isobaric or isothermal)

When the vapour phase composition is also available, the Gibbs-Duhem quality test can be used to validate the data. This method is based on the observation that, at given pressure and temperature, the composition of both phases is known.

For systems with high deviation from ideality and at a relatively low pressure, a hetero-geneous model with activity coefficients will be used (section 2.2.3.1.A, p. 69):

$$y_i P = P_i^{\sigma} \gamma_i x_i \qquad (3.63)$$

where P_i^{σ} represents the vapour pressure of component i, x_i and y_i are the molar fractions in liquid and vapour phases respectively, and γ_i the activity coefficient in the liquid phase. The activity coefficient behaviour is analysed for each binary system from experimental points:

$$\gamma_i = \frac{y_i P}{P_i^{\sigma} x_i} \qquad (3.64)$$

The Gibbs-Duhem criterion is used to validate equilibrium data [119]. This criterion implies that the following equation must be satisfied:

$$\int_0^1 \ln\left(\frac{\gamma_1}{\gamma_2}\right) dx_1 = 0 \qquad (3.65)$$

If the expression $\ln\left(\gamma_1/\gamma_2\right)$ is plotted as a function of x_1, the resulting curve must have the same area on the positive and on the negative side so the integral is zero.

Figure 3.19 represents graphically the significance of the coherence of the thermodynamic test for a specific binary system. The graph shows that in this case the data are not acceptable.

It is important to note that this behaviour is applicable only with non-reactive mixtures.

A more often used approach is the so-called data reduction. In this case, an empirical model is used to fit the data, as discussed hereafter in section 3.3 (p. 148), and the deviation between experimental and correlated data are analysed. Van Ness *et al.* (1973, 1995 [120, 121]) suggest using the deviation on $\ln \gamma_1/\gamma_2$, thus improving the area test (equation 3.65). This method is used in slightly modified form in the book of Fredenslund *et al.* (1977 [122]).

3.2.2 Liquid-liquid equilibrium data

Liquid-liquid solubility data can be given in the literature without any indication of pressure because the effect of pressure on liquid-liquid immiscibility is often negligible (below 1 MPa).

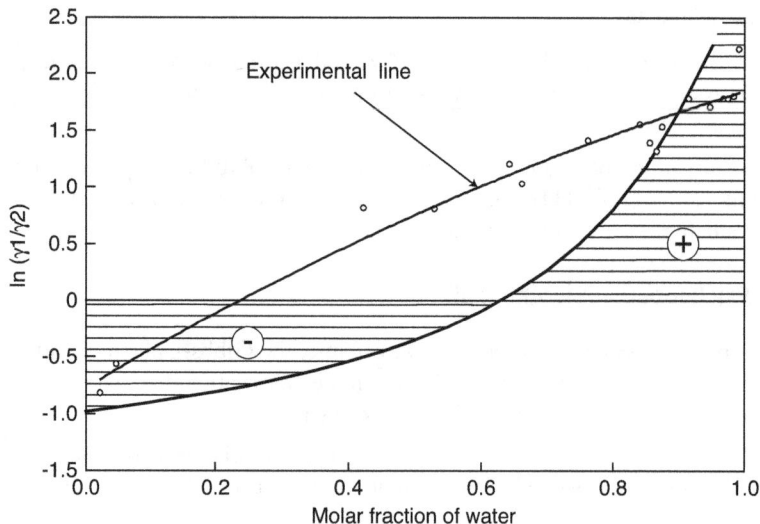

Figure 3.19

Graphical presentation of the thermodynamic coherence for activity coefficient
of a binary mixture. The experimental line (trend line through the experimental
data) does not follow the rule of eq (3.65), in opposition with the line delimit-
ing the hatched zones, which was obtained using the NRTL model.

Either the authors measure at atmospheric pressure (the presence of air is assumed to have no
effect) or their equilibrium cell always contains a vapour phase in which case they actually
work at the system bubble pressure (or, more precisely, its three-phase pressure).

An example of a test that can be performed on liquid-liquid equilibrium data [123] is that
of Othmer and Tobias (1942 [124]). It is based on the observation that for a mixture of three
components of kind 1 (as defined in section 2.1.3.4.B, p. 38) of the liquid-liquid immiscibility curve, the following relationship holds:

$$\ln\left(\frac{1-w_1^\alpha}{w_1^\alpha}\right) = a + b\ln\left(\frac{1-w_2^\beta}{w_2^\beta}\right) \tag{3.66}$$

where indices 1 and 2 refer to the non-mixing components, and α and β to their respective

phases (w refers to weight fractions). The expression shows that $\ln\left(\frac{1-w_1^\alpha}{w_1^\alpha}\right)$ and $\ln\left(\frac{1-w_2^\beta}{w_2^\beta}\right)$

should be on a straight line. Note that the fractions in parentheses can be considered as
"effective" weight fractions of the least soluble component in each phase (*i.e.* not considering
the co-solvent).

Similarly, the test proposed by Hand (1930 [125]) also uses a linearity relation:

$$\ln\left(\frac{w_3^\alpha}{w_1^\alpha}\right) = a' + b'\ln\left(\frac{w_3^\beta}{w_2^\beta}\right) \tag{3.67}$$

where this time the "effective" concentration of the co-solvent, *i.e.* the component that is entirely soluble in both 1 and 2 (here index 3) is analysed.

As an example, these rules can be applied on the 2-2-4 trimethyl-pentane (1)+ water (2) + ethanol (3) + mixture (figure 3.20):

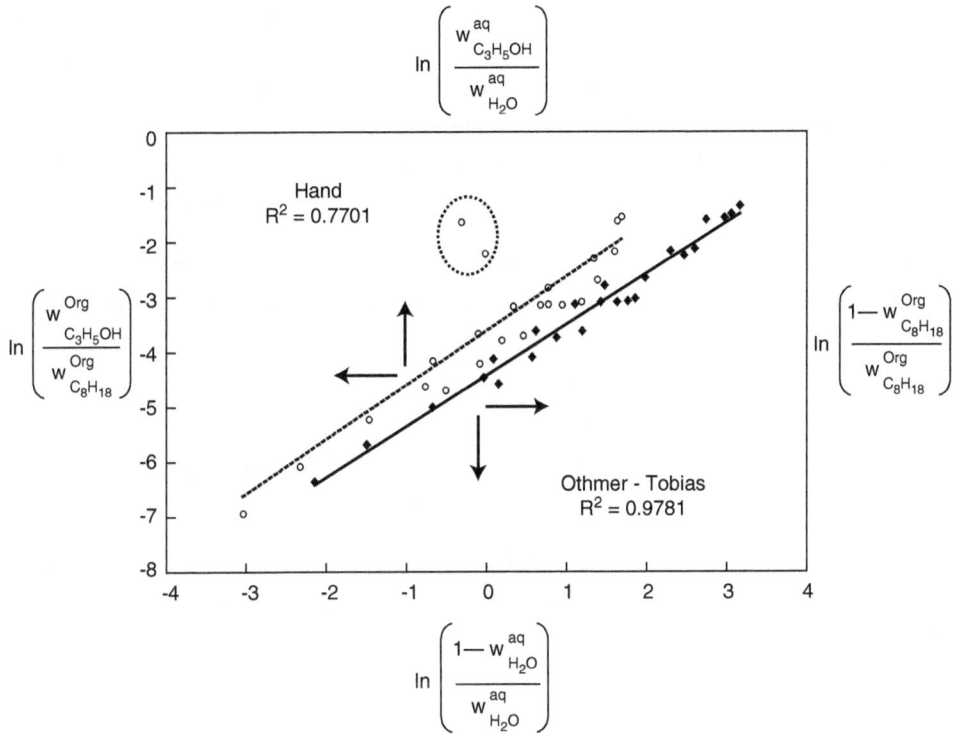

$$\ln \left(\frac{w^{aq}_{C_3H_5OH}}{w^{aq}_{H_2O}} \right)$$

$$\ln \left(\frac{w^{Org}_{C_3H_5OH}}{w^{Org}_{C_8H_{18}}} \right)$$

Hand
$R^2 = 0.7701$

Othmer - Tobias
$R^2 = 0.9781$

$$\ln \left(\frac{1 - w^{Org}_{C_8H_{18}}}{w^{Org}_{C_8H_{18}}} \right)$$

$$\ln \left(\frac{1 - w^{aq}_{H_2O}}{w^{aq}_{H_2O}} \right)$$

Figure 3.20

Othmer-Tobias and Hand methods for screening data, using water + ethanol + 2-2-4 trimethylpentane liquid-liquid equilibrium data.

The Othmer-Tobias relationship yields a very good regression coefficient ($R^2 = 0.9781$). In this plot, all points seem to be accurate. Nevertheless, in Hand's coordinates, two points (in the circle in the upper central part of the figure) lie very far of the main trend (dotted line). This information is translated numerically by a poor regression coefficient ($R^2 = 0.7701$). These two techniques must be used to complement each other to obtain satisfactory results.

3.3 DATA REGRESSION

As already stressed in the introductory chapter, a thermodynamic model always comes with parameters. Among various approaches, one of the best ways to determine model parameters is to fit them on valid and representative experimental data. Several examples of the construction of thermodynamic models will be given in the case studies associated with this book. At this point, it may simply be useful to state that a stepwise approach is recommended:

- first, determine pure component parameters (generally on pure component data),
- second, determine binary parameters (generally on binary data),
- if required by the model, ternary parameters come last.

In this section, we will briefly discuss the most important items in data regression. More details can be found in other documents [126]. The basic mathematics of parameter regression is well developed in Ledanois et al. (2000 [127]) and Chapra and Canale (2006 [128]). We will here focus on some simple issues related to thermodynamic problems.

Parameter-fitting involves minimising a function called the "objective function", whose shape is as follows:

$$F_{Obj} = \frac{1}{n_{exp}} \sum_{i}^{n_{exp}} \left[w_i \left(X_i^{\text{exp}} - X_i^{cal}(C, \mathbf{p}) \right) \right]^2 \qquad (3.68)$$

where X may be any physical property that can be measured experimentally or calculated by the chosen model, considering the experimental conditions C (temperature, pressure, compositions, etc) and a set of numerical parameters $\mathbf{p}$ that is to be regressed. The parameter values that will minimise this function will differ, depending on the experimental property selected.

The factor w_i in equation (3.68) is the weighting factor associated with the experimental data point i. The value of n_{exp} is the number of experimental data points used.

3.3.1 General guidelines

The procedure for parameter regression can be summarised schematically as shown in figure 3.21. It contains several steps that are discussed below and illustrated using two examples in section 3.3.2 (p. 155).

3.3.1.1 Choosing the model

The first guideline is obvious, but essential, especially if the model is to be used in extrapolation (which is often the case): use a model that is adapted to the problem (see below "models recommendation", chapter 4). The resulting model, however good it may be for reproducing the data that have been used in the optimisation, may be inappropriate to describe the other desired properties. Generally, we can state that the more the model is based on physical principles, the better it extrapolates.

3.3.1.2 The parameters to be adjusted

All models have a number of parameters, as will be stressed in section 3.4 (p. 160) of this chapter (so-called unary parameters, related to pure components; binary parameters,

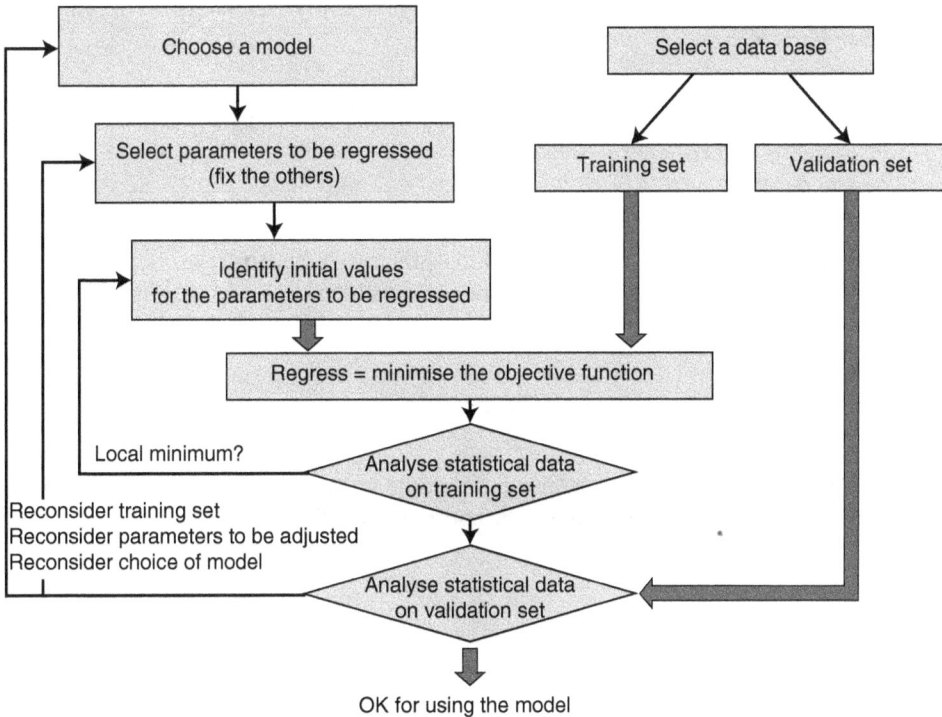

Figure 3.21

Procedure for regressing and validating a model.

describing interaction between components, and sometimes even ternary parameters). It is recommended to use as few adjustable parameters as possible (the **p** vector in equation (3.68)). It may be tempting to use many, as a better fit can certainly be obtained. However, the range of validity of this fit may be extremely limited in extrapolation. In addition, parameters are sometimes correlated with each other. In this case, several sets may provide equivalent results. Finding the best set under the conditions where the model is to be used may be difficult. As an example, the use of the NRTL model is shown in section 3.3.2.2 (p. 156).

3.3.1.3 The data

The quality of the database used is the main guarantee of success of the final model. It is therefore vitally important to select suitable data.

Whenever possible, it is best to select the experimental data that are as close as possible to the properties that need to be described, both in terms of data type (bubble pressure, composition, enthalpy, etc.) and in terms of experimental conditions.

If enough data ara available, the database is generally split into a *training set* and a *validation set*. The training set is used for the actual regression (in the objective function), while the validation set is used at a second stage, to evaluate whether the model developed is suitable for extrapolation.

Two different kinds of data may have to be manipulated:

- different measurements related to identical physical conditions (thus evaluating repeatability and accuracy of the data),
- measurements related to different conditions, that must be compared with results of a model.

It is good to start defining some statistical tools related to the first case (weighted mean, variance and standard deviations). These same tools will then be applied to the deviations between model and data.

A. Measurements related to identical conditions

a. Weighted mean

The weighted mean is a generalisation of the arithmic mean with a specific weight w_i associated with each experimental point. The corresponding formula is:

$$\overline{X} = \frac{\sum_{i}^{n_{exp}} w_i X_i}{\sum_{i}^{n_{exp}} w_i} \tag{3.69}$$

b. Variance and standard deviation

The variance is a property discussed at length in all statistical calculations. It describes the dispersion of the (Gaussian) distribution from a mean value. For a discrete sample, the variance, σ^2, corresponds to the sum of the square deviations between each value X_i of weight w_i and the mean, $\overline{X}$.

$$\sigma^2 = \frac{\sum_{i}^{n_{exp}} w_i \left(X_i - \overline{X}\right)^2}{\sum_{i}^{n_{exp}} w_i} \tag{3.70}$$

In many cases, the standard deviation (SD) is preferred. It has the same physical magnitude as the variable X. Two versions are commonly used; one corresponding to the Root Mean Square Deviation (RMSD), which is the maximum likelihood estimate when the population is normally distributed, and one corresponding to the unbiased measurement of the same variable (n_{exp} is the number of experimental points with non-zero weights).

$$RMSD = \sqrt{\frac{\sum_{i}^{n_{exp}} w_i \left(X_i - \overline{X}\right)^2}{\sum_{i}^{n_{exp}} w_i}} \tag{3.71}$$

$$SD = \sqrt{\frac{\sum_{i}^{n_{exp}} w_i \left(X_i - \overline{X} \right)^2}{\frac{n_{exp} - 1}{n_{exp}} \sum_{i}^{n_{exp}} w_i}}$$

(3.72)

B. Measurements related to different conditions

When the objective is to compare the experimental values with the results calculated through a model, tools are used for evaluating the distance between the data and the calculations. The following quantities are most often employed, with X representing the property used as a basis for comparison. They are calculated from the residue, or the deviation for each individual data point i and have the same dimensions as the experimental variable:

$$X_i^{dev} = X_i^{cal}\left(C, p \right) - X_i^{exp}$$

(3.73)

It may happen, however, that the residue is measured as a relative difference between the experimental and calculated values (which can be expressed in a dimensionless form).

$$X_i^{dev} = \frac{X_i^{cal}\left(C, p \right) - X_i^{exp}}{X_i^{exp}}$$

(3.74)

In particular, when X refers to a variable with a very large range of order of magnitude (several decades), the latter formula is preferred to give the same opportunity to all points to "participate" to fitting. When the range of values is small, equations (3.73) and (3.74) are almost similar. If the values may span both positive or negative values, equation (3.73) should be used.

a. Standard deviation

When the problem is concerned with comparison of experimental data and a model, the standard deviation becomes:

$$SD = \sqrt{\frac{\sum_{i}^{n_{exp}} w_i \left(X_i^{dev} \right)^2}{\frac{n_{exp} - 1}{n_{exp}} \sum_{i}^{n_{exp}} w_i}}$$

(3.75)

b. Absolute average deviation

For practical reasons, most authors prefer to use the absolute average deviation, defined as:

$$AAD = \frac{\sum_{i}^{n_{exp}} w_i \left| X_i^{dev} \right|}{\sum_{i}^{n_{exp}} w_i}$$

(3.76)

The AAD is often used as an estimate of the average uncertainty of the model. This can be done provided that the data used for this comparison are representative of the process conditions.

c. Bias

The SD and the AAD both indicate whether or not the average calculation is large. They yield only positive numbers. They do not show whether the model tends to over- or to under-predict the data values. The use of the bias can help. It is defined in the same way as the AAD, omitting the absolute sign:

$$BIAS = \frac{\sum\limits_{i}^{n_{exp}} w_i X_i^{dev}}{\sum\limits_{i}^{n_{exp}} w_i} \tag{3.77}$$

On average, the bias should be zero, since positive deviations should cancel negative deviations. Yet, in practice, this is never the case. A positive bias means that the model over-predicts the data, and the opposite is true for a negative bias. If the absolute value of the bias is equal to the AAD, all the deviations have the same sign.

d. Maximum deviation

The maximum deviation is the largest of all deviations:

$$MAX\ DEV = \max_{i}\left(\left| X_i^{dev} \right| \right) \tag{3.78}$$

It may be worthwhile checking this value, since it may help identify experimental outliers, which should be removed from the database.

3.3.1.4 The objective function

It is often possible to combine several types of data in the same objective function. This is done routinely when fitting pure component parameters for equations of state, where both vapour pressure and liquid volume at saturation are used. For complex equations, this may not be enough, as illustrated for SAFT by Clark et al. [129], and other types of data, such as heat capacity and second virial coefficients, may provide additional, independent information.

When *TPx* **vapour-liquid equilibrium** data are of interest, the bubble pressure (or the bubble temperature) is generally used for regression. The compositions of the liquid phase are rarely used in the objective function for binary mixtures. In the following equation, the weighting factor defined previously is taken as one:

$$F_{Obj} = \frac{1}{n_{exp}} \sum\limits_{i}^{n_{exp}} \left(\frac{P_i^{cal}(C,\mathbf{p}) - P_i^{exp}}{P_i^{exp}} \right)^2 \tag{3.79}$$

When *TPxy* data are given (*i.e.* the vapour composition is also given), a second term can be added to the above equation. In this case, the distribution coefficients or possibly the logarithm of the distribution coefficient (as discussed by Lopez *et al.* [130]) can be used as an alternative:

$$F_{Obj} = \frac{1}{n_{exp}} \sum_i^{n_{exp}} \left(\sum_j^{nc} \frac{K_{i,j}^{cal}(C, \mathbf{p}) - K_{i,j}^{exp}}{K_{i,j}^{exp}} \right)^2 \tag{3.80}$$

In this case, the distribution coefficient of all components is used, which explains why a double sum is needed: one on each data point and one on each component. It seems important to take the relative deviation (as in equation 3.79), otherwise, the values for heavy components (distribution coefficient below one) will be negligible compared with that of the light components (distribution coefficient higher than one). It is also worth noting that the distribution coefficient can be calculated in different ways: either a bubble pressure calculation is performed, in which case the pressure used for estimating this quantity is the calcu-

lated pressure, or the experimental pressure is used (*i.e.* $K_{i,j}^{cal}(C, \mathbf{p}) = \dfrac{\varphi_{i,j}^{L,cal}(C^L, \mathbf{p})}{\varphi_{i,j}^{V,cal}(C^V, \mathbf{p})}$). In the

latter case, the computer time required is much smaller.

For **liquid-liquid equilibria**, pressures cannot be used in the objective function, since these types of equilibria are almost insensitive to it. It is therefore recommended, if both phase compositions are available, to fit on distribution coefficients, as in (3.80), where K_j is defined as the ratio of mole fractions in both liquid phases.

The conditions (C) for liquid-liquid calculations are such that the model calculates phase split. Temperature and pressure are those of the experimental data (if pressure is not provided, some value at or above the bubble pressure can be taken). Composition is most

conveniently chosen at the middle of the tie-line, i.e. $x_j = \dfrac{x_j^{\alpha, exp} + x_j^{\beta, exp}}{2}$ where $x_j^{\alpha, exp}$ and

$x_j^{\beta, exp}$ are the experimental liquid molar fraction of component j in phases α and β.

If only one phase composition is provided we must rely on Gibbs phase rule, which indicates that *for a binary mixture*, the phase compositions are unique at given pressure and temperature. In this case, the conditions for equilibrium calculations corresponding to the data point i must be chosen arbitrarily such that a liquid-liquid phase split is calculated. More complex choices must be made for multicomponent mixtures.

The objective function can then be chosen as:

$$F_{Obj} = \frac{1}{n_{exp}} \sum_i^{n_{exp}} \left(\sum_j^{nc} \left(x_{i,j}^{cal}(C, \mathbf{p}) - x_{i,j}^{exp} \right)^2 \right) \tag{3.81}$$

However, we must bear in mind that the relative importance of the deviation between 0.95 and 0.96 mole fractions is considered equivalent to the difference between 0.01 and 0.02. Yet, in the former case, the difference corresponds to a 1% error, while in the latter case it is a factor of 2 (100%). Consequently, it may be better in the objective function to consider only the component with the lowest solubility (hence, not sum over all components).

Note that it may occur that in the chosen conditions (C), and with the chosen parameters ($\mathbf{p}$), no phase split at all is calculated (trivial solution). The algorithm should be able to deal with this type of situations.

When accurate temperature extrapolation is required, it may be useful to include heat of mixing data in the regression database. This is well illustrated in the second example that follows.

3.3.1.5 The weighting factor

The deviation used in the objective function is frequently normalised with the number of data points. However, each data point should be considered with a weighting factor that depends on its experimental uncertainty (w_i). Unfortunately, very often no uncertainty is known. The same factor is then taken for all data. Most often, the uncertainty is assumed proportional to the

inverse of the data value squared, in which case $w_i = \left(X_i^{\exp} \right)^{-2}$. This corresponds to the sum of the square relative errors.

$$F_{Obj} = \frac{1}{n_{exp}} \sum_{i}^{n_{exp}} \left(\frac{X_i^{cal}(C,\mathbf{p}) - X_i^{\exp}}{X_i^{\exp}} \right)^2 \tag{3.82}$$

3.3.1.6 The initial values

Finally, it is important to make sure that the resulting parameters values are independent of the initial values used in the optimisation procedure. This objective function is often highly non-linear, and as such, it may yield many local optima (Marquardt 1963 [131]). This is why it is important to have a reasonable first guess of the parameter values (e.g. zero, or any physically meaningful value).

It may be useful as a first step to linearise the model: try to write the problem in terms of a linear combination of elemental functions. If the expression $X_i^{cal}(C,\mathbf{p})$ can be linearised

as $X_i^{cal}(C,\mathbf{p}) = \sum_{j=0}^{m} p_j f_j(C_i)$, the coefficients can be obtained using only matrix operations.

A rectangular matrix based on the $n+1$ points with $m+1$ parameters ($m<n$) can be constructed:

$$[\mathbf{F}] = \begin{pmatrix} f_0(C_0) & f_1(C_0) & \cdot & f_j(C_0) & \cdot & f_m(C_0) \\ f_0(C_1) & f_1(C_1) & \cdot & f_j(C_1) & \cdot & f_m(C_1) \\ \cdot & \cdot & \cdot & \cdot & \cdot & \cdot \\ f_0(C_i) & f_1(C_i) & \cdot & f_j(C_i) & \cdot & f_m(C_i) \\ \cdot & \cdot & \cdot & \cdot & \cdot & \cdot \\ f_0(C_n) & f_1(C_n) & \cdot & f_j(C_n) & \cdot & f_m(C_n) \end{pmatrix} \tag{3.83}$$

The least squares method indicates that the parameters $\begin{bmatrix} \mathbf{p} \end{bmatrix}$ are solutions of:

$$\left(\begin{bmatrix} \mathbf{F} \end{bmatrix}^T \begin{bmatrix} \mathbf{F} \end{bmatrix} \right) \begin{bmatrix} \mathbf{p} \end{bmatrix} = \begin{bmatrix} \mathbf{F} \end{bmatrix}^T \begin{bmatrix} \mathbf{X}^{\exp} \end{bmatrix} \tag{3.84}$$

where $\begin{bmatrix} \mathbf{F} \end{bmatrix}$ is a $(n+1) \times (m+1)$ matrix, $\left(\begin{bmatrix} \mathbf{F} \end{bmatrix}^T \begin{bmatrix} \mathbf{F} \end{bmatrix} \right)$ is a $(m+1) \times (m+1)$ matrix, $\begin{bmatrix} \mathbf{X}^{\exp} \end{bmatrix}$

is a $(n+1)$ column vector and $\begin{bmatrix} \mathbf{F} \end{bmatrix}^T \begin{bmatrix} \mathbf{X}^{\exp} \end{bmatrix}$ is a $(m+1)$ column vector. The solution of the

[P] vector is obtained from eq (3.84) as $\begin{bmatrix} \mathbf{p} \end{bmatrix} = \left(\begin{bmatrix} \mathbf{F} \end{bmatrix}^T \begin{bmatrix} \mathbf{F} \end{bmatrix} \right)^{-1} \left(\begin{bmatrix} \mathbf{F} \end{bmatrix}^T \begin{bmatrix} \mathbf{X}^{\exp} \end{bmatrix} \right)$

3.3.1.7 Optimisation algorithm

The system is generally far from linear however. In this case, a non-linear optimisation algorithm must be used. Many such algorithms exist (from simple Newton methods, to simplex, Nelder-Mead, etc.) but one of the best method for the type of problem under consideration is that of Marquardt-Levenberg ([131, 132]) which was especially designed for this purpose.

3.3.1.8 The resulting uncertainties on the parameters

To obtain an idea of the uncertainty on the regressed parameters and of their possible inter-correlation, it is recommended to analyse the Hessian matrix (second derivatives of the objective function with respect to the parameters), assuming a linear behaviour close to the optimal result.

3.3.2 Detailed examples

3.3.2.1 Vapour pressure fit

Example 3.10 Find the parameters to fit the vapour pressure of ethyl oleate

Experimental data on the vapour pressure of ethyl oleate have been obtained experimentally using saturation and synthetic method techniques. The results are given in the table 3.16 .

Table 3.16 Experimental data for vapour pressure of ethyl oleate

Temperature (°C)	80.902	100.89	120.78	139.53	161.26	181.4	202.31
Vapour pressure (Pa)	0.35	2.06	9.15	31.58	131.59	416.46	1112.08
Temperature (°C)	118.93	129.04	139.10	149.14	150.08	179.06	
Vapour pressure (Pa)	8.68	18.16	36.85	63.16	118.17	348.16	

What are the best parameters to fit these data with the DIPPR equation:

$$\ln P^\sigma = A + \frac{B}{T} + C \ln T + D T^E$$

Analysis:

The given properties are vapour pressure as a function of temperature all along the saturation curve.

Component is an ester with very few data published.

Phases are vapour and liquid in the range of 350 K to 475 K.

Solution:

The objective function must first be selected. Different residues can be constructed:

• absolute pressure difference,
• relative pressure difference,
• absolute log(pressure) difference,
• relative log(pressure) difference.

The selected equation is not linear, so a numerical solver is required to minimise the root mean square error. In the solution, the Excel solver is used. Error minimisation is carried out with various initial values so as not to remain trapped in a local minimum. Obviously, in this specific case, many local solutions may exist.

When looking closer at the DIPPR equation, it appears that, except for the parameter E, all other are linear combination parameters. Hence, using a fixed value for E, the others can be found using (3.84). When this set of values is used as initial guess for the minimisation procedure, the solution offers excellent stability.

This example is discussed on the website:
http://books.ifpenergiesnouvelles.fr/ebooks/thermodynamics

3.3.2.2 VLE fit

Example 3.11 Fitting of binary interactions parameters for the water + MEA mixture with the NRTL model

The vapour-liquid equilibrium of water + monoethanolamine (MEA) is discussed here. This example is of great practical use for CO_2 capture from exhaust gases of power stations. Amine losses in the process can be evaluated through accurate knowledge of the vapour-liquid equilibrium behaviour.

The model chosen here is the NRTL activity coefficient model, presented in section 3.4.2.2.C, p. 177. The vapour-liquid equilibrium calculation is then a heterogeneous symmetric approach as discussed in section 2.2.3.1.B (p. 68): the vapour pressures (pure component liquid fugacities) of the pure components are calculated using a correlation that has been validated and that will not be further analysed here.

The parameters to be adjusted can then be identified: the NRTL model contains three adjustable parameters for every pair of components (τ_{ij}, τ_{ji} and $\alpha_{ij} = \alpha_{ji}$). It is well known that if temperature extrapolations are needed, which is the case here, these parameters should be considered as a function of temperature, thus doubling the number of adjustable parameters. In order to limit their number, we decided in this case to fix $\alpha_{ij} = \alpha_{ji} = 0.2$. The other two parameters are functions of temperature according to:

$$\tau_{ij} = A_{ij} + B_{ij}/T$$

There is a large vapour-liquid database for the binary system under consideration. It is summarised in table 3.17 . This database has been split into a training set and a validation set. For the purpose of this example, several combinations are chosen in the training set (fit 1-4):

Table 3.17   Availability of experimental data for the mixture Water + MEA and choice of training sets in fits 1 through 4

Data set	Temperature (K)	Data type	Number of data points	Source	Fit 1	Fit 2	Fit 3	Fit 4
a	298.15	TPx (VLE)	15	[135]	X	X	X	X
b	323.15	TPx (VLE)	13	[136]			X	X
c	298.15	H^E	67	[135, 137, 138]		X		X
d	363.15	TPx (VLE)	20	[139, 140]				X
e	308.15-573.15	TPx (VLE)	130	various [141]				X

The resulting parameters can be presented in the following regression table (3.18). Four different regressions have been performed, yielding four different parameter sets. The objective function of the different regressions contained various types of data, as indicated by the bold values in the regression table.

The objective function for sets *a*, *b*, *d* and *e* (*TPx* data type) is $\dfrac{1}{N}\sum\left|\dfrac{p^{exp}-p^{calc}}{p^{exp}}\right|$.

The objective function for set *c* (H^E) is $\dfrac{1}{N}\sum\left|\dfrac{H^{exp}-H^{calc}}{H^{exp}}\right|$.

Table 3.18   Average absolute deviations (AAD, in%) of the regressions (values in bold have been used in the objective function to calculate each fit)

Parameters	Fit 1	Fit 2	Fit 3	Fit 4
A_{12}	1.279	− 1.894	8.049	− 0.01599
B_{12}	–	419.1	− 2019	7.561
A_{21}	− 1.774	3.869	− 2.479	1.693
B_{21}	–	− 1338	210.3	− 828.4
Dataset	AAD (%)	AAD (%)	AAD (%)	AAD (%)
a	**4.6**	7.1	**4.6**	5.2
b	7.1	5.2	**2.1**	**2.9**
c	*	**2.8**	11.5	**3.9**
d	9.3	2.9	2.8	**2.6**
e	13.1	6.6	7.3	**5.7**

*: In the case of fit 1, no mixing enthalpy calculations can be performed since the parameters are independent of temperature.

All four fits lead to a good representation of the equilibrium data at 25 °C (dataset a), which is always included in the training set. The best result for this dataset is obtained with fit 1 that has been obtained using these data only. Yet, extrapolation to other temperatures using the parameter set from fit 1 provides relatively poor results.

It is interesting to note that fit 3 which uses data at 298.15 K and 333.15 K is unable to represent the heat of mixing correctly (dataset c: AAD of 11.5%).

Obviously the best results are obtained in fit 4, where the four datasets a, b, c and d have been used. Simultaneously, no data is available for validation.

Note nevertheless that fit 2 that uses heat of mixing data (dataset b), and only one VLE isotherm (dataset a) is able to correctly represent all other isotherms: it provides a representation close to that of fit 4.

The comparison between calculations and data for one isotherm is presented graphically in figure 3.22.

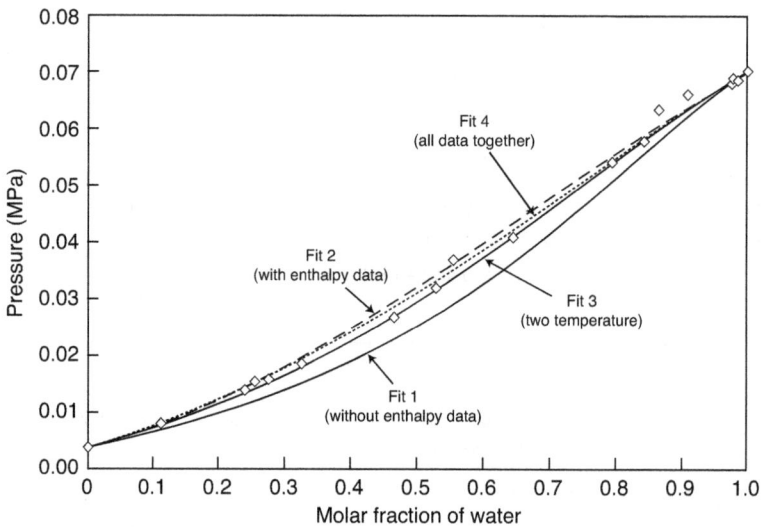

Figure 3.22

VLE behaviour of the water + monoethanolamine binary mixture at 363.15 K and model behaviour using the various fits of table 3.18.

A similar observation has already been made by Renon *et al.* (1971 [142]) who concluded that the use of mixing enthalpies greatly improves the capacity of the model for temperature extrapolations.

A second conclusion of this example is related to the numerical values of the parameters. The NRTL model is known to lead to energy parameters that are strongly inter-correlated. Figure 3.23 shows that parameters τ_{21} and τ_{12} at 298 K are on a single line. Parameter sets 1 and 3 are almost identical in this representation, and all sets are correlated. The uncertainties associated with the parameters are also shown: they are smaller with fits 2 and 4, *i.e.* using mixing enthalpy data.

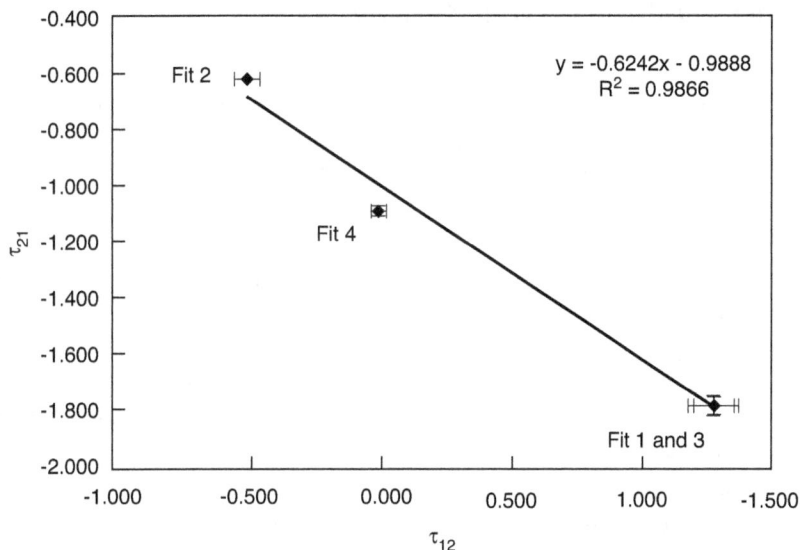

Figure 3.23

Numerical values of the parameter τ_{21} and τ_{12} sets at 298.15 K.

This example used NRTL, but the same conclusion can be drawn for other models using the thermodynamic relationships.

This example is discussed on the website:
http://books.ifpenergiesnouvelles.fr/ebooks/thermodynamics

3.3.3 Conclusion

The quality and the type of the experimental data available are the most important factors to ensure good regression. They may be used for two purposes: either as a basis for tuning the parameters, as shown above, or to evaluate the quality of the model calculations.

Use of combined data types can improve the predictive capacity of the model.

In the absence of experimental data, as mentioned above, molecular simulation may provide additional information ("pseudo-experimental" data). However, we must bear in mind that, while this approach is based on a physical picture of the material, it remains a numerical approximation of reality [41].

Within simulators, the parameters may originate from different sources: databases, correlations or group contribution methods or direct data fitting. As a part of the analysis of a thermodynamic problem, it is essential to know where the parameters originate from.

3.4   MODELS FOR THE MIXTURE PROPERTIES

So far, components have been considered individually in this chapter, *i.e.* in an environment containing only molecules of the same kind. The behaviour of molecules depends however to a very large extent on their immediate environment. When closely surrounded by molecules of a different kind, especially in the liquid phase, a non-volatile molecule may become volatile (since repelled by the others), or vice versa.

This section aims at providing guidelines to the reader on how mixtures may behave, and what tools (models) are available to describe this behaviour.

In the first point we will show, using simple models, how the phase diagram of a binary mixture can be modified by the molecular interactions. The concept of ideal mixture and activity coefficient will be used to show when azeotropic behaviour or liquid-liquid phase split can be observed.

In the second point, the molecular interactions themselves will be discussed, and the corresponding activity coefficient models presented.

Equations of state (EoS) are a different family of models that can be used to describe the fluid phase behaviour, both for pure components and for mixtures. Here, molecular interactions are discussed using the concept of density, which is absent from the activity coefficient models. The third point will focus on EoS construction, from both a molecular and a practical point of view.

In some cases, it is important to be able to model infinite dilution properties (e.g. Henry constants) directly. A short section on this topic will follow.

3.4.1   Prediction of some phase diagrams using the infinite dilution activity coefficients

The main feature of the ideal mixture defined in section 2.2.2.2 (p. 60) is that the bubble pressure curve in a *Pxy* diagram is a straight line. The low pressure behaviour of the ideal mixture can be simplified by Raoult's law:

$$y_i P = x_i P_i^\sigma \tag{3.85}$$

Even though a mixture of non-polar components is not exactly an ideal mixture, most hydrocarbon mixtures will in fact follow this rule closely. More generally, this type of simple behaviour is encountered when the molecules in the mixtures are alike.

Departure from this ideal behaviour is generally found in the liquid phase (we will focus on this case). It is described by an activity coefficient γ_i, as follows (the exact definition is provided in section 2.2.2.2.B, p. 61):

$$y_i P = \gamma_i x_i P_i^\sigma \tag{3.86}$$

The "measure" of the non-ideality of a liquid phase is its activity coefficient. As shown with the index i in equation (3.86), each component has its own activity coefficient. Its value depends on the mixture composition as well as on temperature (generally, the pressure dependence is neglected). In order to quantify the behaviour of a mixture, the infinite dilution activity coefficient can be used. This is the value that the activity coefficient takes when component i is infinitely diluted in a solvent:

$$\lim_{\substack{x_i \to 0 \\ x_s \to 1}} \gamma_i = \gamma_i^{\infty} \tag{3.87}$$

Under these conditions, the interactions between component i and its surrounding solvent are strongest, and depending on the value of γ_i^{∞}, the behaviour of component i may change dramatically from its pure component behaviour.

In order to illustrate how equation (3.86) results in different types of phase diagram, the relatively simple expression provided by the Margules equation [1] for a binary mixture of components "1" and "2" will be used. This equation is written as:

$$\frac{g^E}{RT} = x_1 x_2 \left(A_{21} x_1 + A_{12} x_2 \right) \tag{3.88}$$

or equivalently as:

$$\ln \gamma_1 = x_2^2 \left(A_{12} + 2 \left(A_{21} - A_{12} \right) x_1 \right) \tag{3.89}$$

and

$$\ln \gamma_2 = x_1^2 \left(A_{21} + 2 \left(A_{12} - A_{21} \right) x_2 \right) \tag{3.90}$$

The infinite dilution activity coefficients are then easily found from (3.87):

$$\ln \gamma_1^{\infty} = A_{12} \tag{3.91}$$

and

$$\ln \gamma_2^{\infty} = A_{21} \tag{3.92}$$

The activity coefficient expresses the ratio between the actual volatility of a component and its theoretical volatility if it was an ideal mixture. This ratio can become very large at infinite dilution. In the figure 3.24, *n*-hexane diluted in benzene is shown to be twice as volatile as if it were pure.

---

1. The list of all commonly used activity coefficient models, and their domain of use is provided in section 3.4.2, p. 171 of this chapter. Margules is rarely used, but is convenient here for its simplicity, and because of the very straightforward relationship between its parameters and the infinite dilution activity coefficients.

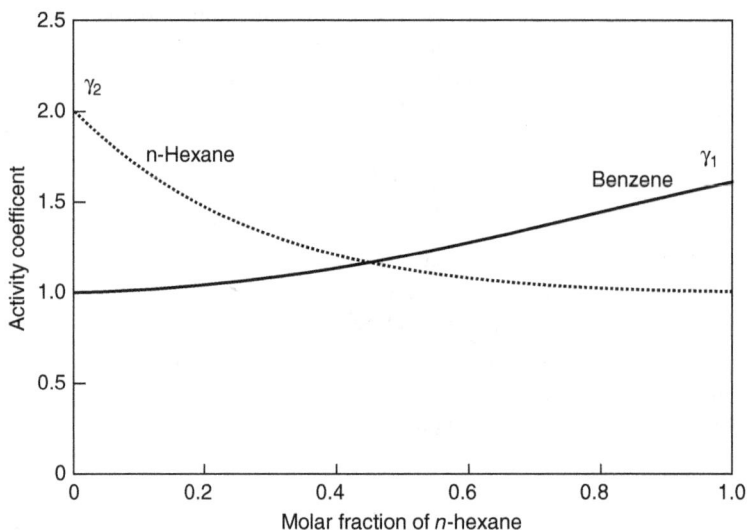

Figure 3.24

The activity coefficient as a function of composition for *n*-hexane + benzene mixture at 303.2 K, according to the Margules equation.

3.4.1.1 Positive and negative deviations from ideality

While the activity coefficient is always positive, its logarithm may be larger or smaller than zero. If $\ln \gamma_1 > 0$ and $\ln \gamma_2 > 0$, equation (3.86) indicates that the partial pressure of component (1) is larger than that of the ideal mixture, in equation (3.85). The opposite is true if $\ln \gamma_1 < 0$ and $\ln \gamma_2 < 0$. It is clear that $\ln \gamma_1 = \ln \gamma_2 = 0$ corresponds to the Raoult law presented in chapter 2. This leads to the relationship between the activity coefficient and the *Pxy* diagram, illustrated in figure 3.25.

Positive deviations ($\ln \gamma_1 > 0$) can be understood on the microscopic scale as if the molecules were preferentially attracted to molecules of the same kind. When another kind of molecule is added, they prefer leaving the liquid phase (partial pressure higher than ideal). For example, alcohols will form hydrogen bonds with other alcohol molecules (auto-association). When non-polar molecules are added, the hydrogen bonds are broken and the alcohol becomes more volatile.

On the contrary, negative deviations ($\ln \gamma_1 < 0$) result from the molecules being attracted preferentially to molecules of the opposite kind. As a result, they will stay in the liquid phase longer (*i.e.* at lower pressure) than if the mixture was ideal. Their partial pressure is smaller. This occurs when cross-association interactions are larger than auto-association as in alcohol + amine mixtures, for example.

Figure 3.25 illustrates how the activity coefficient behaviour affects the vapour-liquid equilibrium behaviour for binary mixtures.

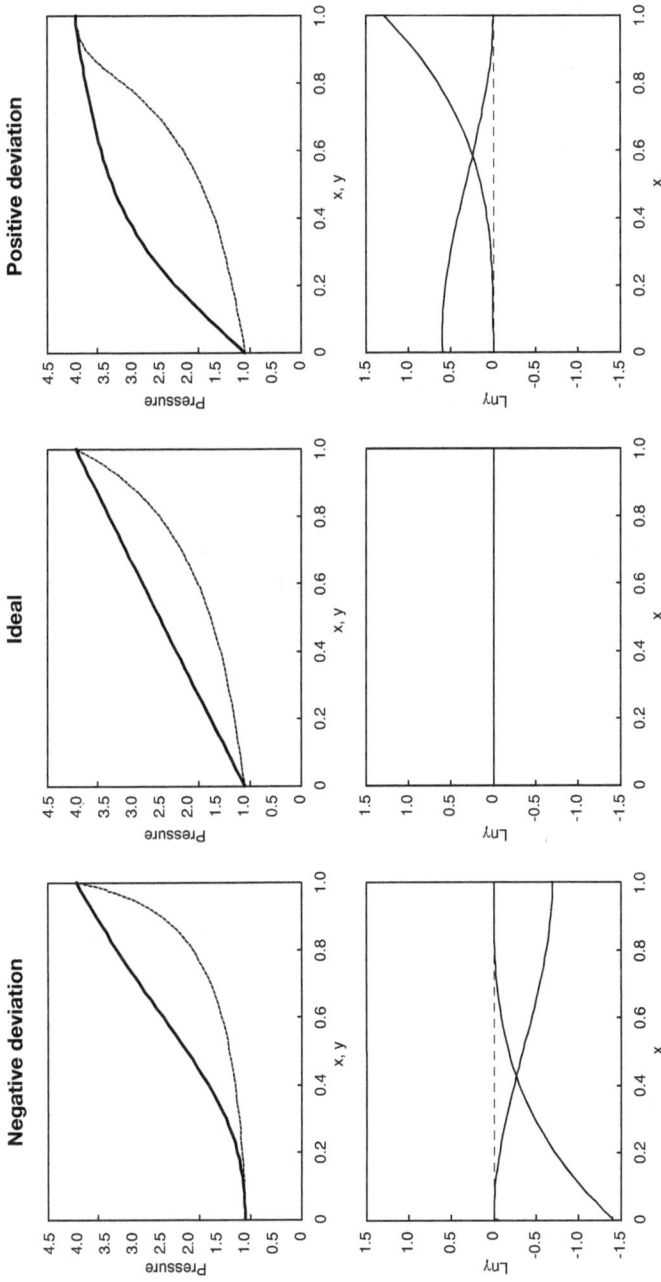

Figure 3.25

Phase diagram of negative or positive deviations from ideal behaviour.

Non-ideality means that a component may behave very differently in a liquid solvent than it would if it were alone. The volatility, or tendency to go into the vapour phase, may be strongly modified. This may be a very useful feature for separating closely boiling mixtures. Examples of applications are found in extractive distillation processes (example 3.12).

3.4.1.2 Azeotropy

Azeotropic behaviour is defined as that for which the liquid and vapour phases have the same composition:

$$x_i = y_i \tag{3.93}$$

for all components i. Hence, from (3.86):

$$P = \gamma_i P_i^\sigma \tag{3.94}$$

And, since the azeotropic pressure is identical for the two components:

$$\gamma_1 P_1^\sigma = \gamma_2 P_2^\sigma \tag{3.95}$$

equation (3.95) can be written as:

$$\frac{\gamma_1}{\gamma_2} = \frac{P_2^\sigma}{P_1^\sigma} \tag{3.96}$$

where we can see that the right-hand side of the equation is independent of composition, while the left-hand side is a continuous function that will vary with composition between the values of $\dfrac{1}{\gamma_2^\infty}$ (for $x_1 = 1$) and γ_1^∞ (for $x_1 = 0$). As a result, we observe that if

$$\frac{\gamma_1^\infty}{1} < \frac{P_2^\sigma}{P_1^\sigma} < \frac{1}{\gamma_2^\infty} \quad \text{or if} \quad \frac{\gamma_1^\infty}{1} > \frac{P_2^\sigma}{P_1^\sigma} > \frac{1}{\gamma_2^\infty} \text{ , then equation (3.96) must be true for some interme-}$$

diate composition.

Example 3.12 Separation of *n*-butane from 1,3-butadiene at 333.15 K using vapour-liquid equilibrium

Find a component that will enhance the separation between these two components.

Analysis:

The physical property used for separation by vapour-liquid equilibrium is the volatility of the components. The vapour pressures of the pure components are the most fundamental piece of information related to volatilities. They can be found in many databases. For example, at 333.15 K, the vapour pressures of the two components are $P_{nC4}^\sigma = 729$ kPa and $P_{nBTD}^\sigma = 639$ kPa respectively (source DIPPR).

These two values are fairly close to each other. Taking into account non-idealities, we can determine (using the NRTL activity coefficient model, for example) that the infinite dilution activity coefficients are very close to one ($\gamma_{nC4}^\infty \approx \gamma_{BTD}^\infty = 1.1$), indicating an almost ideal mixture. The separation of the two components by distillation will be very difficult.

In order to improve the separation, a solvent can be found in which the activity coefficients of both components are quite different. This solvent should, in addition, remain in the liquid phase to be effective, in other words have a low vapour pressure.

Solution:

Using a large database of vapour pressure data, we find that acetonitrile (CH3CN), for example, has a boiling pressure at 333.15 K of 49.6 kPa. This component has a lower vapour pressure than the time components to be separated and will therefore preferentially be found in the liquid phase.

The two activity coefficients at infinite dilution of butane and butadiene in acetonitrile are respectively $\gamma_{nC4}^{\infty} = 12.7$ and $\gamma_{BTD}^{\infty} = 4.5$.

In other words, 1,3-butadiene forms a much less non-ideal mixture with acetonitrile than n-butane. Butane has in fact no polar nature at all, while the double bonds of the butadiene result in a polarity that makes it more similar to acetonitrile. This behaviour is made visible in figure 3.26, which shows that the resulting distribution coefficient of n-butane and 1,3-butadiene are both large (*i.e.* they are preferentially in the vapour phase) and different (*i.e.* they can now be separated). The acetonitrile distribution coefficient is small (<1) indicating that it remains preferentially in the liquid phase.

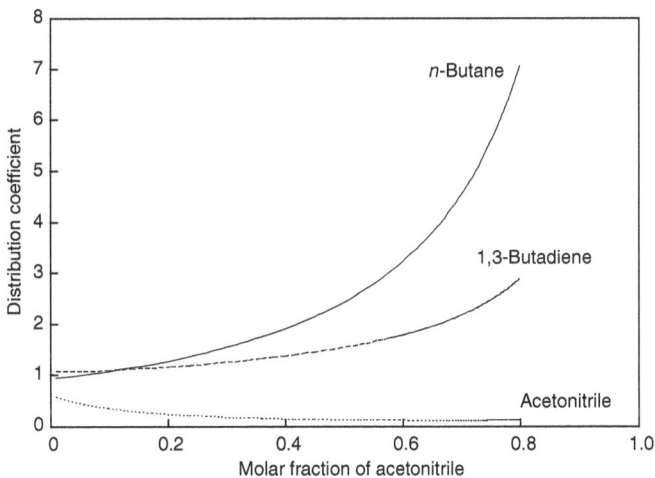

Figure 3.26

Acetonitrile as an extractive solvent for the n-butane + 1,3-butadiene mixture. The plot shows the distribution coefficient of the components at 333.15 K at the bubble pressure.

This example is discussed on the website:
http://books.ifpenergiesnouvelles.fr/ebooks/thermodynamics

In distillation, the relative volatility α_{12} between the two components is the quantity that will help the process engineer determine how easy the separation will be. It is defined as:

$$\alpha_{12} = \frac{\gamma_1 P_1^{\sigma}}{\gamma_2 P_2^{\sigma}}$$

(3.97)

The relative volatility changes with temperature and composition. At the azeotropic point, $\alpha_{12} = 1$, meaning that no further separation can be achieved since the volatilities of the two components are identical.

> According to the criterion derived from expression (3.96), it appears that an azeo-trope can be found in either of two cases:
> • either the second component has a vapour pressure very close to that of the first component, in which case the activity coefficients need not be very large for an azeotrope to be found,
> • or the components form a strongly non-ideal mixture.

Hydrocarbon mixtures can satisfy the first criterion, for example a mixture of benzene (normal boiling temperature 353.25 K) with cyclohexane (normal boiling temperature 353.85 K). An example of the second criterion is found in the case of *n*-hexane and meth-anol mixture (figure 3.27).

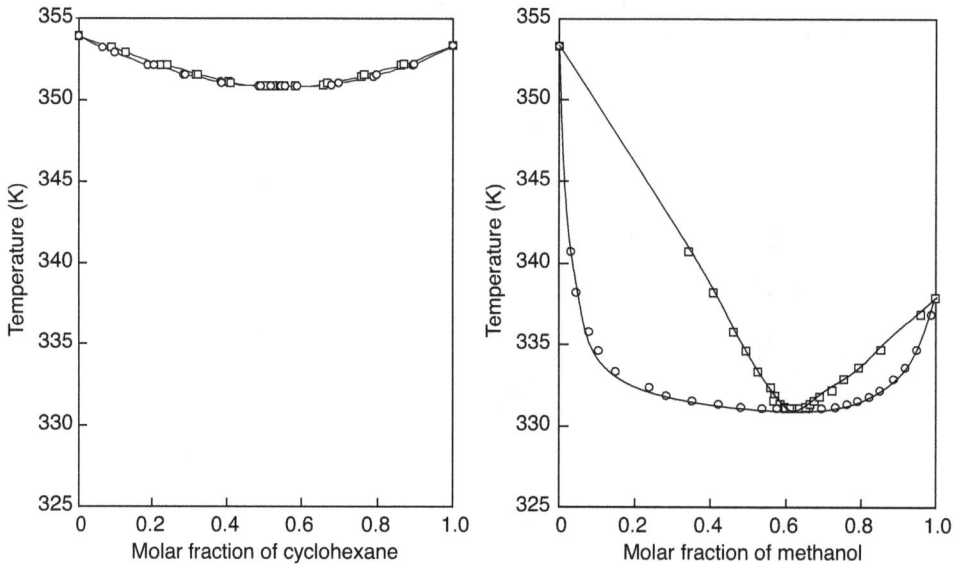

Figure 3.27

Benzene azeotrope at atmospheric pressure with 2 different components [141, 143].

3.4.1.3 Liquid-liquid equilibria

Liquid phase split occurs when the tendency of the individual components to remain with the same kind becomes so strong that their stability is greater in individual phases. Liquid-liquid phase split can be calculated using either of two methods:

- either the chemical potentials of each component in each phase are equal. The chemi-cal potentials may be replaced by fugacities, which, at low pressure, are written using the excess approach (see section 2.2.2.2, p. 60):

$$x_i^{\alpha}\gamma_i^{\alpha}\left(T,\mathbf{x}^{\alpha}\right)P_i^{\sigma}\left(T\right)=x_i^{\beta}\gamma_i^{\beta}\left(T,\mathbf{x}^{\beta}\right)P_i^{\sigma}\left(T\right) \qquad (3.98)$$

Note that it is essential to make a clear difference between the components (i, j, etc.) and the phases (α, β). When phase split is very strong (e.g. for water-hydrocarbon mixtures), water is frequently identified with the aqueous phase. This may be convenient but is strictly speaking wrong, and may lead to confusion, since water is also present (in small quantities) in the organic phase. Equation (3.98) can be simplified: the vapour pressure has no effect on the liquid-liquid phase split:

$$x_i^{\alpha}\gamma_i^{\alpha}\left(T,\mathbf{x}^{\alpha}\right)=x_i^{\beta}\gamma_i^{\beta}\left(T,\mathbf{x}^{\beta}\right) \qquad (3.99)$$

• or the minimum in the Gibbs energy is calculated. It is written as:

$$G=G^{M,id}+G^{E}=\left(\sum N_i g_i^*+RT\sum N_i \ln x_i\right)+RT\sum N_i \ln \gamma_i \qquad (3.100)$$

where the value of the pure component Gibbs energy, g_i^*, depends on the reference state selected (most conveniently chosen as $g_i^*=0$ for the pure i, at the same pressure and temperature as the system, and in the liquid state). Therefore, for an ideal mixture, equation (3.100) becomes:

$$G=G^{M,id}=RT\sum N_i \ln x_i \qquad (3.101)$$

which is represented for a binary mixture in figure 3.28 as a solid line (Liquid Gibbs energy).

When the mixture shows non-ideal behaviour, the Gibbs free energy curve may have two inflection points, as presented in figure 3.28. In this case, liquid phase split is observed. Using the Margules equation with one parameter (3.88) with $A_{12}=A_{21}$, it can be shown that this occurs when:

$$A_{21}=A_{12}=\ln \gamma_1^{\infty}>2 \text{ , or } \gamma_1^{\infty}>7.5 \qquad (3.102)$$

The upper limit proposed in equation (3.102) is a result of the use of the Margules equation. It does not imply that above this infinite dilution activity coefficient phase split always occurs. There are in fact numerous examples with much larger values of γ_1^{∞} without phase split.

The main point to remember is that the larger the value of γ_1^{∞}, the higher the risk of liquid phase split.

A straight line has been drawn on figure 3.28 to show that the two compositions at equilibrium (x_1^{α} and x_1^{β}) are located on the tangent plane (bold, straight line) to the Gibbs energy line. It is in fact possible to demonstrate that the intersection of the tangent to the $G(x)$ curve with the pure component axes is the chemical potential (which is related to the fugacities, as was discussed in chapter 2). Hence, locating the representative points of the two liquid phases is equivalent to stating that their chemical potentials (or fugacities) are equal in both phases (as in equation 3.99). In between these two limits (x_1^{α} and x_1^{β}) the liquids are immiscible.

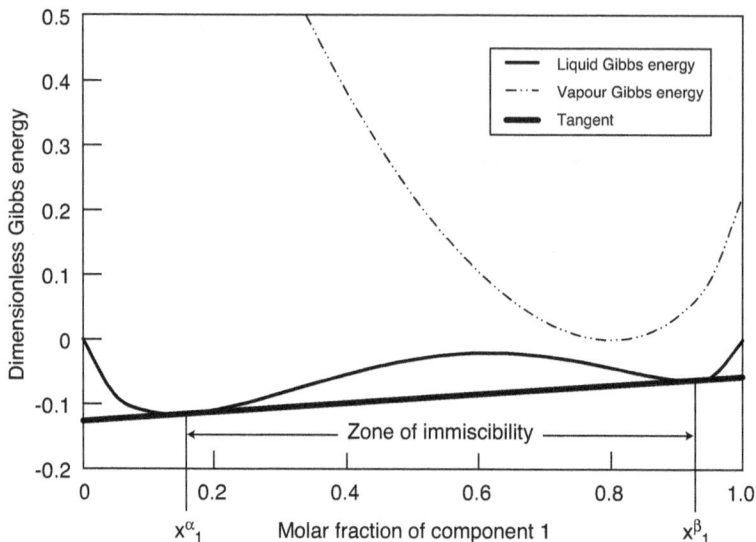

Figure 3.28

Dimensionless Gibbs energy of a liquid-liquid equilibrium: the dashed line represents the ideal mixing in the vapour phase. The solid line is the total Gibbs energy calculated from equation (3.100).

To calculate the phase equilibrium, the algorithm must find the lowest tangent that touches the surface of all phases present (tangent plane algorithm [144-146]). However, since the liquid phase surface is generally much flatter than the vapour phase surface (shown in figure 3.28 with a dashed line), we can see that finding a liquid-liquid tangent is much more difficult than finding a vapour-liquid tangent. It may also happen that the algorithm finds a vapour-liquid equilibrium when in actual fact the liquid-liquid solution is more stable.

Some examples of infinite dilution activity coefficients are shown in table 3.19. All the examples show positive deviations. Note that there is a distinction between solvents and solutes: infinite dilution is not symmetric. The order of magnitudes of the numbers vary greatly (remember that these numbers tell us how the solute volatility changes compared to its pure component volatility): water in methanol is twice as volatile as when it is pure; the volatility of n-hexane in methanol is 17 times greater than pure and the volatility of hexane in water is increased by a factor 30000. The larger the number, the more likely the mixture will split in two liquid phases.

For strongly dissimilar mixtures (water + hydrocarbon), the infinite dilution activity coefficients provide an indication of the mole fraction solubility: using equation (3.99), and considering that the other phase is (almost) pure, we find:

yielding $x_i^\alpha \gamma_i^\alpha = 1$ or $x_i^\alpha = 1/\gamma_i^\alpha$

From table 3.19, we can thus read that at 25 °C the hexane solubility in water is $1/30000 = 33$ ppm. The water solubility in hexane is $1/2000 = 500$ ppm.

Table 3.19 Infinite dilution activity coefficient of various solutes in various solvents, at 298.15 K

Solute	Water	nC_4	nC_6	nC_8	Sulfolane	Methanol	MTBE	Acetonitrile	Benzene
Water	1	2100	2000	1500	3.2	2			240
nC_4	21600	1			30	16		15	
nC_6	30000		1		70	17		27	2.2
nC_8	700000			1	160	58		55	2.3
Sulfolane			1400		1				6
Methanol	1.8				2.1	1	3.7	3.1	
MTBE	115				7.4	6	1		
Acetonitrile	11		28	31				1	3.3
Benzene	2500		1.7	1.5	2.1	7		2.9	1

Example 3.13 Draw the phase diagram of the binary mixture of water and butanol at 373.15 K

Problem statement:

The water (1) + 1-butanol (2) mixture can be modelled using the Margules activity coefficient representation (equations 3.89 and 3.90) with following parameters $A_{12} = 1.3863$ and $A_{21} = 3.0445$ (dimensionless).

The vapour pressures of the pure components are at this temperature:

P_1^σ = 101.33 kPa

P_2^σ = 51.89 kPa

Analysis:

A complete isothermal phase diagram for a binary system must be constructed. Therefore, no feed composition is required. Only phase equilibria need to be calculated. However, we do not know in advance what type of phase equilibria may be found, and all possibilities must be checked.

Vapour-liquid equilibrium always occurs if the temperature is between the triple point temperature and the critical temperature of any one of the components. Using the margules equation, we find:

$\gamma_2^\infty = \exp(A_{12}) = 4$

$\gamma_1^\infty = \exp(A_{21}) = 21$

With the information provided, and using equation (3.96), we may even state that the vapour-liquid equilibrium will present an azeotropic behaviour:

$$\frac{1}{\gamma_2^\infty} = 0.25 < \frac{P_2^\sigma}{P_1^\sigma} = 0.512 < \frac{\gamma_1^\infty}{1} = 21$$

When drawing the Gibbs energy surface as in figure 3.28, an inflection point appears, indicating that a liquid-liquid phase split will occur. In combination with the azeotropic behaviour, this may lead to a heteroazeotrope.

The triple point temperatures of both components are too low to generate a risk of a solid phase appearing at 373.15 K.

Solution:

The pressure is well below 500 kPa so the corrected Raoult's law (3.86) can be used to calculate the vapour-liquid equilibrium. It requires vapour pressure (provided) and activity coefficients.

For the activity coefficients, the choice of the Margules model is proposed here in the problem statement, but any other two parameter activity coefficient model could have been used.

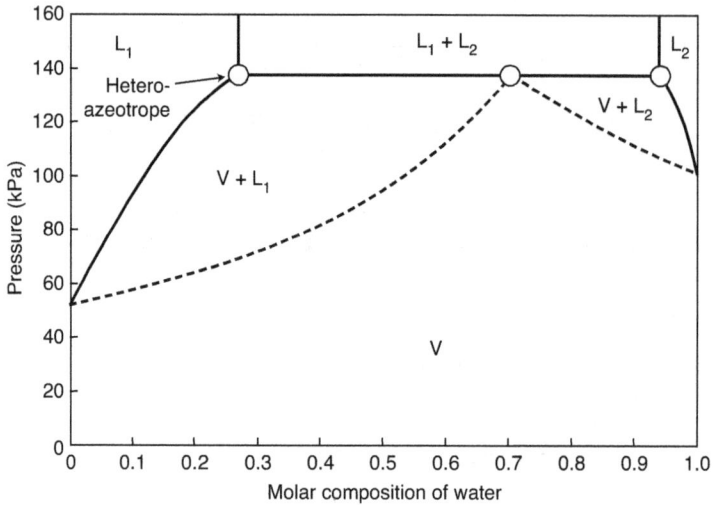

Figure 3.29

Pxy diagram for the water + 1-butanol mixture at 100 °C (373.15 K).

Figure 3.29 shows at 373.15 K the full isothermal hetero-azeotropic phase diagram of the water + 1-butanol mixture. Note that the liquid-liquid region is presented by a vertical slice. This is a result of the assumption that the activity coefficient is independent of pressure. As a first approximation, it is a good solution, but the graph cannot be extended to pressures larger than 1 MPa.

This example is discussed on the website:
http://books.ifpenergiesnouvelles.fr/ebooks/thermodynamics

3.4.1.4 Conclusion

We have shown how the simultaneous evaluation of vapour pressures and activity coefficients could help identify the type of phase diagram that may be expected. The combination of the criteria (3.96) for azeotropy and (3.102) for liquid phase split helps understand the conceptual figure 3.30 that locates the type of binary isothermal phase diagram on a two-dimensional plot:

- Along the abscissa, the volatility ratio of the pure components is plotted $(\alpha_{12} = p_1^\sigma / p_2^\sigma)$. To the left, close-boiling mixtures are found. They almost always feature azeotropic behaviour.
- Along the ordinate, the mixture non-ideality is expressed in terms of the logarithm of the infinite dilution activity coefficients (either one since the plot is drawn for symmetric deviations from ideality), the "zero" line corresponds to the ideal behaviour. Below this line, negative deviations from ideality are found; above this line, positive deviations. When the value of $\ln \gamma_1^\infty$ reaches two, the risk of finding liquid-liquid phase split becomes important (once again, remember this is only an approximate value).

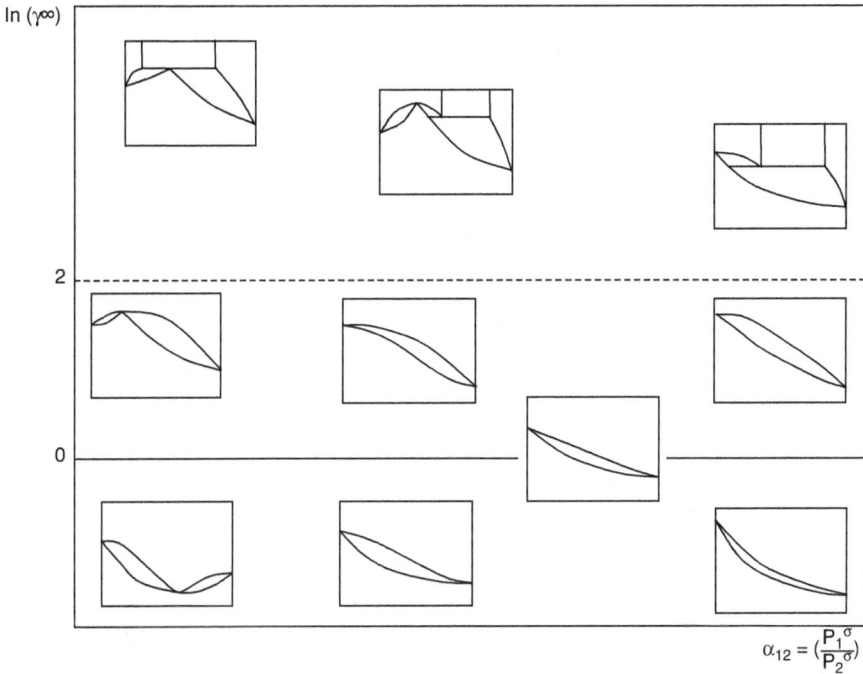

Figure 3.30

Types of *Pxy* diagram as a function of the vapour pressure ratio and activity coefficient at infinite dilution (from [147] – this graph was obtained with a Margules 1 parameter model $A_{12} = A_{21}$).

Reprinted from [147], with permission from Elsevier.

3.4.2 Activity models, or how the molecular structure affects the non-ideal behaviour

The behaviour of a particular compound will be the result of its interactions with the neighbouring molecules. Some are small molecules (hydrogen, methane, ethane, propane, etc.) others very large (naphthalene). Some are linear like *n*-octane, others branched like iso-octane (2,2,3 trimethyl-pentane) or cyclic like benzene (C_6H_6) and cyclohexane (C_6H_{12}). Some have heteroatoms like oxygen in methanol (CH_3OH) and dimethyl-ether

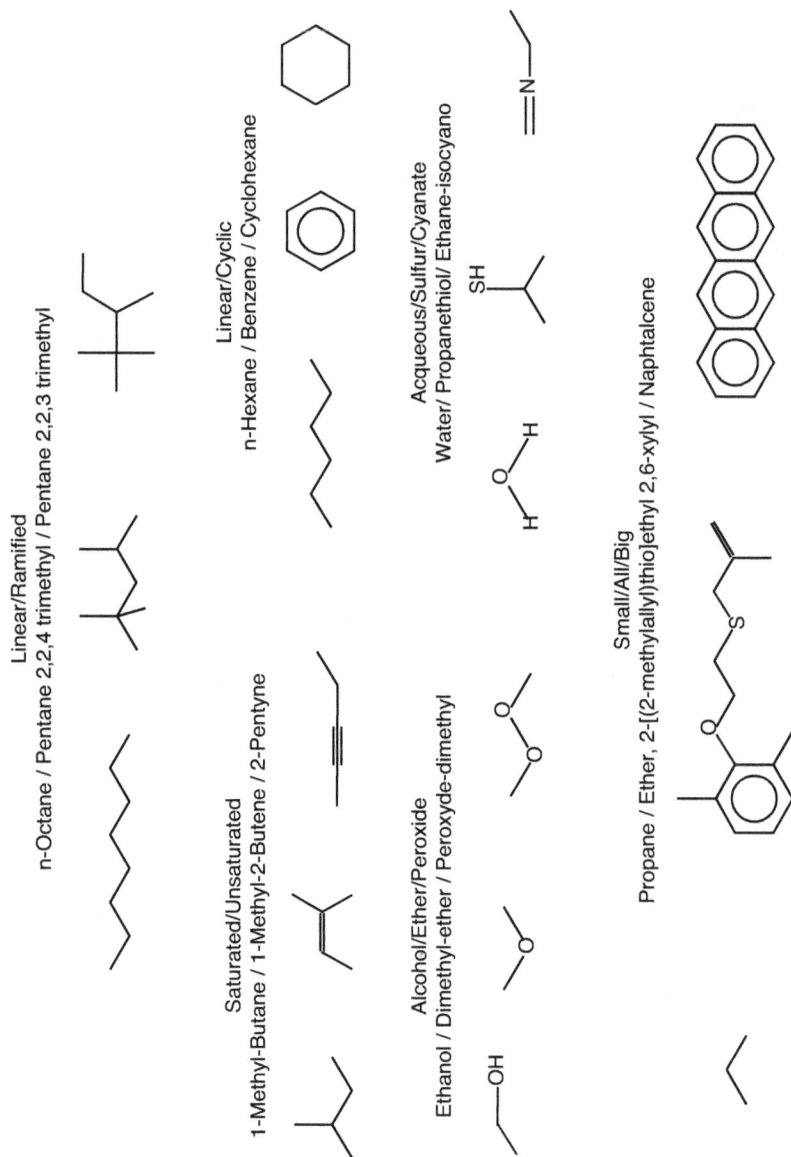

Linear/Ramified
n-Octane / Pentane 2,2,4 trimethyl / Pentane 2,2,3 trimethyl

Linear/Cyclic
n-Hexane / Benzene / Cyclohexane

Saturated/Unsaturated
1-Methyl-Butane / 1-Methyl-2-Butene / 2-Pentyne

Acqueous/Sulfur/Cyanate
Water/ Propanethiol/ Ethane-isocyano

Alcohol/Ether/Peroxide
Ethanol / Dimethyl-ether / Peroxyde-dimethyl

Small/All/Big
Propane / Ether, 2-[(2-methylallyl)thio]ethyl 2,6-xylyl / Naphtalcene

Figure 3.31
Influence of the structure of a molecule in a mixture.

$(CH_3–O–CH_3)$. Some of this diversity is illustrated in figure 3.31. We may therefore understand that some unlike molecules "attract" each other more while others "repel" each other.

Over the years, many authors have investigated these interactions, and this is how the activity models have come to existence. Their main focus was the liquid phase, as this is where the interactions are strongest. This is also why these models generally only consider incompressible fluids, which limits their range of application to low pressure.

> Consequently, unless the parameter values are adapted to work at high pressure, the use of activity coefficient models is generally limited to 1.5 MPa.

It is not the purpose, in this section, to provide an exhaustive list of the existing models. Many excellent textbooks [38, 148-150] have been written for that purpose. The aim of this book is more to guide the practicing engineer when evaluating the reasons for the particular behaviour observed, and lead him or her to the most appropriate choice of model and parameter combination.

It is essential to stress at this stage that, because of the very concept of molecular interactions, use of the activity models assumes a full molecular description of the mixture. The use of petroleum pseudo-components (corresponding states principle) is not appropriate, since the way they are characterised focuses only on properties related to their volatility. In actual fact, only Hildebrand's regular solution theory provides a tool for calculating activity coefficients of petroleum mixtures, and this theory is based on a Van der Waals' type cubic equation of state model.

> Use of activity coefficient models assumes that the pure component behaviour is known. They require molecular parameters as well as binary (or higher) interaction parameters, originating from a database, fitted on binary (or higher) experimental data, or calculated through a group contribution method.

More recently, quantum-mechanical tools have also been used for predicting these parameters [151].

Models that can provide results without interaction parameters are called predictive. The quality of their calculations is obviously limited but, if used on mixtures which follow the theory they are built on, they can provide useful results.

3.4.2.1 Enthalpic *vs.* entropic contributions

The cause of non-ideal behaviour (expressed in terms of excess Gibbs energy, or activity coefficients, as described in section 2.2.2.2, p. 60) can be classified in two major categories: enthalpic, and entropic. This is the direct consequence of the Gibbs energy calculation as a sum of two terms:

$$G^E = H^E - TS^E \tag{3.103}$$

As a result, the activity coefficient itself can be written as the sum of an enthalpic (or "residual" [1]) and entropic (or "combinatorial") contribution:

$$\ln \gamma_i = \ln \gamma_i^{res} + \ln \gamma_i^{comb} \tag{3.104}$$

---

1. The use of the word *residual* is different from the meaning used in section 2.2.2.1.B, p. 55 corresponding to the difference between the real state and the ideal gas state.

Although in most cases the two contributions are present, it will be shown that their behaviour is different and that this classification can improve the understanding of the non-ideal behaviour of a mixture.

3.4.2.2 Enthalpic deviation from ideality

A. Origin of the enthalpic deviation from ideality

In this section, we will consider molecules of similar size and shape. In this case, the interactions among molecules are well known [148] and can be summarised as depicted in figure 3.32.

Ion-dipole interactions are the strongest because the electrostatic charges are explicit and localised. They are sometimes called "hydration" interactions, and form one of the reasons why it is difficult to describe electrolyte mixtures. Their energy is of the order of several hundreds kJ mol^{-1}.

When the charge distribution on two molecules is such that a strongly negatively charged surface can touch a strongly positively charged surface, the molecules will "stick" to each other. This is often called "hydrogen bonding", since the positively charged site is often a hydrogen atom: it is well known that when this atom is located next to an electronegative atom (such as oxygen, but also nitrogen, sulphur, etc.), it will lose its electron and gain a surface charge. The order of magnitude of the energy of hydrogen bonds is 10 to 40 kJ mol^{-1}.

When the surface charge distribution is not strong enough for the oppositely charged ends to stick, it may be enough for a permanent molecular dipole to form. In this case, a "dipole-dipole" interaction is observed: the molecules tend to orient themselves such that their dipole moment is parallel. The energy of such "bonds" is of the order of 5 to 25 kJ mol^{-1}.

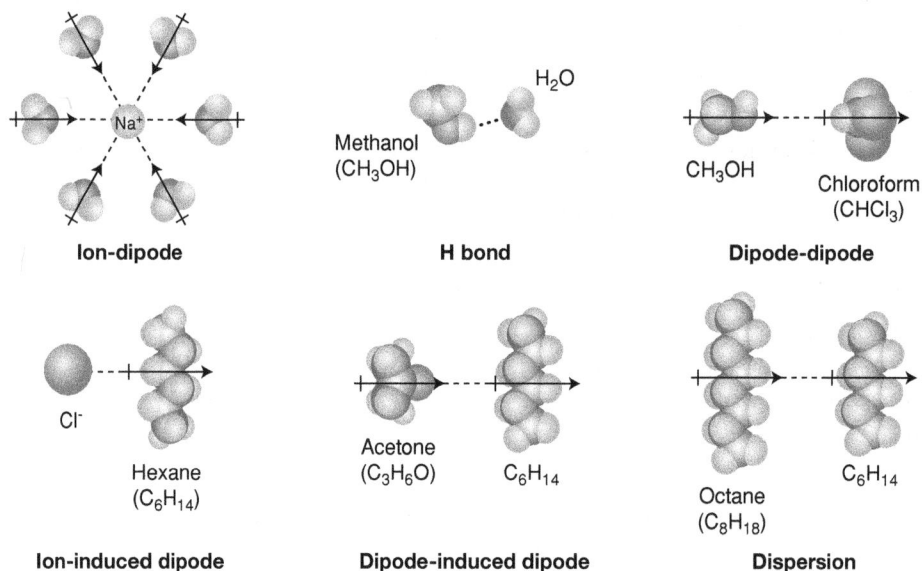

Figure 3.32

Spatial relation between molecules and atoms
(from http://www.science.uwaterloo.ca/~cchieh/cact/c123/intermol.html).

When only one of the two molecules has a permanent dipole (or is ionic), it will disturb the electron cloud of the neighbouring non-polar molecule, thus creating an "induced dipole". When the induction is due to an ion, the strength of this interaction is stronger than when it is due to a dipole. Its strength can reach 2 to 10 kJ mol^{-1}.

Finally, the lowest energy interactions are called "dispersive" or "Van der Waals" interactions. These are a result of oscillations within the electron clouds, thereby giving rise to temporary orientations that result in an attraction between molecules. These forces are described by the London theory. Their strength can be very variable. In general, the heavier is the molecule, the stronger is the Van der Waals force of interaction. For example, the boiling points of inert gases increase as their atomic masses increase due to stronger London dispersion interactions.

Quantum-mechanical tools are able to describe the tendency of each molecule to exhibit the above interactions [151, 152]. Of these, COSMO-RS [153, 154] is probably the most well-known in the engineering community. The COSMO (conductor like screening model) software can be used to visualise the surface charges and thus predict non-ideal behaviour.

Some models designed to describe these types of interaction are discussed below.

B. Scatchard-Hildebrand's regular solution theory

Regular solutions are defined as mixtures where both excess volume and excess entropy (discussed in section 2.2.2.2, p. 60) are zero.

$$\ln \gamma_i = \ln \gamma_i^{res} + \ln \gamma_i^{comb} = \ln \gamma_i^{res} \qquad (3.105)$$

The regular solution model, proposed by Scatchard (1931 [155, 156]) as a result of the work of Hildebrand (1916 [157]), is essentially based on the concepts formulated by Van Laar. The Hildebrand and Wood paper of 1933 [158] followed by the communication of Scatchard and Hildebrand of 1934 [159] provided the definitive basis of this theory. Development of the model is based on the Van der Waals equation of state (section 3.4.3.4, p. 204), with a classical mixing rule between parameters a and b (equations 3.224 and 3.227).

This model is still used today as it is well adapted to non-polar mixtures, such as petroleum mixtures. The Gibbs energy is written as:

$$\frac{g^E}{RT} = \frac{x_1 v_1^* x_2 v_2^*}{RT \left(x_1 v_1^* + x_2 v_2^* \right)} \left(\delta_1 - \delta_2 \right)^2 \qquad (3.106)$$

and the associated activity coefficient is given by:

$$\ln \gamma_1 = \frac{v_1^*}{RT} \left(\delta_1 - \delta_2 \right)^2 \left(\phi_2 \right)^2 \qquad (3.107)$$

introducing the volume fraction ϕ_i defined by:

$$\phi_i = \frac{x_i v_i^*}{\sum_j x_j v_j^*} \qquad (3.108)$$

For multicomponent mixtures, the expression becomes:

$$\frac{g^E}{RT} = \frac{v}{2RT}\sum_{i=1}\sum_{j=1}\phi_i\phi_j\left(\delta_i-\delta_j\right)^2 \tag{3.109}$$

$$\ln\gamma_i = \frac{v_i^*}{RT}\left(\delta_i-\overline{\delta}\right)^2 \tag{3.110}$$

where v_i^* is the liquid molar volume of component i (in m^3 mol^{-1}) and δ_i is the solubility parameter ((J m^{-3})$^{1/2}$). $v=\sum_{i=1}x_iv_i^*$ is the total volume. The solubility parameter, already defined in section 3.1.1.1.C of this chapter (p. 103), is expressed as:

$$\delta_i = \sqrt{\frac{\Delta u_i^\sigma}{v_i^*}} \tag{3.111}$$

where Δu_i^σ is the internal energy of vapourisation of component i. The solubility parameter depends on the temperature, but only its value at 298.15 K is usually taken as the difference between solubility parameters (as in equations 3.106 and 3.109) is almost independent of temperature.

The average solubility parameter is defined as:

$$\overline{\delta} = \sum_{i=1}\phi_i\delta_i \tag{3.112}$$

This method is well adapted for non-polar components. It is very powerful in that it only uses pure component parameters that generally have a physical meaning. It can therefore be considered as predictive. The following limitations must be borne in mind however:
• it only predicts positive deviations from ideality ($\gamma_i > 1$),
• it is not applicable to polar components, that generally show large deviations from ideality.

Using equation (3.110), we may observe that:
• enthalpic deviations from ideality result in positive values of $\ln\gamma_i$ (possibly leading to a positive azeotrope).
• the numerical value of the solubility parameter (available in most databases) provides direct information about the importance of the non-ideality. Mixtures containing components whose solubility parameters have similar values will show a low deviation from ideality, while mixtures containing components whose solubility parameters have very different values will probably behave as strongly non-ideal mixtures.

This theory has also been used successfully by the Yarranton group to predict asphaltene precipitation [160-162].

C. Local composition models and NRTL

The physical foundation of local composition models was laid by Wilson (1964 [163]). It states that the "local composition" (represented by x_{ij}, the molar fraction of i around molecule j) is different from the "global composition" ($x_i \neq x_{ij}$) as shown in figure 3.33. Double indices are now required. Please note that the order of the indices is meaningful, *i.e.* $x_{ji} \neq x_{ij}$.

The mass balance around molecule j is nevertheless retained:

$$\sum_j x_{ij} = 1 \tag{3.113}$$

$$x_{21} = 4/6 \; ; \; x_{11} = 2/6 \qquad\qquad\qquad x_{22} = 3/6 \; ; \; x_{12} = 3/6$$

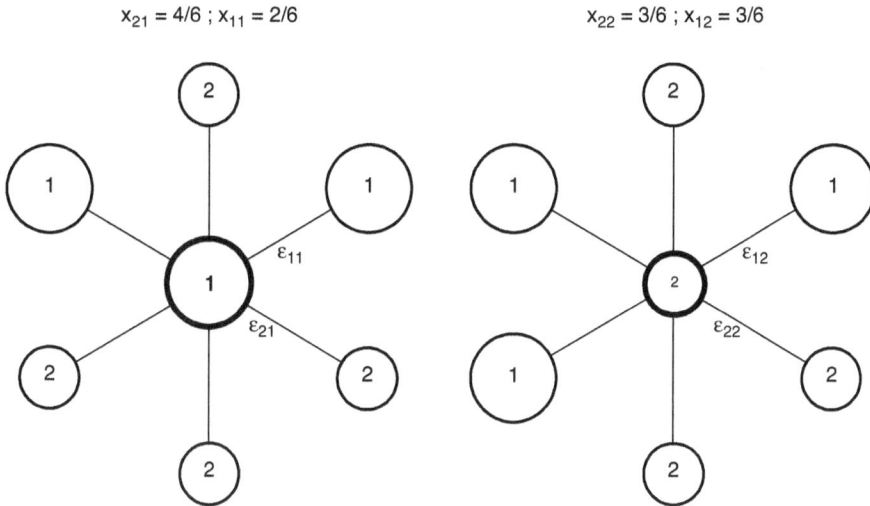

Figure 3.33

Representation of local composition.

The challenge lies in the calculation of this local composition. Using the Boltzmann distribution function, we can state that the probability of finding j close to i (*i.e.* its local mole fraction) is proportional to the exponential of the energy of the i-j bond:

$$x_{ij} \sim x_i \exp\left(-\frac{(\varepsilon_{ij})}{kT}\right) \tag{3.114}$$

using (3.113), we obtain:

$$x_{ij} = \frac{x_i \exp\left(-\frac{\varepsilon_{ij}}{kT}\right)}{\sum_k x_k \exp\left(-\frac{\varepsilon_{kj}}{kT}\right)} \tag{3.115}$$

The theoretical foundation of the NRTL model [142] can be simplified by stating that it is found from the assumption that $g^E = u^E$ (*i.e.* we assume that both v^E and s^E are negligible)[1]. Since the excess internal energy is equal to the internal energy of mixing, we may state:

$$G^E = U^M = \frac{z}{2} \sum_i N_i \sum_j x_{ji} \left(\varepsilon_{ji} - \varepsilon_{ii} \right) \tag{3.116}$$

which is a sum over all central molecules i, of the difference between the energy in the mixture and the energy in the pure component. This sum is multiplied by the number of neighbouring sites, z, divided by two since every molecule is taken once as a central molecule (i), and once as a neighbouring molecule (j). Equation (3.116) is now used to express x_{ij}, and, with

$$\tau_{ji} = \frac{z}{2RT} \left(\varepsilon_{ji} - \varepsilon_{ii} \right) \tag{3.117}$$

and

$$\alpha_{ij} = \alpha_{ji} = \frac{2}{z} \tag{3.118}$$

From (3.116) the NRTL expression is found:

$$g^E = RT \sum_i x_i \sum_j \frac{x_j \exp\left(-\alpha_{ji}\tau_{ji}\right)}{\sum_k x_k \exp\left(-\alpha_{ki}\tau_{ki}\right)} \tau_{ji} \tag{3.119}$$

and using

$$G_{ji} = \exp\left(-\alpha_{ji}\tau_{ji}\right) \tag{3.120}$$

the corresponding activity coefficient expression is obtained as:

$$\ln \gamma_i = \frac{\sum_j \tau_{ji} G_{ji} x_j}{\sum_j G_{ji} x_j} + \sum_j \frac{G_{ij} x_j}{\sum_k G_{ki} x_k} \left(\tau_{ij} - \frac{\sum_k \tau_{kj} G_{kj} x_k}{\sum_k G_{kj} x_k} \right) \tag{3.121}$$

It is important to note here the number of binary parameters. The values of τ_{ij} and τ_{ji} (watch for the dimensions: here they are dimensionless but in some software they are expressed as energy or temperatures depending on the use of RT in the definition equation 3.117) are different and regressed on binary data. Their physical meaning can be inferred from equation (3.117), but generally they are regressed on experimental data.

In addition, $\alpha_{ij} = \alpha_{ji} = \frac{2}{z}$ is typically close to 0.2 since z, the coordination number or number of neighbouring sites, can be considered as being close to 10 in a three dimensional arrangement. It can also be adjusted if necessary however.

For a given temperature, therefore, the NRTL model has a maximum of three binary parameters. However, these parameters may also depend on the temperature. Linear functions are generally used, although this doubles the number of parameters.

---

1. See section 2.2.2.2.B (p. 61) for an explanation of the relationships between these properties.

The large number of parameters available with NRTL makes it very powerful for describing non-ideal mixtures, including different types of properties (vapour-liquid equilibria, liquid-liquid equilibria or mixing enthalpies). However, it may be dangerous to leave too many adjustable parameters as correlations may exist between them (see example 3.11 in section 3.3.2.2, p. 156).

D. Other approaches

From the discussion above, it is clear that one of the main contributions to the non-ideal behaviour of a mixture is the hydrogen bonding behaviour. This phenomenon can be described using a quasi-chemical approach, as comprehensively described by Elliott [164].

The highly successful Wertheim theory [165, 166 which will be discussed in section 3.4.3.2 (p. 192) regarding equations of state, can be simplified to correct the liquid phase activity coefficient. This has been done by Mengarelli *et al.* and Ferreira *et al.* [167, 168], who combined it with UNIFAC (see section 3.4.2.4.C, p. 184), thereby enabling them to predictively calculate (after re-fitting all the UNIFAC parameters), both cross- and self-associating components as various sugar types.

Today, the development of equations of state is such that their ability to describe strongly non-ideal mixtures in a semi-predictive manner has greatly improved. The Cubic Plus Association (CPA), Statistical Association Fluid Theory (SAFT) or its polar version and the Non Random Hydrogen Bonding (NRHB) equations of state, all of which are further discussed in section 3.4.3.5 of this chapter (p. 216), should be examined.

Although strictly speaking not an activity coefficient model, COSMO-RS [154] also calculates activity coefficients. It is very well adapted as a predictive tool for describing enthalpic deviations from ideality.

3.4.2.3 Entropic deviation from ideality (athermal solutions)

A. Origin of entropic contribution

The entropic deviation from ideality is caused by differences in size or shape of the molecules. Its contribution to the excess Gibbs energy, as shown in equation (3.103), is negative: for athermal solutions (solutions whose excess enthalpy is zero), the excess Gibbs energy is

$$G^E = 0 - TS^E \tag{3.122}$$

indicating that for positive excess entropies, the deviation from ideality will be negative (possible formation of negative azeotropes).

$$\ln \gamma_i = \ln \gamma_i^{res} + \ln \gamma_i^{comb} = \ln \gamma_i^{comb} \tag{3.123}$$

The models developed for athermal solutions are therefore based on the Boltzmann definition of entropy:

$$S = R \ln W \tag{3.124}$$

where R is the universal gas constant and W is the number of different configurations that the system may have. The challenge encountered by the developers of these models is how to determine these configurations.

The mixing entropy for a binary mixture is calculated as

$$S^M = R\left(\ln W_{12} - \ln W_{11} - \ln W_{22}\right) \tag{3.125}$$

where W_{12} are the number of configurations in the mixtures, while W_{11} and W_{22} are the number of configurations in the pure components 1 and 2 respectively. The excess entropy is readily found by subtracting the ideal mixture entropy (see section 2.2.2.2.B, p. 61):

$$s^E = s^M - s^{M,id} = s^M - R\sum_i x_i \ln x_i \tag{3.126}$$

B. Flory model

Although quantitatively not accurate enough, the Flory [169] model offers a simple way of understanding the effect of differences in molecular volume. It is predictive, requiring only pure component molecular properties (the volume) as input and is used for monomer + polymer mixtures.

If a lattice-like structure is considered for the liquid mixture, the molecules are characterised by the number of adjacent lattice-sites they occupy. For example, a 1-mer + *m*-mer mixture will have a mixing entropy that can be calculated as:

$$S^M = -R\left(N_1 \ln \frac{N_1}{N_1 + mN_2} + N_2 \ln \frac{mN_2}{N_1 + mN_2}\right) \tag{3.127}$$

where N_1 and N_2 are the number of 1-mer and *m*-mer molecules. Note that the ratios within the logarithms are the volume fractions of the species considered:

$$\frac{m_i N_i}{\sum_j m_j N_j} = \frac{v_i x_i}{\sum_j v_j x_j} = \phi_i \tag{3.128}$$

When the sizes of the molecules are different, non-ideal behaviour will be observed as a result of excess entropy. This is the case in particular when polymers are mixed with small solvent molecules. In this case the Flory model is a very convenient tool for evaluating the effect of the difference in molar volume. Constructed using the statistical significance of the entropy of mixing, its shape is as follows:

$$g^E = -Ts^E = RT\sum_i x_i \ln \frac{\phi_i}{x_i} \tag{3.129}$$

where ϕ_i represents the volume fraction of component *i*, according to (3.128). The Flory model can also be written as:

$$\ln \gamma_i = \ln \frac{\phi_i}{x_i} + 1 - \frac{\phi_i}{x_i} \tag{3.130}$$

This effect is called "entropic" and leads to a negative deviation from ideality (activity coefficient less than one).

The Flory equation (3.130) discussed here is unable to take into account all entropic effects, primarily since it only considers size effects (it uses the molar volume as sole parameter), while in actual fact the molecular shapes also have an influence on non-ideality.

C. Staverman-Guggenheim

Flory only uses information on molecular volumes. The molecules may have similar volumes but different shapes, however, which may also result in an entropic deviation. This observation led Staverman [170] and Guggenheim [171] to propose an expression which incorporates a different pure component property, the molecular surface area, q_i. The molar volume parameter is here expressed as r_i, to indicate that it should now be considered as a model parameter rather than as a molecular property. The equation can now be expressed as follows:

$$g^E = RT \left(\sum_i x_i \ln \frac{\phi_i}{x_i} + \frac{z}{2} \sum_i q_i x_i \ln \frac{\theta_i}{\phi_i} \right) \tag{3.131}$$

The activity coefficient using this equation is:

$$\ln \gamma_i = \ln \frac{\phi_i}{x_i} + 1 - \frac{\phi_i}{x_i} - \frac{z}{2} \left(\ln \frac{\phi_i}{\theta_i} + 1 - \frac{\phi_i}{\theta_i} \right) \tag{3.132}$$

and the surface fraction θ_i and the volume fraction ϕ_i of molecule are:

$$\theta_i = \frac{x_i q_i}{\sum_j x_j q_j} \quad \text{and} \quad \phi_i = \frac{x_i r_i}{\sum_j x_j r_j} \tag{3.133}$$

An additional parameter is found here, z, the coordination number of the liquid lattice. It corresponds to the number of nearest neighbours of any particular molecule. For a three-dimensional lattice, this value is usually taken as $z = 10$.

This model is adapted to mixtures without mixing enthalpy, but with different shapes and sizes.

The Staverman-Guggenheim equation offers the same main advantage of only containing pure component parameters that can be found in many databases. It is therefore purely predictive, with the resulting advantages and disadvantages (poor accuracy). Consequently, it is almost never used as is, but rather as the combinatorial part of the UNIQUAC model, which contains additional binary parameters (see hereafter 3.4.2.4.B, p. 183).

D. Other models

As already mentioned, the entropic effect leads to negative deviations from ideality, which favours mixing in the liquid phase. Yet, it is well known that polymer-solvent systems exhibit phase split phenomena, with a Lower Critical Solution Temperature (LCST). This phase split phenomenon is discussed in section 4.2.1.1 (p. 265). It is explained, for a fully athermal system, by what is called "free volume" [38, 172]: it has been shown [173] that the fraction of free or empty volume is much larger for a long chain molecule (polymer) than for a pure component consisting of small molecules. This difference increases with increasing temperatures. As a result, a negative excess entropy is observed for the mixture (the number of combinations is reduced upon mixing), which eventually leads to a positive Gibbs excess energy (equation (3.122)).

Several models based on this concept have been proposed (entropic free volume [174], GC-Flory equation [175, 176]) but they should in fact be in the category of equations of state since they depend on pressure. They are all predictive in the sense that they are based on the group contribution concept. In the same category, the group contribution lattice fluid equation of state will be briefly discussed in section 3.4.3.6 of this chapter (p. 219).

3.4.2.4 Mixed enthalpic and entropic deviation from ideality

The mixed models are able to predict both positive and negative deviations from ideality. It has been shown that the empiric Margules method, or its more extended Redlich-Kister [177] equivalent are able to describe both kinds of deviation. The lack of physical meaning of these equations has limited their use however.

Based on the observation that the Gibbs excess energy is the sum of an enthalpic and entropic term (3.103), several models have been proposed that contain a sum of these two terms. Note that in principle, any combination of a combinatorial term with a residual term is feasible.

A. Flory-Huggins

The Flory-Huggins equation was proposed almost simultaneously by Huggins (1941 [178]) and Flory (1941-1942 [179, 180]).

In polymer applications molecular sizes are very different, but it is not correct to state that the mixing enthalpy with the solvents is zero. The athermal Flory model can therefore be combined with the regular solution theories using:

$$g^E = h^E - Ts^E \tag{3.134}$$

to produce the following equation:

$$\frac{g^E}{RT} = \frac{v}{2RT}\sum_{i=1}^{n}\sum_{j=1}^{n}\phi_i\phi_j\left(\delta_i - \delta_j\right)^2 + \sum_i x_i \ln\frac{\phi_i}{x_i} \tag{3.135}$$

The convention is to call the energy term "residual" and the entropic term "combinatorial". The activity coefficient is:

$$\ln\gamma_i = \frac{v_i^*}{RT}\left(\delta_i - \overline{\delta}\right)^2 + \ln\left(\frac{\phi_i}{x_i}\right) + 1 - \frac{\phi_i}{x_i} \tag{3.136}$$

Often, the square difference between the solubility parameters (this parameter is unknown for polymers) is replaced by an adjustable parameter κ, which yields for a binary mixture:

$$\frac{g^E}{RT} = \kappa\left(x_1 + \frac{v_1^*}{v_2^*}x_2\right)\phi_1\phi_2 + \left(x_1 \ln\frac{\phi_1}{x_1} + x_2 \ln\frac{\phi_2}{x_2}\right) \tag{3.137}$$

The Flory-Huggins model, expressed as in equation (3.136) is fully predictive, which makes it an interesting tool for evaluating the trends observed in the non-ideality of mixtures when both entropic and enthalpic effects are present. Nevertheless, its accuracy is generally very low.

Example 3.14 Isothermal phase diagram using the Flory-Huggins activity coefficient model

The Flory-Huggins model is applied to the hexane + benzene mixture.

This example is discussed on the website:
http://books.ifpenergiesnouvelles.fr/ebooks/thermodynamics

B. Universal Quasi-Chemical (UNIQUAC)

The UNIQUAC [182] model is based on the approach presented above, consisting of the sum of a combinatorial and a residual contribution. The combinatorial contribution is taken directly from the Staverman-Guggenheim expression. The result is as follows:

$$g^{E.comb} = RT \left(\sum_i x_i \ln \frac{\phi_i}{x_i} + \frac{z}{2} \sum_i q_i x_i \ln \frac{\theta_i}{\phi_i} \right) \tag{3.138}$$

This term is predictive in that it does not contain any binary parameter. The residual contribution is taken from a Wilson-type equation which has not been presented in this work:

$$g^{E.res} = -RT \sum_i x_i q_i \ln \left(\sum_j \tau_{ji} \theta_j \right) \tag{3.139}$$

This residual term is inspired by Flory's expression, where the volumes and volume fractions have been replaced with surfaces and surface fractions. Once again the parameter q refers to the molecular surface, and θ_j is the surface fraction of molecule j. The binary parameter τ_{ji} is a local composition type binary interaction parameter. It is asymmetric since $\tau_{ji} \neq \tau_{ij}$. If these parameters are temperature-dependent, the number of binary parameters can increase to four. The τ_{ji} parameter is related to a characteristic energy interaction between i and j.

$$\tau_{ji} = \exp \left(-\frac{\varepsilon_{ji} - \varepsilon_{ii}}{RT} \right) \tag{3.140}$$

The activity coefficients corresponding to the above excess Gibbs energy equations can be calculated from (3.104):

$$\ln \gamma_i = \ln \gamma_i^{comb} + \ln \gamma_i^{res} \tag{3.141}$$

where the combinatorial part is that of Staverman-Guggenheim (eq. 3.132), written as:

$$\ln \gamma_i^{comb} = \ln \frac{\phi_i}{x_i} + 1 - \frac{\phi_i}{x_i} - \frac{z}{2} \left(\ln \frac{\phi_i}{\theta_i} + 1 - \frac{\phi_i}{\theta_i} \right) \tag{3.142}$$

and the residual part is

$$\ln \gamma_i^{res} = q_i \left[1 - \ln \left(\sum_j \theta_j \tau_{ji} \right) - \sum_j \frac{\theta_j \tau_{ij}}{\sum_k \theta_k \tau_{kj}} \right] \tag{3.143}$$

The UNIQUAC model contains the same features as NRTL, but it is somewhat more powerful as it takes explicitly into account the entropic contribution. It is also adapted to calculate liquid-liquid equilibria and may be able to calculate mixing enthalpies if adequate temperature dependent parameters are used.

The UNIQUAC parameters are generally fitted on experimental data, but proposals have been made for using quantum-mechanical calculations in order to use it as a predictive model.

C. UNIFAC

The UNIFAC equation (Fredenslund *et al.*, 1975 [81]) for G^E model is directly inspired by the UNIQUAC equation [sum of equations (3.142) and (3.143)], but it is designed as a group contribution model, and as such, it is a predictive model. Each molecule is split into its composing chemical groups (as in section 3.1.2.2, p. 124).

In the combinatorial part of the equation (3.142) the parameters r_i and q_i associated with the volume fraction and surface fraction respectively (from equation 3.133) are obtained by the basic rules:

$$r_i = \sum_{k=1} \upsilon_{kj} R_k \quad \text{and} \quad q_i = \sum_{k=1} \upsilon_{ki} Q_k \tag{3.144}$$

υ_{ki} represents the number of groups of type k in molecule i.

The residual part of the activity coefficients is given by:

$$\ln \gamma_i^{res} = \sum_k^{groups} \upsilon_{ki} \left(\ln \Gamma_k - \ln \Gamma_{ki} \right) \tag{3.145}$$

where Γ_k is the "activity coefficient" of group k in the mixture, while Γ_{ki} is the activity coefficient of this same group in the pure component i. The difference of the two terms is used to validate the basic rule that in the limit of pure i, the molecular activity coefficient γ_i^{res} is equal to 1. The group activity coefficients are calculated using the same approach as for UNIQUAC (equation 3.143):

$$\ln \Gamma_k = Q_k \left[1 - \ln \left(\sum_m^{groups} \tau_{mk} \Theta_m \right) - \sum_m^{groups} \frac{\Theta_m \tau_{km}}{\sum_n^{groups} \Theta_k \tau_{nm}} \right] \tag{3.146}$$

where

$$\Theta_m = \frac{Q_m X_m}{\sum_n^{groups} Q_n X_n} \tag{3.147}$$

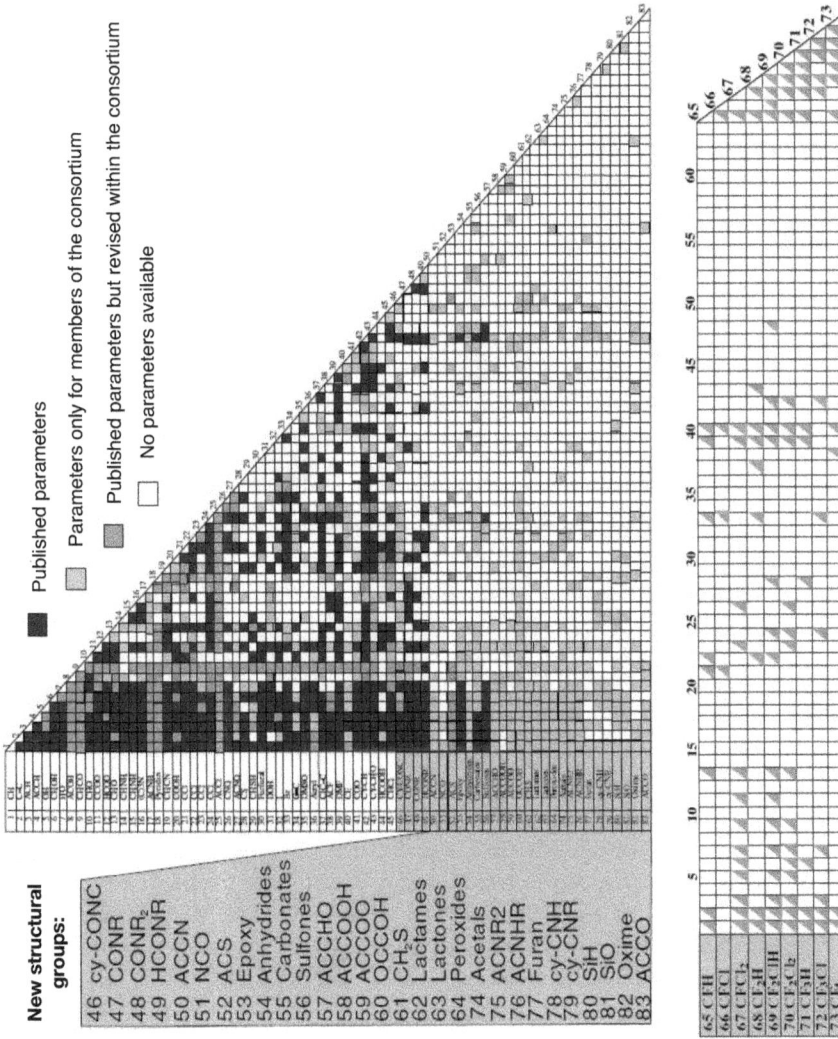

Figure 3.34

Unifac published groups from Gmehling 2009 [56].
Reprinted from [56], with permission from Elsevier.

is the group volume fraction, and

$$X_m = \frac{\overset{moles}{\underset{i}{\sum}} \upsilon_{mi} x_i}{\overset{moles\ groups}{\underset{i}{\sum}\underset{k}{\sum}} \upsilon_{ki} x_i}$$

(3.148)

is the group mole fraction. τ_{mk} has the same meaning as in the UNIQUAC model.

As a result of the success encountered by the UNIFAC model, several extensions have been proposed [182-189]. The most well-known modification is that proposed by Weidlich and Gmehling [190, 191], often called Dortmund – UNIFAC. Both the combinational part (eq. 3.146) and the temperature dependence of the parameters have been changed in order to improve predictions of vapour-liquid equilibria, infinite dilution activity coefficients and heats of mixing.

The new combinatorial part is written as:

$$\ln \gamma_i^{comb} = \ln\frac{\phi_i{}'}{x_i} + 1 - \frac{\phi_i{}'}{x_i} - \frac{z}{2}q_i\left(\ln\frac{\phi_i}{\theta_i} + 1 - \frac{\phi_i}{\theta_i}\right)$$

(3.149)

where

$$\phi_i{}' = \frac{x_i r_i^c}{\sum_j x_j r_j^c}$$

(3.150)

and c depends on the version ($c = 2/3$ for the Lyngby version and $c = 3/4$ for the Dortmund version).

The process simulators today propose a large number of alternative UNIFAC versions. The main difference between these versions is the parameter table that is used. New parameters are continually added to the database. The latest parameters are available from the UNIFAC consortium at www.unifac.org. The last published groups are given in the figure 3.34.

3.4.2.5 Electrolyte models

The presence of dissociated salts or more generally electrolyte species in a solution results in large non-idealities. An additional difficulty in this case is that the reference state that has been used so far (pure component at the same pressure and temperature as the mixture) cannot be used for the simple reason that electrolytes cannot exist in their pure state. Hence, the asymmetric convention is used (see the asymmetric convention in section 2.2.3.1, p. 67). The activity coefficients, defined as

$$\gamma_i = \frac{f_i}{f_i^{ref}}$$

(3.151)

uses as reference state, for ionic species the infinite dilution in the solvent, and for the solvent (generally water) the pure component. In addition, since ions cannot exist in the absence of their counter-ion, their individual activity coefficients cannot be measured. The mean ionic activity coefficient is therefore defined as:

$$\hat{\gamma}_{\pm} = (\gamma_{+}^{\nu_{+}} \gamma_{-}^{\nu_{-}})^{1/(\nu_{+}+\nu_{-})} \tag{3.152}$$

where ν_{+} and ν_{-} are the stoichiometric coefficients of the salt (ν_{-} for the anion and ν_{+} for the cation). A very clear account of the difficulties that arise are discussed by Prausnitz *et al.* [148].

The appropriate models are discussed in detail by more specialised handbooks, for example Zemaitis *et al.* [193] or Rafal *et al.* [194]. Here, we simply mention the construction of these models as a sum of terms, as in (3.104).

$$\ln \gamma_i = \ln \gamma_i^{lr} + \ln \gamma_i^{sr} \tag{3.153}$$

The first term refers to the "long range" interactions and the second term to the "short range" interactions. For this second term, authors have proposed to use any of the previously mentioned models, for example an e-NRTL model [195], an e-UNIQUAC model [196], or (taking advantage of the group contributions) an e-UNIFAC model [197, 198]. The model proposed by Pitzer [199], based on a theoretical development, and including ternary interaction parameters, nevertheless remains one of the most well-known and most accurate models up to very high salt concentrations. Parameters for many ionic species exist for this model [193, 200].

For all these models the "long range" Coulomb interactions are described using the model proposed by Debye and Hückel (1923) [201, 202]. They assume that the electrostatic interactions, can be described using point charges surrounded by an electron cloud of opposite charge. Using the Poisson and Boltzmann equations it is then possible to calculate the excess energy and thus the activity coefficient of the charged species predictively (*i.e.* without adjustable parameter) as a function of the ionic strength, $I = \dfrac{1}{2} \sum_{i}^{ions} m_i z_i^2$ (m_i is the ion molality in mol kg^{-1}, and z_i is the ionic charge):

$$\ln \gamma_{\pm}^{(m)} = -|z_+ z_-| A^{DH} \sqrt{I} \tag{3.154}$$

This expression provides an accurate limiting law, but must be corrected for an ionic strength up to 0.1 mol kg^{-1} [203]:

$$\ln \gamma_{\pm}^{(m)} = -\frac{|z_+ z_-| A^{DH} \sqrt{I}}{1 + a B^{DH} \sqrt{I}} \tag{3.155}$$

where a can be related to the ionic, closest approach diameter, and $B^{DH} = e N_{Av} \sqrt{\dfrac{8\pi d_s}{\varepsilon RT}}$. This is often called the "Pitzer-Debye-Hückel" model.

The superscript (m) indicates that this activity coefficient is on a molality basis (for more details, see [148]) and

$$A^{DH} = \frac{\sqrt{2d_s}}{8\pi} \frac{N_{Av}^2 e^3}{(\varepsilon RT)^{3/2}} \tag{3.156}$$

where d_s stands for the solvent density (in kg m^{-3}), e is the electronic charge ($e = 1.60218 \times 10^{-19}$C), ε is the permittivity of the solvent ($\varepsilon = \varepsilon_0 \varepsilon_r$ where $\varepsilon_0 = 8.85419 \times 10^{-12}$ C^2 N^{-1} m^{-2} and ε_r is the relative permittivity), R is the ideal gas constant ant T the temperature in kelvin.

For concentrations up to 1 mol kg^{-1}, a linear, empirical improvement can be used [193]:

$$\ln \gamma_{\pm}^{(m)} = -\frac{\left|z_+ z_-\right| A^{DH} \sqrt{I}}{1 + a B^{DH} \sqrt{I}} + bI \tag{3.157}$$

where b is an adjustable parameter.

3.4.2.6   Conclusion on activity coefficient models

Activity coefficient models are very convenient tools as they express directly the deviation from the ideal behaviour. Using this feature, the relationship between the shape of the phase diagrams and the deviation from ideality was shown in section 3.4.1 (p. 160). They are constructed in such a way as to take into account separately and explicitly the entropic and enthalpic contributions to this deviation. An additional contribution for electrolyte solutions is also available.

In addition, as also summarised in table 3.20, the distinction has been made between "predictive" and "correlative" models. The former require only pure component parameters, but can only be used to identify trends, the latter allow accurate calculations, but necessitate a large database for adjusting parameters.

Table 3.20   Overview of the most important activity coefficient models

	Predictive	**Correlative**
Enthalpic	Scatchard-Hildebrand	NRTL Margules Van Laar
Entropic	Flory Staverman-Guggenheim	
Enthalpic + entropic	UNIFAC Flory-Huggins	Wilson UNIQUAC (Flory-Huggins)*

* The Flory Huggins model is presented here as predictive, but it is very often used with a correlated parameter for solvent-polymer mixtures.

The models of Margules, Van Laar and Wilson are mentioned but have not been discussed in this text; we refer to classical textbooks as [38].

The *Ab-Initio* calculation tool COSMO-RS [153, 154], now increasingly used in industrial applications with complex chemicals, may be of great interest for calculating activity coefficients directly. It is sometimes considered as an activity coefficient model, although it is not used directly, but rather to predict NRTL or UNIQUAC parameters that can subsequently be used in process simulation.

The use of activity coefficient models has several limitations, however:

- They are only valid for liquid or condensed phases. Another model is required for vapour-liquid equilibrium calculations. Consequently, these models cannot be used to calculate vapour-liquid critical points.

- The assumption that the pure components exist in the liquid phase is not compulsory, but whenever this is not the case (one of the components is supercritical), the use of these models becomes more complex: the asymmetric convention must then be used and the model is only valid within a limited composition range.
- It is assumed that the pure component properties (or at least the properties of all components in their reference states) are known. Independent models should be used for these. Most often, the models described in section 3.1 of this chapter (p. 102) are used.
- It has been stressed on several occasions that activity coefficient models cannot be used over a large pressure range: it is generally accepted that the activity coefficient is independent of pressure below 1.5 MPa. Above this pressure, another approach should be used.
- All phenomena due to density variation effects require the phase volume as input parameter, and cannot therefore be described by activity coefficient models. This is particularly the case for the liquid-liquid phase split that occurs in monomer + polymer solutions.

For all these reasons, a clear trend exists nowadays to increasingly use equations of state for describing phase behaviour and phase properties.

3.4.3 Equations of State, EoS (all fluid phases)

Activity models focus on unlike interactions and therefore assume that the pure component behaviour of each compound is known. In addition, they can only be used for dense phases, where these interactions are strong. Yet, it is clear that from a molecular point of view, molecules of the same kind also interact with each other. Theories which take into account all types of interaction and which consider the effect of density on these interactions are called equations of state. They allow the calculation of pure as well as mixture properties as a function of the density. Their application range therefore covers all fluid phase conditions and they can be used to calculate many more properties (residual properties, as discussed in section 2.2.2.1, p. 52). Equations of States were born with the works of Boyle (1660), Mariotte (1787) and Gay-Lussac (1802), some of the pioneers in this field, but the very first important EoS is attributed to Van der Waals (1873 [204]).

Good reviews of the state of the art of equations of state exist [205-207], as well as good textbooks with many details [208]. Only a few are discussed here, the equations most used for industrial purposes. Equations of state with an increasingly physical background will most probably be more readily available in the future, however. This is why some of their underlying physical principles will also be presented briefly.

3.4.3.1 Introduction on the use of equations of state

An equation of state can be expressed in various ways. Often, it is considered as a relationship between pressure, volume and temperature. In this case, a distinction is made between pressure explicit EoS and volume explicit EoS. Volume explicit equations of state are unfortunately unable to describe multiple phases at a fixed pressure (there is a unique volume for every pressure). They are therefore of little practical use (except for the virial equation that can only be used for the vapour phase) and will not be further discussed here.

Pressure explicit equations of state can be regarded as originating from an expression of the Helmholtz energy as a function of temperature, phase volume and phase composition:

$$A(T,V,N) \tag{3.158}$$

It is important to stress that these are phase properties, not system properties: if a system splits into several phases, the equation of state must be applied separately for each phase (see section 2.2.3, p. 63 for more details).

A second point to stress is that pressure is not an independent variable in the equation of state expression (3.158). Instead, it must be calculated from the thermodynamic relationship (see section 2.2.1.1.C, p. 46):

$$P(T,V,N) = -\frac{\partial A}{\partial V}\bigg|_{T,N} \tag{3.159}$$

Figure 3.35 illustrates the typical shape of equation (3.159) when pressure is plotted *versus* volume at a given temperature. Point A shows the *PV* behaviour in the vapour: compressibility is large; on the other side of the figure, point F shows the *PV* behaviour in the liquid phase: compressibility is low. Between these two points (more precisely between points B and E), the equation of state shows a behaviour that does not reflect the physical equilibrium (*i.e.* the straight line). Rather, it shows that, for a given pressure located between the pressures of points D and C, three possible volumes can be found. Calculating the phase volume, knowing the phase pressure, is called "solving" the equation of state. The smallest of all physically signifi-cant volumes (*i.e.* larger than the molecular hard sphere volume) is called the liquid volume. The largest volume is called the vapour volume. The other root(s) are not considered.

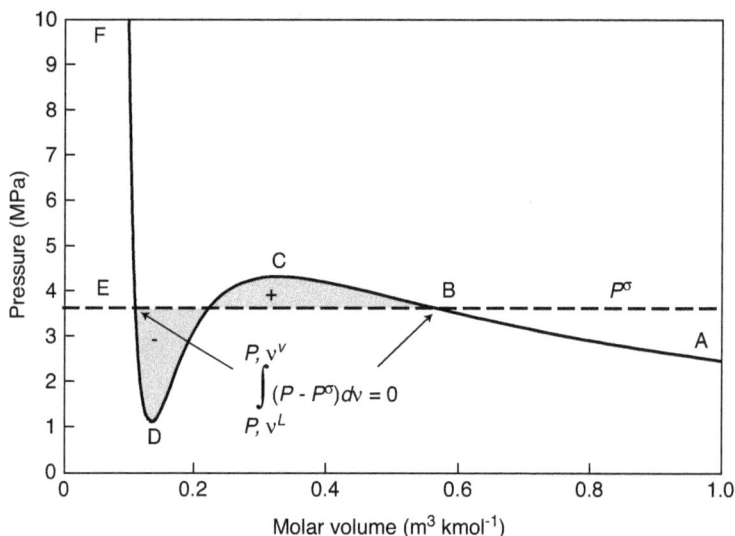

Figure 3.35

Behaviour of a pressure *versus* volume isotherm with an EoS (sub-critical).

In addition to pressure, the second most important property that can be calculated from an equation of state is the chemical potential:

$$\mu_i\left(T,V,\mathrm{N}\right) = \frac{\partial A}{\partial N_i}\bigg|_{T,V,N_{j\neq i}} \tag{3.160}$$

As explained earlier (section 2.2.2.1.B, p. 55), the equation of state describes the deviation of the real fluid behaviour with respect to the ideal gas. These deviations are called the residual properties (in contrast with the excess properties that require excess Gibbs energy models). For phase equilibrium calculations, the residual chemical potential $\mu_i^{res}\left(T,P,N\right)$ is also expressed as a fugacity coefficient φ_i. It was shown that it is calculated using the equation of state from:

$$\mu_i^{res}\left(T,P,\mathrm{N}\right) = \frac{\partial A}{\partial N_i}\bigg|_{T,V,N_{j\neq i}} - \frac{\partial A^{\#}}{\partial N_i}\bigg|_{T,V^{\#}=\frac{RT}{P},N_{j\neq i}} = RT \ln \varphi_i$$

$$= \int_{\infty}^{V}\left(-\frac{\partial P}{\partial N_i}\bigg|_{T,V,N_j} + \frac{RT}{V}\right)dV - RT \ln Z \tag{3.161}$$

Example 3.15 Use of an equation of state for pure component vapour pressure calculation

Application:

What equations must be solved to calculate a pure component vapour pressure using an equation of state? What are the consequences on the PV plot?

Solution:

Two conditions must be satisfied simultaneously:

1. To reach mechanical equilibrium the pressure must be identical in both phases. This must be expressed explicitly, since the equation of state is pressure explicit. Hence, from (3.159), we state:

$$P^L = -\frac{\partial A}{\partial V}\left(V^L,T\right) = -\frac{\partial A}{\partial V}\left(V^V,T\right) = P^V \tag{3.162}$$

2. Chemical equilibrium is reached when the Gibbs energies are equal in both phases (chemical potentials are equal to the molar Gibbs energies for pure components):

$$g(V^L) = \frac{\partial A(V^L)}{\partial N}\bigg|_{T,V^L} = \frac{\partial A(V^V)}{\partial N}\bigg|_{T,V^V} = g(V^V) \tag{3.163}$$

This yields a set of two equations with two unknowns (V^L and V^V) which can be solved.

The PV plot of the isotherm considered, as calculated with the equation of state, is shown in figure 3.35. This purely mathematical behaviour can be analysed as follows in terms of the physical picture it should represent.

It is clear that at high molar volume (point A), the equation should tend towards the ideal gas behaviour. When the volume is decreased, the pressure increases, which corresponds to the compressibility of the gas phase (point B). However, it is known that the behaviour depicted between point C and point D, *i.e.* for a negative isothermal compressibility coefficient (volume increases when pressure increases) is physically incorrect. A vapour-liquid equilibrium should in fact be observed, represented by a horizontal line on this diagram, since the saturated liquid and the saturated vapour volume are at the same pressure.

The label (E) shows the saturated liquid, and when the fluid is still compressed from that point on, pressure will increase very rapidly while the volume decreases very little (point F). Note that the liquid volume does not asymptote to zero when the pressure reaches an infinite value. This is due to the incompressible molar volume. The limiting volume is called the covolume and is noted *b*.

Equation (3.162) can easily be made visible on the figure 3.35, as it corresponds to the horizontal line shown (liquid and vapour molar volumes should yield equal pressure). In order to visualise equation (3.163), it should first be written in a different form: if liquid and vapour Gibbs energy are equal, then the integral of the Gibbs energy between liquid and vapour volume must be zero:

$$\int_{v^V}^{v^L} dg = 0 \qquad (3.164)$$

Using the expressions of chapter 2, we can deduce that (Maxwell criterion):

$$\int_{P,v^V}^{P,v^L} v dP = 0 \qquad (3.165)$$

This is exactly the integral under the v(P) curve. If the horizontal line represents vapour liquid equilibrium, equation (3.163) tells us that the area below the curve must equal the area on top of the curve in order to show vapour-liquid equilibrium. It can also be written as:

$$\int_{v^V}^{v^L} \left(P - P^\sigma \right) dv = 0 \qquad (3.166)$$

This is known as Maxwell's relation for vapour-liquid equilibrium.

This example is discussed on the website:
http://books.ifpenergiesnouvelles.fr/ebooks/thermodynamics

3.4.3.2 Molecular basis for Equations of State

Without providing considerable detail on the molecular construction of equations of state, it is useful to state, in the same way as Elliott [164], that it is based on two types of assumption: one considering the interaction energy between molecules, the other regarding the shape of the molecules. The classification is not unlike that used to distinguish between enthalpic and entropic contributions to the Gibbs Excess Energy in section 3.4.2.1 (p. 173).

The statistical mechanical theories for developing equations of state generally aim at providing a Helmholtz energy function, as expressed previously in (3.158). Making a number of assumptions, it can be shown that the expression can be written as a sum of contributions which are more or less independent from each other and which refer to different molecular interactions:

$$A(T,V,\mathbf{N}) = A^{\#}(T,V,\mathbf{N}) + A^{rep}(T,V,\mathbf{N}) + A^{att}(T,V,\mathbf{N}) + A^{pol}(T,V,\mathbf{N}) + ... \qquad (3.167)$$

The first of these terms refers to the ideal gas, which will not be discussed further. Each of the other potential contributions (repulsion, attraction, polar…) will be introduced briefly.

A. The fundamental molecular interactions

Assuming spherical segments [209], several models have been developed to describe intermolecular interactions. The most well-known are depicted on figure 3.36 which shows the interaction energy as a function of the distance between the segments. Positive interaction means that the segments repel each other, negative interaction means that they attract each other.

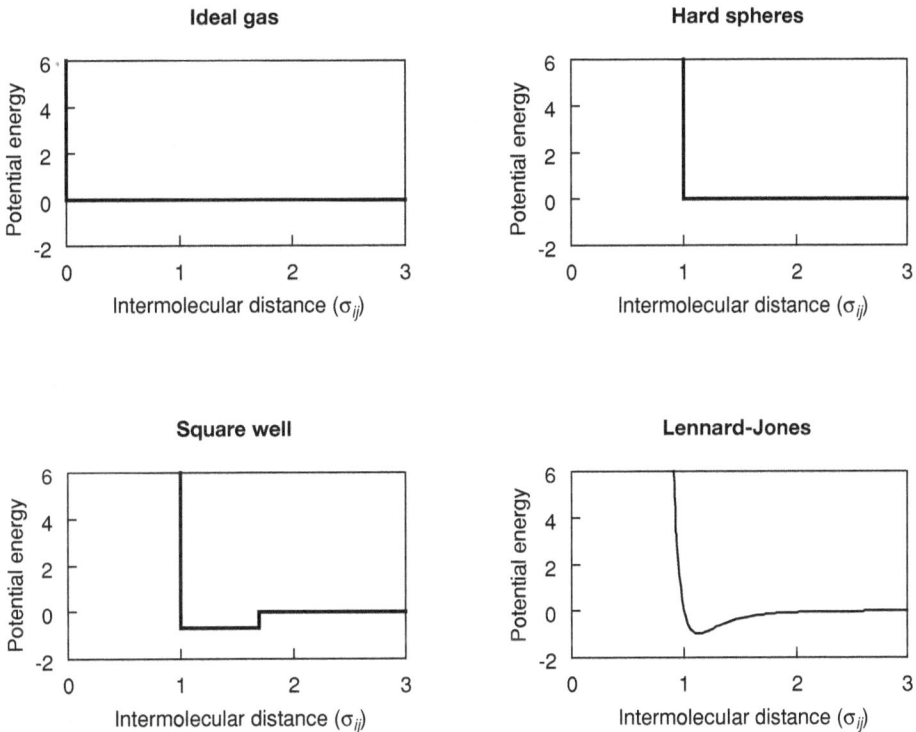

Figure 3.36

Molecular interactions (attraction and repulsion) of different models.

Four models are shown:

- ideal gas: the segments do not interact at all and have no volume;
- hard spheres: the segments do not interact, but have a volume: at distance $r = \sigma_{ij}$, the interaction becomes infinite, indicating that the segments cannot interpenetrate;
- square well: the segments do not interact beyond a certain range, that is described by an additional parameter $r > \lambda\sigma_{ij}$. Between σ_{ij} and $\lambda\sigma_{ij}$, their interaction energy has a negative value ($- \varepsilon_{ij}$) indicating that the segments attract each other;

- Lennard-Jones: the most physical model. It is generally written as a sum of a positive (repulsive) and negative (attractive) terms, with power-laws with respect to the inverse distance, often taken as 12 and 6:

$$u_{ij}(r) = 4\varepsilon_{ij}\left[\left(\frac{\sigma_{ij}}{r_{ij}}\right)^{12} - \left(\frac{\sigma_{ij}}{r_{ij}}\right)^{6}\right]$$ (3.168)

The origin of the attractive potential is not further discussed here, since it was already mentioned in section 3.4.2.2 (p. 174). The major difference between the construction of Gibbs Excess energy models and equation of state models is that now the distance between the molecules is considered. Hence, assuming that only pair-interactions are important, the residual energy (*i.e.* the energy excluding the kinetic energy, which is the ideal gas contribution) is written as the integral of this interaction energy $u(r)$ times the probability of finding another molecule at the distance r from the first one:

$$U = \frac{1}{2}\sum_i N_i \sum_j \frac{N_j}{V}\int_0^\infty u_{ij}(r_{ij})g_{ij}(r_{ij})4\pi r_{ij}^2 dr_{ij}$$ (3.169)

The radial distribution function $g_{ij}(r_{ij})$ expresses the probability of finding a molecule j at a distance r_{ij} from molecule i. This function depends on the potential energy model chosen, as well as on density and temperature. Since no exact expression exists for solving (3.169), numerous investigations have been conducted aiming at finding reasonable simplifications.

If the pair-interaction energy can be written as the sum of a repulsive and an attractive contribution, the resulting Helmholtz energy equations will show the same feature:

$$A(T,V,\mathbf{N}) = A^{rep}(T,V,\mathbf{N}) + A^{attr}(T,V,\mathbf{N})$$ (3.170)

a. Repulsive contribution

Several authors have worked on the repulsive term, by considering the hard-sphere model. Probably the most known equation, and certainly the most used of these results from the work of Carnahan and Starling [210], later extended to mixtures by Mansoori *et al.* [211]:

$$A^{rep} = RT\frac{6V}{\pi N_{Av}}\left[\left(\frac{\zeta_2^3}{\zeta_3^2} - \zeta_0\right)\ln(1-\zeta_3) + \frac{3\zeta_1\zeta_2}{1-\zeta_3} + \frac{\zeta_2^3}{\zeta_3(1-\zeta_3)^2}\right]$$ (3.171)

with: $\zeta_n = \frac{\pi}{6}\frac{N}{V}\sum_i x_i d_i^n$ and d is the hard sphere diameter.

However, an "excluded volume" theory can also be used to explain the Van der Waals repulsive term (it corresponds to the ideal gas equation with the assumption that the available volume is reduced with the hard sphere volume of the molecules, b) [148]:

$$A^{rep} = RT \sum_i N_i \ln \left(V - \sum_i N_i b_i \right) \tag{3.172}$$

b. Attractive or dispersive contribution

Concerning the attractive term, the theories are much more complex. Alder *et al.* [212] have proposed a fit on molecular simulation data, which has been combined to produce the Carnahan and Starling model [210].

The most frequently used approach in molecular thermodynamics is the perturbation theory (often called TPT, for Thermodynamic Perturbation Theory), as developed by Barker and Henderson [213]. They express the deviation with respect to a reference fluid (generally the hard sphere fluid) as a mathematical expansion, which may be truncated at the first or at the second order. The SAFT equations of state, which will be mentioned in more detail, are based on this development.

It is interesting to see that, using the first order thermodynamic perturbation theory, assuming a mean field (the radial distribution function is a constant, $g_{ij}(r_{ij}) = 1$), and

taking an attractive potential energy $u_{ij}(r_{ij}) = -4\varepsilon_{ij}\left(\dfrac{\sigma_{ij}}{r_{ij}}\right)^6$ as suggested by equation

(3.168), the attractive pressure can be calculated (see for example Elliott *et al.* [164]:

$$P^{att} = -\frac{\partial A}{\partial V}\bigg|_T = \frac{8\pi}{3}\frac{\sigma_{ij}^3 \varepsilon_{ij}}{(V/N_{Av})^2} \tag{3.173}$$

Expression (3.173) is directly related to Van der Waals' attractive term: $P^{att} = -a_{ij}/v^2$,

with the parameter $a_{ij} = 8\pi(N_{Av})^2 \sigma_{ij}^3 \varepsilon_{ij}/3$, thus related to fundamental molecular properties (the Avogadro number N_{Av} is needed in order to transform the molecular properties into molar properties).

> All equations of state are constructed using a minimum of two terms. One describes the molecular repulsion and contains a parameter that reflects the molecular volume. The second describes the attraction or dispersion, and contains a parameter that reflects the intermolecular potential well.

B. Chemical association

The interactions among segments may, in some cases, be much stronger than indicated by the Lennard-Jones equation. Possibly the most effective source of non-ideality in a fluid mixture is related to its tendency to form multimers (as in the case of hydrogen bonding). The most well-known example is that of acetic acid, which forms dimers even in the ideal gas state. Several approaches have been developed to describe this phenomenon:

- the pseudo-chemical theory (introduced in the Sanchez-Lacombe equation of state by Panayiotou *et al.* [214, 215]),
- the associative equations of state (initially proposed by Heidemann & Prausnitz [216], and very nicely summarised and extended by Elliott [164]),
- the newly developed Wertheim-based association term (best described by Jackson *et al.* [217] and later developed in the various SAFT theories [218, 219]).

Economou *et al.* [220] have compared these different models and concluded that they were essentially equivalent. We will focus here on the latter (Wertheim) since it now ranks amongst the most widely used in engineering models (SAFT, GCA and CPA).

It is important to notice that the association occurs between sites rather than molecules. The sites are labelled below as A_i and B_j, the capital letter referring to the site and the lower case index to the molecule. Let us consider the association as a chemical reaction. The equilibrium constant can then be written as:

$$\Delta^{A_i B_j} = \frac{\left[A_i B_j \right]}{\left[A_i \right]\left[B_j \right]} \tag{3.174}$$

where the square brackets designate a molar concentration of either bonds ($\left[A_i B_j \right]$) or free sites ($\left[A_i \right]$). The equilibrium constant is computed using, for example [217]:

$$\Delta^{A_i B_j} = D_{ij}^3\, g_{ij}\left(v\right) \kappa^{A_i B_j} \left[\exp\left(\frac{\varepsilon^{A_i B_j}}{kT} \right) - 1 \right] \tag{3.175}$$

where D_{ij} is the hard-sphere distance between components i and j, g_{ij} is the radial distribution function at close contact between the two molecules, which is a function of the system density; $\kappa^{A_i B_j}$ (dimensionless interaction volume) and $\varepsilon^{A_i B_j}$ (interaction energy) are adjustable parameters for the association.

When using this concept, the user must determine the number of sites on each molecule and how they will interact with sites of another molecule (*i.e.* what parameter should be used for $\kappa^{A_i B_j}$ and $\varepsilon^{A_i B_j}$: these parameters are in fact site-site parameters, although often considered as pure component parameters).

Several types of association can be identified on figure 3.37:

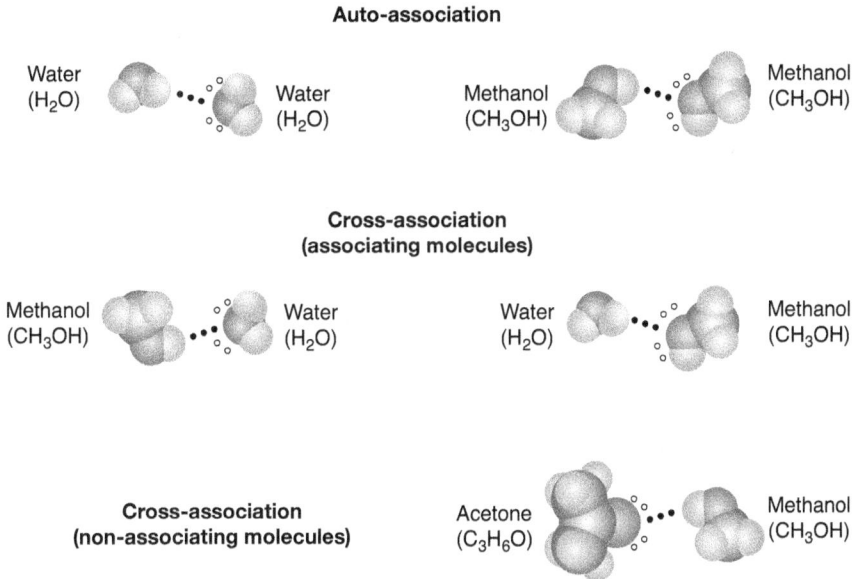

Auto-association

Cross-association
(associating molecules)

Cross-association
(non-associating molecules)

Figure 3.37

Possible types of auto- or cross-association between different molecules.

a. Auto- or self-association

Huang and Radosz [219] have described a number of such possible auto-association schemes, and labelled them with a number (number of association sites on each molecule) and a letter (that make it possible to distinguish between donor sites and acceptor sites). The most well-known are 1A (one site that associates with any other site, typically used for acids), 2B (two sites, one donor, one acceptor, typically used for alcohols), 3B (3 sites, two donors, one acceptor, sometimes used for water), and 4C (two donors, two acceptors sites, alternative choice for water).

b. Cross-association between associating molecules

When two different associating molecules coexist in the same mixture, both auto- and cross-association can be observed. The interaction parameters for cross-association between sites are often determined using so-called combination rules, which can be either arithmetic or geometric averages:

$\varepsilon^{A_i B_j} = \sqrt{\varepsilon^{A_i} \varepsilon^{B_j}}$	(3.176)	$\kappa^{A_i B_j} = \sqrt{\kappa^{A_i} \kappa^{B_j}}$	(3.177)
$\varepsilon^{A_i B_j} = \dfrac{\varepsilon^{A_i} + \varepsilon^{B_j}}{2}$	(3.178)	$\kappa^{A_i B_j} = \dfrac{\kappa^{A_i} + \kappa^{B_j}}{2}$	(3.179)

Elliott's rule also has sometimes been recommended:

$$\Delta^{A_i B_j} = \sqrt{\Delta^{A_i} \Delta^{B_j}}$$

(3.180)

c. Cross-association between non-associating molecules (solvation)

It has been proposed to use association to explain non-ideal behaviour between molecules that do not auto-associate. In this case, none of the combination rules discussed above can be used, because the association parameters are not pure component parameters. In this case, they must be considered as binary interaction parameters. This phenomenon is sometimes called solvation.

C. Polarity

Some molecules may not be considered associating, because the interactions between electron donor and electron acceptor sites are not strong enough. Their electronic structure, however, is such that they contain a native dipole or quadrupole (typically, esters, ketones, but also carbon dioxide, etc.). Several equation of state contributions have been proposed to describe this phenomenon (Gubbins and Twu [221], Kraska and Gubbins [222-223], Jog and Chapman [224], Karakatsani and Economou [225]). They all use as a basis a Padé approximant expression, as initially proposed by Stell [226]:

$$A^{polar} = A_2^{polar} \left[\frac{1}{1 - A_3^{polar} / A_2^{polar}} \right]$$

(3.181)

where A_2^{polar} represents the second order perturbation and A_3^{polar} the third order perturbation. The parameters involved in this expression are the polar moments of the molecule (dipole or quadrupole; no higher order has yet been used). In the approach initially proposed by Jog and Chapman [224] for the SAFT theory, and further developed by Nguyen-Huynh *et al.* [209], an additional parameter is used to represent the polar fraction of the chain molecule.

The use of this term can be very powerful in that it makes the theory predictive for polar mixtures. It must be used with great caution, however, since the additional parameters are highly correlated with the dispersive parameters.

D. The molecular shape

It is clear that molecules cannot generally be considered as spherical. The above spherically-symmetric approximations must therefore be corrected with a non-sphericity parameter, in particular when mixtures of small and large molecules are considered, as for example in polymer-solvent mixtures. Several types of theories exist [227].

1. Prigogine's [228] theory takes into account the fact that internal degrees of freedom of the molecule may also be density dependent. Consequently, this number of degrees of freedom, c, appears in the equation of state as a third parameter in addition to the segment diameter σ, and the dispersive energy ε. The equation of state most often quoted and developed directly from this theory is the Perturbed Hard Chain Theory (PHCT), by Prausnitz and co-workers [229, 230].

2. The Wertheim association theory [166], discussed above can be used in a very elegant manner to predict the behaviour of chain molecules. This was first implemented by Chapman [218, 231] and is now widely recognised in the various versions of the SAFT equation of state [207]. Their major contribution was to consider a chain molecule as a mixture of segments with infinite association strength. By doing so and using a Lennard-Jones type potential interaction, they write (for non-associating molecules):

$$A^{res} = m\left(A_0^{hs} + A_0^{disp}\right) + A^{chain} \tag{3.182}$$

where m is the number of segments in the molecule, A_0^{hs} and A_0^{disp} are the segment hard sphere (repulsion) and dispersion (attraction) terms and A^{chain} is the contribution of chain-forming to the Helmholtz free energy. It is worth noting that in reality, m is often taken as an adjustable parameter that can take any real value.

3. An alternative theory was developed based on Flory's entropic contribution (see 3.4.2.3.B, p. 180). Sanchez and co-workers introduced empty lattice sites in the equation and thus proposed an equation of state that is directly tuned to polymer-solvent systems. An elegant account of the principles of this equation is provided in [232]. Equations of this type, now extensively developed in Panayiotou's group [233, 234], are known as Lattice-Fluid equations of state.

3.4.3.3 Virial equations of state

The virial expansion, introduced in 1901 by Heike Kamerlingh Onnes, is a generalisation of the ideal gas law using a development in power of $1/v = N/V$ (virial, from the Latin *vir*, genitive *viris*, means force [235]):

$$\frac{PV}{NRT} = 1 + \frac{B(T)N}{V} + \frac{C(T)N^2}{V^2} + \frac{D(T)N^3}{V^3} + ... \tag{3.183}$$

A. The virial coefficients

The virial coefficients are the multiplicative factors of each term of the expansion. The expression is most often written in terms of the compressibility factor:

$$\frac{P(T,V,N)V}{NRT} = Z(T,V,N) = 1 + \frac{B(T)}{v} + \frac{C(T)}{v^2} + \frac{D(T)}{v^3} + ... \tag{3.184}$$

The coefficients are called the second virial coefficient (B), the third virial coefficient (C) and so on. Theoretically, they refer to collision integrals between two components, three components, and so on [236]. Useable expressions have only been developed for the second virial coefficient (see section 3.1.1.2.G, p. 120).

In a mixture, since the second virial coefficient concerns interactions between two molecules it is possible to show rigorously that the mixing rule for this parameter is

$$B = \sum_i \sum_j y_i y_j B_{ij} \tag{3.185}$$

where a binary interaction parameter B_{ij} expresses the interactions between unlike molecules i and j. A method useful in chemical engineering for associating systems has been published by Hayden and O'Connell (1975) [237].

The use of an infinite series expansion is obviously not possible for engineering calculations. Several methods therefore exist for approximating the equation.

B. The truncated pressure virial expansion

The simplest form of the virial equation of state is as follows:

$$Z = 1 + \frac{BP}{RT} \tag{3.186}$$

which is equivalent to (only the volume explicit equation will be discussed here):

$$v = \frac{RT}{P} + B \tag{3.187}$$

The Z vs. P plot is a straight line. This expression is very simple to use and, if limited to low pressure (< 1.5 MPa) vapours, provides a good indication of the gas phase properties.

Using the expressions presented previously, the most important residual properties can be computed as:

$$h^{res}(T, P) = \left(B - T \frac{dB}{dT} \right) P \tag{3.188}$$

and

$$\frac{g^{res}(T, P)}{RT} = \ln \varphi = \frac{BP}{RT} \tag{3.189}$$

Then, the fugacity coefficient of a component i in a mixture may be written as:

$$\ln \varphi_i = \frac{\left(2 \sum_{j=1}^{n} y_j B_{ij} - B \right) P}{RT} \tag{3.190}$$

C. The modified virial equations (so-called pseudo-experimental equations)

In order to be able to use a virial equation of state for the liquid phase, it should be truncated after many more than two terms. As discussed by Jacobsen *et al.* [238] and as first proposed by Benedict, Webb and Rubin [239] (BWR), an exponential term is also required for a correct representation of the critical point region. Since then, many extensions of the original BWR equation have been proposed.

> Considering the complexity of the equations that are thus derived from the virial expression, they are almost never used (nor recommended) for phase equilibrium calculations. However, they are of great interest for calculating residual properties, in particular volume, enthalpy, entropy and heat capacity, but also Joule-Thomson coefficients, speed of sound, etc. Their large number of parameters makes them suitable for fitting on large amounts of data with good accuracy (not predictive).

a. Benedict Webb Rubin (BWR)

Benedict, Webb and Rubin (1940 [239]) use a virial expansion up to the fifth term, further corrected with a term whose purpose is to approximate the inflection point of the critical isotherm as closely as possible. The expression shown here is the one presented by Lee and Kesler (1975 [19]):

$$Z = \frac{Pv}{RT} = 1 + \frac{G_1}{v_r} + \frac{G_2}{v_r^2} + \frac{G_3}{v_r^5} + \frac{c_4}{T_r^3 v_r^2} \left(1 + v_r^{-2}\right) \exp\left(-v_r^{-2}\right) \qquad (3.191)$$

where

$$G_1 = b_1 - \frac{b_2}{T} - \frac{b_3}{T^2} - \frac{b_4}{T^3}$$

$$G_2 = c_1 - \frac{c_2}{T} - \frac{c_3}{T^2} \qquad (3.192)$$

$$G_3 = d_1 - \frac{d_2}{T}$$

where v_r is the reduced volume ($v_r = v/v_c$).

The ten parameters b_i, c_i and d_i, must be known. They have no physical meaning, but are regressed on PVT and vapour pressure data of the pure components.

An extensive compilation of constants for the BWR equation of state was published by Cooper and Goldfrank [240]. Benedict *et al.* have extended the equation to mixtures [241]. This equation is quite correct on densities, but errors on caloric properties can reach 10%.

b. Starling (1973)-BWR equation of state

In order to make the BWR equation predictive, several authors have proposed a corresponding states approach for determining the 11 equation parameters (including v_c).

Starling's method [242] is relatively accurate for low molar mass non-polar fluids. The parameters for pure components are generalised as functions of the component acentric factor, critical temperature, and critical density. The mixing rules for the eleven mixture parameters are similar to the mixing rules used for the BWR equation. Using a binary interaction parameter, it can predict mixture properties for light hydrocarbons very accurately when experimental data covering entire ranges are available. For this reason, it has been widely used in the Liquefied Natural Gas industry. Nevertheless, it has difficulties describing the critical region.

c. Soave (1995, 1999)-BWR equation

Soave [243, 244] uses a four-parameter corresponding states principle (he uses Z_c in addition to T_c, P_c and acentric factor), to determine an improved BWR equation that describes the full phase diagram (including the critical region) relatively accurately, but is limited to hydrocarbons. For mixtures, instead of defining mixing rules on the parameters, the author proposes mixing rules on the critical parameters that include an adjustable interaction parameter.

d. Lee-Kesler (1975)

Probably the most well-known predictive extension of the BWR equation of state is that of Lee and Kesler [19]. They used a method inspired directly from the three parameters corresponding states principle (introduced in section 2.2.2.1.C, p. 57).

Consequently, the various thermodynamic functions of a "real" fluid are obtained as an interpolation between a "simple" fluid (methane, superscript $^{(0)}$) and a "reference" fluid (n-octane, superscript $^{(r)}$) using the acentric factor ω as the interpolation parameter. At the time of the publication, the acentric factor of methane was known to have a value of almost $\omega^{(0)} = 0$, while the acentric factor of n-octane was $\omega^{(r)} = 0.3978$.

The equation proposed by Lee and Kesler for the compressibility factor of any fluid is then:

$$Z\left(T_r, P_r\right) = Z^{(0)}\left(T_r, P_r\right) + \frac{\omega}{\omega^{(r)}}\left(Z^{(r)}\left(T_r, P_r\right) - Z^{(0)}\left(T_r, P_r\right)\right) \tag{3.193}$$

For equilibrium purposes, the fugacity-pressure ratio is obtained using the same kind of expression:

$$\ln\left(\frac{f}{P}\right)\left(T_r, P_r\right) = \ln\left(\frac{f}{P}\right)^{(0)}\left(T_r, P_r\right) + \frac{\omega}{\omega^{(r)}}\left[\ln\left(\frac{f}{P}\right)^{(r)}\left(T_r, P_r\right) - \ln\left(\frac{f}{P}\right)^{(0)}\left(T_r, P_r\right)\right] \tag{3.194}$$

To complete the calculations, enthalpy and entropy, in their adimensional form are similarly:

$$\frac{h^{res}}{RT_c}\left(T_r, P_r\right) = \left(\frac{h^{res}}{RT_c}\right)^{(0)}\left(T_r, P_r\right) + \frac{\omega}{\omega^{(r)}}\left[\left(\frac{h^{res}}{RT_c}\right)^{(r)}\left(T_r, P_r\right) - \left(\frac{h^{res}}{RT_c}\right)^{(0)}\left(T_r, P_r\right)\right] \tag{3.195}$$

and

$$\frac{s^{res}}{R}\left(T_r, P_r\right) = \left(\frac{s^{res}}{R}\right)^{(0)}\left(T_r, P_r\right) + \frac{\omega}{\omega^{(r)}}\left[\left(\frac{s^{res}}{R}\right)^{(r)}\left(T_r, P_r\right) - \left(\frac{s^{res}}{R}\right)^{(0)}\left(T_r, P_r\right)\right] \tag{3.196}$$

The four expressions presented above apply to pure components. An extension to multicomponents has been proposed by the authors using mixing rules on the corresponding states parameters [19].

$$\omega_m = \sum x_i \omega_i \tag{3.197}$$

$$Z_{cm} = 0.2905 - 0.085\omega_m \tag{3.198}$$

$$P_{cm} = Z_{cm}\frac{RT_{cm}}{v_{cm}} \tag{3.199}$$

$$v_{cm} = \sum_i \sum_j x_i x_j \left(\frac{v_{ci}^{1/3} + v_{cj}^{1/3}}{2}\right)^3 \tag{3.200}$$

The mixture critical temperature, required in (3.199) is calculated as follows:

$$T_{cm} = \frac{1}{v_{cm}^\eta} \sum_i \sum_j x_i x_j \left(\frac{v_{ci}^{1/3} + v_{cj}^{1/3}}{2} \right)^{3\eta} \sqrt{T_{ci}T_{cj}} \tag{3.201}$$

where the empirical parameter η, initially equal to one, has been introduced by Plocker (1978 [245]) in order to improve the mixing enthalpy predictions.

The Lee-Kesler-Plocker method is available in many commercial simulators, and is appropriate for calculating single phase residual properties. Its accuracy was evaluated [246] and it was found that the original Lee-Kesler method provides accurate results for density throughout the phase diagram (the worst results are close to the critical point, of the order of 5% for small hydrocarbons) and that it could be used for n-alkanes up to C_{20}. The derivative properties (heat capacities) have also been found relatively well-represented for simple hydrocarbons [247].

e. MBWR

The concept of using a multi-parameter equation of state in order to calculate as accurately as possible the full thermodynamic surface of single-phase fluids has further been developed by Jacobsen and Stewart [248] who propose a modified BWR equation (MBWR). They express their equation as follows $\left(v_r = v/v_c \right)$:

$$P = \sum_{n=1}^{9} a_n v_r^{-n} + \exp(-v_r^{-2}) \sum_{n=10}^{15} a_n v_r^{(17-2n)} \tag{3.202}$$

where the a_n are temperature dependent parameters adjusted on a large number of pure component properties.

This concept has been further improved by Wagner and co-workers [249, 250]. In order to better approximate the fluid behaviour close to the critical point, they add terms with $\exp\left(-v_r^{-4}\right)$:

$$\frac{P}{RT} = \sum_{n=1}^{13} a_n v_r^{-r_n} T_r^{s_n} + \exp\left(-v_r^{-2}\right) \sum_{n=14}^{24} a_n v_r^{-r_n} T_r^{s_n} + \exp\left(-v_r^{-4}\right) \sum_{n=25}^{32} a_n v_r^{-r_n} T_r^{s_n} \tag{3.203}$$

where the a_n r_n and s_n parameters are all fitted on experimental data. Using a large number of parameters, they manage to reproduce the experimental density to within 0.2% over a very large range of pressure and temperature.

These equations are of great use for specific, very well-known components. IUPAC has used them for developing specific equations for methane, ethane, oxygen, nitrogen, CO_2, many refrigerants, and quite a number of simple hydrocarbons, in addition to the well-known steam tables. More details on this topic can be found on http:/www.nist.gov.

3.4.3.4 The cubic equations of state

A. Mathematical expressions

Due to their simplicity of use, versatility and accuracy, a large number of cubic equations of state have been developed. It is impossible to avoid mentioning the first one, since it led to a real breakthrough in the understanding of fluid behaviour. It was published in 1873 by Van der Waals [251]:

$$P = \frac{RT}{v-b} - \frac{a}{v^2} \tag{3.204}$$

In this equation, the first term corresponds to the repulsive part of the pressure and the second to the attractive part. For this novel point of view, Van der Waals was awarded the Nobel Prize in 1910. The cohesion term (attractive part) was introduced as depending on temperature by Clausius (1880-1881 [252, 253]) but including a third parameter and later by Berthelot in 1903 [254] for a two-parameter EoS. These publications have included the cohesion function (or "alpha" function $a(T) = a_c \alpha(T)$) with $\alpha(T) = 1/T$ in this case. The most frequently used cubic equations are that of Redlich and Kwong (1949 [255] $\alpha(T) = 1/\sqrt{T}$), adapted by Soave (traditionally called SRK) [256] in 1972, with $\alpha(T) = \left[1 + m\left(1 - \sqrt{T}\right)\right]^2$:

$$P = \frac{RT}{v-b} - \frac{a_c \alpha(T)}{v(v+b)} \tag{3.205}$$

and that of Peng and Robinson (traditionally called PR) [257, 258]:

$$P = \frac{RT}{v-b} - \frac{a(T)}{v^2 + 2bv - b^2} \tag{3.206}$$

using the same cohesion function as Soave.

The shape of the physical isotherms (i.e. without the unstable loop that was shown in figure 3.35) is illustrated in figure 3.38. At low temperature $(T < T_c)$, it shows three distinct volumes at a given pressure (thus yielding a horizontal equilibrium line as explained in example 3.15). At high temperatures $(T > T_c)$, monotonic behaviour is observed. The limiting isotherm is called the critical isotherm with an inflection point at $T = T_c$.

These equations are called cubic because when solving for the volume at fixed pressure and temperature, an equation of the third order in v must be solved. All basic cubic equations can be written using a common shape as:

$$P = \frac{RT}{v-b} - \frac{a(T)}{(v-r_1 b)(v-r_2 b)} = \frac{RT}{v-b} - \frac{a(T)}{\left(v^2 + ubv + wb^2\right)} \tag{3.207}$$

where r_1 and r_2 (or u and w) can take many different values (table 3.21).

Many other versions of cubic EoS derived from the RK and PR equations exist (several hundreds). It is not the purpose of this book to cite all of them. Reviews from Abott (1979 [259]), Sandler and Orbey (1995 [260]), Sengers et al. (2000 [208]), Wei and Sadus (2000 [205]), and Valderrama (2003 [206]) provide all the information necessary for further investigation.

Table 3.21   Parameters to be used for cubic equations of state

Parameter	Van der Waals 1873	Redlich-Kwong (Soave) 1949 (1972)	Peng-Robinson 1976 (1978)
r_1	0	0	$-1-\sqrt{2}$
r_2	0	-1	$-1+\sqrt{2}$
u	0	1	2
w	0	0	-1
Ω_a	27/64	0.427480*	0.457235*
Ω_b	1/8	0.08664*	0.077796*
Z_c	3/8	1/3	0.30740*

* Values truncated with six decimal places.

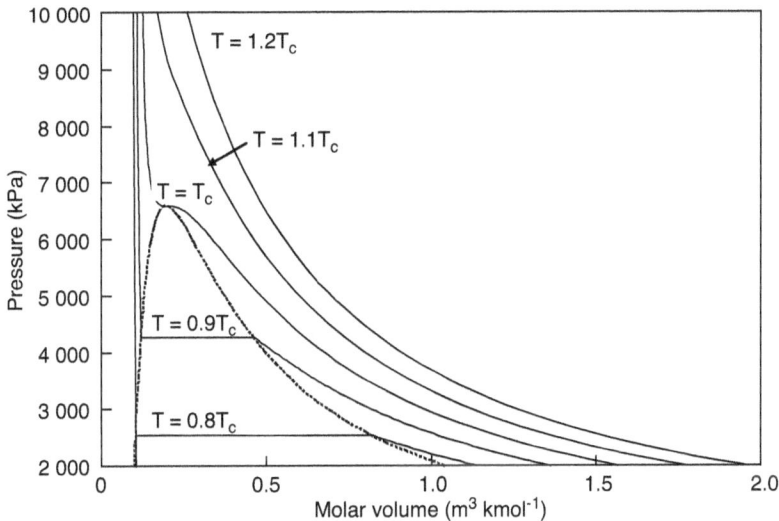

Figure 3.38

Pv representation of different isotherms according to a cubic EoS.

Using the fundamental relations (table 2.13, section 2.2.2.1.B, p. 55), the residual properties are calculated as a function of volume. If pressure appears in the equations, the assumption is made that it is calculated using the equation of state, at the volume provided.

The fugacity coefficient, required for the phase equilibrium calculation, is then found using (3.161):

$$\ln \varphi_i = -\ln \frac{P(v-b)}{RT} + \frac{b_i}{b}(Z-1) + \frac{a}{bRT}\left(\frac{2\sum_j a_{ij}z_j}{a} - \frac{b_i}{b} \right) U\left(v,b,r_1,r_2\right) \qquad (3.208)$$

where $a_{ij} = \sqrt{a_i a_j}\left(1-k_{ij}\right)$ and z_j is the molar fraction of component j in the phase considered. Parameters a_i and b_i describe pure component i while a and b are for the mixture parameters. The k_{ij} is a **binary interaction parameter** (BIP). Z is the compressibility factor. Note that for pure components, the equation simplifies as:

$$\frac{g^{res}(T,P)}{RT} = -\ln\frac{P(v-b)}{RT} + \frac{a(T)}{bRT}U\left(v,b,r_1,r_2\right)+Z-1 \tag{3.209}$$

The residual enthalpy is similarly calculated as:

$$\frac{h^{res}(T,P)}{RT} = \frac{1}{RT}\frac{a(T)-T\dfrac{da(T)}{dT}}{b}U\left(v,b,r_1,r_2\right)+Z-1 \tag{3.210}$$

In all of these equations, the function U depends on the equation chosen, as shown in table 3.22:

Table 3.22 U function for the main representatives of the cubic EoS

	Van der Waals	Redlich Kwong	Peng Robinson
$U\left(v,b,r_1,r_2\right)$	$-\dfrac{b}{v}$	$\ln\left(\dfrac{v}{v+b}\right)$	$-\dfrac{1}{2\sqrt{2}}\ln\left(\dfrac{v+b\left(1+\sqrt{2}\right)}{v+b\left(1-\sqrt{2}\right)}\right)$

Note that the temperature derivative of $a(T)$ is required for the residual enthalpy. The residual heat capacity expression uses the second derivative. As a result, these properties will be even more sensitive to the choice of the "alpha" function.

Some versions of cubic EoS have three or four parameters. This allows a better description of the critical locus (for example Fuller 1976 [276], Patel and Teja 1982 [277] and its Valderrama's modification [278]) for pure components but its use with mixtures is more complex and is not available in all simulators.

B. Pure component parameter values

Cubic equations of state use two parameters. Parameter b is the covolume [261], the molar volume occupied by the fluid at very high pressure. It essentially describes the liquid volume. The vapour pressure is highly sensitive to parameter a, which accounts for the attractive interactions. These parameters are generally calculated using the critical constraint (that imposes that for the critical isotherm, the critical pressure is found at the inflexion point of the Pv curve, see figure 3.38):

$$\left.\frac{dP}{dv}\right|_{T=T_c} = \left.\frac{d^2P}{dv^2}\right|_{T=T_c} = 0 \tag{3.211}$$

Solving for a and b, one finds:

$$a_c = \Omega_a\frac{R^2T_c^2}{P_c} \tag{3.212}$$

which is corrected with the alpha function at temperatures other than T_c ($a(T) = a_c\alpha(T)$).

and

$$b = \Omega_b \frac{RT_c}{P_c} \qquad (3.213)$$

Parameters Ω_a and Ω_b are fixed values that depend upon the equation. They are provided in table 3.21. Bian *et al.* (1992 [262]) point out that the number of significant digits may be important near the critical point.

As an alternative to the corresponding states approach (equations (3.212) and (3.213)), Coniglio *et al.* (2000 [62]) proposed a method for calculating these parameters from group contributions, so it can be used directly with complex molecular components without having to calculate critical properties first (which often have no physical significance).

A correct calculation of the vapour pressure requires a correct temperature dependence of $\alpha(T)$ (further called "alpha" or cohesion function). Several equations exist; only a few are mentioned here.

a. The Soave function

The original function, proposed by Soave, is:

$$\alpha(T) = \left[1 + m\left(1 - \sqrt{T/T_c}\right)\right]^2 \qquad (3.214)$$

where the parameter m is calculated as a function of the acentric factor (defined in section 3.1.1.1.E, p. 105). The equation is usually a polynomial function:

$$m = M_0 + M_1\omega + M_2\omega^2 + M_3\omega^3 \qquad (3.215)$$

where the M_i coefficients are fitted so that the acentric factor definition is matched (equation 3.2). Various values for M_i have been published and some are given in table 3.23. The acentric factor range for which the equation is accurate is illustrated in figure 3.39.

Table 3.23   Various coefficients of m for the "alpha" (cohesion) function (3.215)

	M_0	M_1	M_2	M_3	ω range
SRK					
Soave 1972 [256]	0.48	1.574	− 0.176		0-0.5
Grabosky and Daubert 1978 [263]	0.48508	1.55171	− 0.15613		≤ 0.91
Ledanois *et al.* 1998 [264]	0.478972559	1.576809191	− 0.187219516	0.020424946	− 0.8-2.4
PR					
Peng and Robinson 1976 [257]	0.37464	1.54226	− 0.26992		0-0.5
Robinson and Peng 1978 [258]	0.379642	1.487503	− 0.164423	0.016666	0.2-2.0
Ledanois *et al.* 1998 [264]	0.378710697	1.487972964	− 0.16754831	0.017169486	− 0.5-2.1

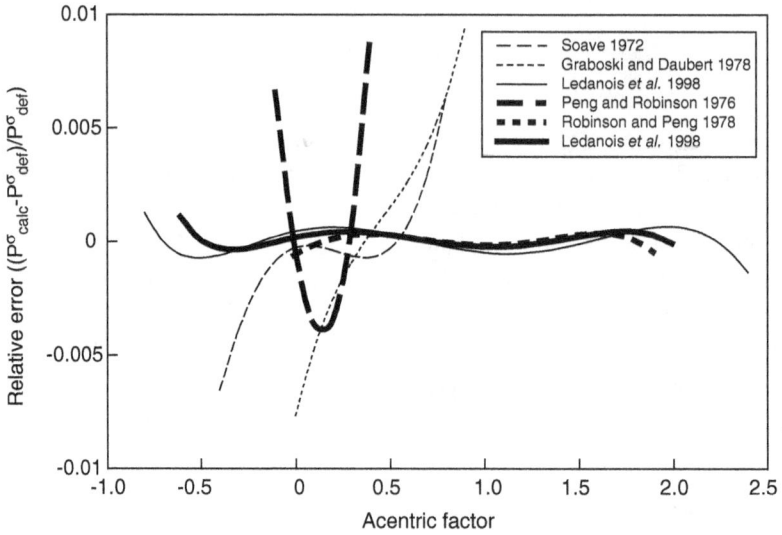

Figure 3.39

Error in the vapour pressure prediction with SRK (fine) and PR (bold) EoS at
$T_r = 0.7$ [264].

b. Boston and Mathias

Boston and Mathias (1980 [265]) improve the $\alpha(T)$ behaviour for supercritical conditions.
For the SRK EoS, they propose to use the original Soave equation below the critical point
and a new extrapolation above the critical temperature.

$$\text{for } T > T_c: \quad \alpha(T) = \left[\exp\left(c\left(1 - \left[T/T_c \right]^d \right) \right) \right]^2 \tag{3.216}$$

where c and d are calculated from the acentric factor.

c. Mathias and Copeman

Mathias and Copeman (1983 [266]) extended the Soave method:

$$\alpha(T) = \left[1 + c_1\left(1 - \sqrt{T/T_c} \right) + c_2\left(1 - \sqrt{T/T_c} \right)^2 + c_3\left(1 - \sqrt{T/T_c} \right)^3 \right]^2 \tag{3.217}$$

No general expression for the parameters of this equation (c_1, c_2, c_3) exist. They must be
fitted to experimental data.

d. Twu

Twu et al. (1991 [267] 1995 [268, 269]) propose an alternative method:

$$\alpha(T) = T_r^{N(M-1)} \exp\left[L\left(1 - T_r^{NM} \right) \right] \tag{3.218}$$

that allows an improved prediction of vapour pressure and of derivative properties over a large temperature range. For the original equation, the constants must be fitted for each component. In order to make this method truly predictive, the corresponding states principle can be used with:

$$\alpha(T) = \alpha^{(0)} + \omega\left(\alpha^{(1)} - \alpha^{(0)}\right) \tag{3.219}$$

where $\alpha^{(0)}$ and $\alpha^{(1)}$ are calculated from equation (3.218) using the parameters of table 3.24.

The parameters L, M, N are different for the SRK and the PR EoS. Extrapolation above the critical point requires the use of specific parameters.

Table 3.24 Parameters values for the generalised Twu correlation [268, 269]

	SRK				PR			
	$T_r \leq 1$		$T_r > 1$		$T_r \leq 1$		$T_r > 1$	
Parameters	$\alpha^{(0)}$	$\alpha^{(1)}$	$\alpha^{(0)}$	$\alpha^{(1)}$	$\alpha^{(0)}$	$\alpha^{(1)}$	$\alpha^{(0)}$	$\alpha^{(1)}$
L	0.141599	0.500315	0.441411	0.032580	0.125283	0.511614	0.401219	0.024955
M	0.919422	0.799457	6.500018	1.289098	0.911807	0.784054	4.963070	1.248089
N	2.496441	3.291790	-0.2	-8	1.948150	2.812520	-0.2	-8

C. Predictive capacity of the cubic EoS for pure components

Like all equation of state, cubic equations can be used to calculate volume, vapour pressure and all residual properties.

a. Volume prediction

As an illustration of the predictive capacity of the most common cubic EoS (SRK and PR), Figure 3.40 shows the deviation of the liquid saturated volume as a function of temperature for a few members of the *n*-alkane family.

It is observed that:

- The liquid molar volumes are more accurate at low reduced temperature. This observation is true for all equations of state: the critical volume cannot be described correctly by an analytical equation.
- The PR equation shows a smaller deviation on liquid molar volumes than the SRK equation for molecules close to C_6. SRK shows better results for very small molecules.
- The alpha function has (almost) no effect on the volume calculation.

The liquid molar volume predictions are generally very bad. Cubic equations of state are not used for calculating liquid volumes except with a volume translation (see section 3.4.3.4.D, p. 211).

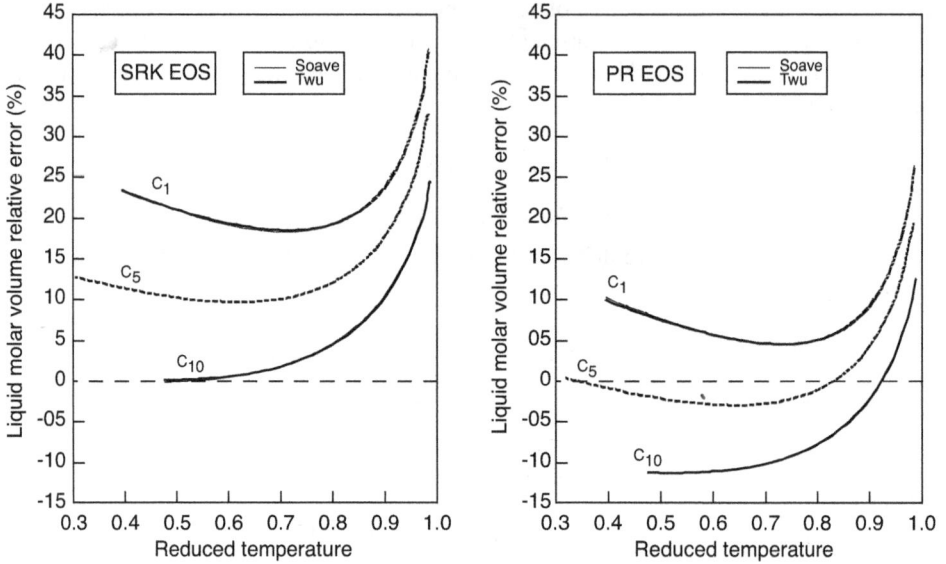

Figure 3.40

Deviation on the liquid molar volume of some *n*-alkanes as a function of temperature, using the Soave-Redlich-Kwong (SRK, left) or Peng-Robinson (PR, right) EoS.

One of the limitation of the cubic EoS is the error in the prediction of the critical volume as pointed out by Valderrama and Alfaro (2000 [272]). Some improvements have been made using a volume translation technique (Péneloux *et al.* 1982 [273]) for the SRK EoS. The same technique has also been proposed by Mathias *et al.* (1989 [274]) or Tsai and Chen (1998 [275]) for the PR EoS (see point D below).

b. Vapour pressure

The correct calculation of the pure component vapour pressure is essential before vapour-liquid phase equilibrium calculation of mixtures can be performed. Figure 3.41 shows the quality of the predictions for different components and different alpha functions.

For vapour pressure calculations, the predictions are most accurate at higher reduced temperatures. This is not surprising since the parameters have been fitted on the critical point. At the reduced temperature of 0.7, the vapour pressure is correctly calculated as well. At low reduced temperature, the calculation can deviate quite significantly. The Twu cohesion (alpha) function allows for a slightly better temperature extrapolation than the Soave function.

> The quality of the vapour pressure calculation will depend entirely on the shape of the $\alpha(T)$ function: the use of a function other than Soave's function will provide an entirely different deviation plot.

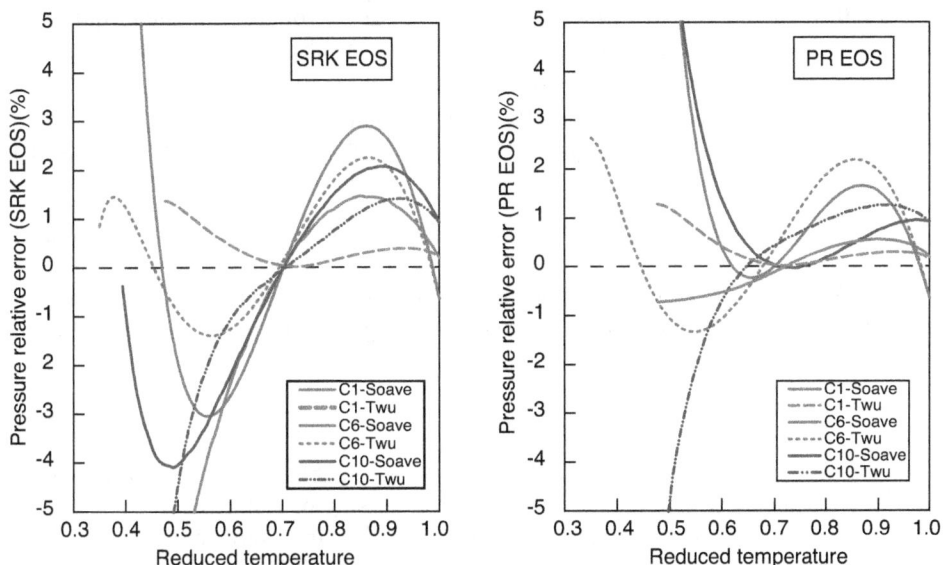

Figure 3.41

Deviation on the vapour pressure of *n*-alkanes as a function of temperature, using the Soave-Redlich-Kwong (SRK, left) or Peng-Robinson (PR, right) EoS.

c. Residual properties

A recent (2009) review of Neau *et al.* [270-271] give a complete analysis of advantages and disadvantages of the various methods. They study the behaviour of the first and second derivative of "alpha" with respect to the temperature (it appears in the calculation of enthalpy and heat capacity) concluding that the original Soave cohesion function is applicable in all industrial applications. They have observed that both the Boston and Mathias and the Twu generalised "alpha" function may lead to unphysical behaviour for the heat capacity in supercritical conditions.

D. Volume translation for cubic EoS

Considering the very bad predictions of the liquid molar volumes using a cubic equation of state, Martin [279] and later Péneloux and coworkers [273] have proposed to correct the equation of state volume by translating it with a fixed value:

$$v = v^{eos} - \sum_i x_i c_i \qquad (3.220)$$

where v^{eos} is the volume resulting from the EoS calculation and a new parameter, c_i, appears for each component. This parameter can be temperature dependent, and can be either directly fitted on experimental liquid volumes, or calculated using another model. It is worth noting that this volume translation has no effect on phase equilibrium calculation, on the condition that a homogeneous method is used (*i.e.* the same, volume translated, equation of state is used for both liquid and vapour fugacity coefficient calculation). A recent discussion of the use of volume translations is provided by Frey *et al.* [280].

Péneloux initially proposed to use a Racket-type method (see section 3.1.1.2.D, p. 114) to calculate the volume translation using the corresponding states principle (for the SRK EoS):

$$c_i = 0.40768 \frac{RT_{ci}}{P_{ci}} \left(0.29441 - Z_{RA} \right)$$
(3.221)

where the Rackett compressibility factor can be calculated using equation (3.19), if needed. Twu [281] recently discusses this approach and suggests a similar equation for the Peng-Robinson EoS:

$$c_i = 0.406501 \frac{RT_{ci}}{P_{ci}} \left(0.260484 - Z_{RA} \right)$$
(3.222)

Several temperature-dependent correlations have been published, but great care must be taken when using them because they can result in crossing of isotherms thus yielding an unphysical negative thermal expansion coefficient [282]. In a further discussion by de Sant'Ana [283] it is recommended to use high pressure volumetric data in order to avoid the inconsistency close to the critical point. For petroleum fluids, and the PR EoS, the recommended method is [282]:

$$c_i \left(T \right) = \frac{\left(0.023 - 0.00056 M_w \right) T + \left(-34.5 + 0.4666 M_w \right)}{10^6}$$
(3.223)

where M_w is the molar mass (in g mol^{-1}) of the component.

So far, the volume translation concept has found very limited use in process simulators [206]. They are much more used in other applications, as for example reservoir simulation.

E. Cubic equations of state for mixtures

The "one-fluid" hypothesis states that a mixture behaves in exactly the same way as a pure component. As a result, the mathematical expression of the cubic equation of state is identical and can be written for example as in (3.207). The difference lies in the calculation of the parameters. The usual approach consists in using mixing rules based on the pure component parameters.

It is essential to bear in mind that the quality of the cubic EoS predictions for mixtures phase behaviour lies essentially in the mixing rule. This is the reason why so many developments are related to this issue [206, 284].

a. *The covolume (b) parameter mixing rule*

The mixing rule generally applied is a simple linear combination usually named after Lorentz and Berthelot:

$$b_m = \sum_i x_i b_i$$
(3.224)

It may also be written as:

$$b_m = \sum_i \sum_j x_i x_j b_{ij}$$

(3.225)

with

$$b_{ij} = \frac{b_i + b_j}{2}$$

(3.226)

This rule is generally sufficient and no interactions parameters are used.

b. The a parameter mixing rules

The mixing rule on the *a* parameter can have many different shapes. The quality of the phase equilibrium calculation greatly depends on the choice of this mixing rule.

The classical mixing rule

The most common rule is:

$$a = \sum_i \sum_j x_i x_j a_{ij}$$

(3.227)

where

$$a_{ij} = \sqrt{a_i a_j}\left(1 - k_{ij}\right)$$

(3.228)

Here, k_{ij} is an empirical parameter. Note that $k_{ij} = k_{ji}$. This parameter may be temperature dependent, using:

$$k_{ij} = k_{ij}^{(0)} + k_{ij}^{(1)}T + k_{ij}^{(2)}T^2$$

(3.229)

or

$$k_{ij} = k_{ij}^{(0)} + \frac{k_{ij}^{(1)}}{T} + \frac{k_{ij}^{(2)}}{T^2}$$

(3.230)

For hydrocarbon mixtures of similar molar mass, $k_{ij} = 0$ may generally be assumed. When the mixture contains molecules of different sizes or from different families, it is recommended to fit the parameter on experimental phase equilibrium data. Chapter 4 discusses the choice of the optimal k_{ij} value. Most simulators contain an extensive k_{ij} database, but it is good to always validate the result on experimental data.

Some authors have proposed a method to calculate the k_{ij} parameter based on the Hudson and McCoubrey [285] equation. In particular, we can quote Coutinho *et al.* [286], who suggest:

$$k_{ij} = 1 - \left(\frac{2\sqrt{b_i b_j}}{b_i + b_j}\right)^{(n/3-2)}$$

(3.231)

The parameter n in this equation varies however depending on the symmetry of the mixture. They suggest using $n = 6$ for very asymmetric mixtures, but n may reach very large values for systems whose ratio of covolumes is close to one.

Recently, Jaubert's group [65] have proposed a group contribution method for calculating the k_{ij} of the Peng-Robinson EoS. The thus redesigned equation is called "PPR78". They suggest:

$$
k_{ij}(T) = \frac{-\dfrac{1}{2}\left[\displaystyle\sum_{k=1}^{N_g}\sum_{l=1}^{N_g}\left(\alpha_{ik}-\alpha_{jk}\right)\left(\alpha_{il}-\alpha_{jl}\right)A_{kl}\left(\dfrac{298.15}{T/K}\right)^{\left(\frac{B_{kl}}{A_{kl}}-1\right)}\right]-\left[\dfrac{\sqrt{a_i(T)}}{b_i}-\dfrac{\sqrt{a_j(T)}}{b_j}\right]^2}{2\dfrac{\sqrt{a_i(T)a_j(T)}}{b_i b_j}}
\tag{3.232}
$$

where α_{ik} is the fraction of group k in molecule i; $A_{kl} = A_{lk}$ and $B_{kl} = B_{lk}$ are interaction parameters between groups. This method has proven very useful for VLE calculation of hydrocarbon mixtures as well as mixtures containing CO_2, H_2S and N_2.

The Simsci mixing rule

For complex mixtures, Twu and co-workers (1994 [287]), based on an original work of Panagiotopoulos and Reid (1986 [288]), proposed to differentiate between k_{ij} and k_{ji}, using the following relationship, introducing a new parameter, $c_{ij} = c_{ji}$:

$$
a_m = \sum_i\sum_j x_i x_j \sqrt{a_i a_j}\left(1 - k_{ij} + \left(k_{ij} - k_{ji}\right)\left(\frac{x_i}{x_i + x_j}\right)^{c_{ij}}\right)
\tag{3.233}
$$

The usual choice is $c_{ij} = 1$, in which case this mixing rule proposes a continuous transition between the value of k_{ij} and k_{ji}, in proportion to the molar ratio of components i and j. In the limiting case when $k_{ij} = k_{ji}$, the classical mixing rule (3.227) is recovered.

It should be noted that the Simsci mixing rule is not invariant when a component is split into two or more identical fractions [289], although this problem has rarely practical consequences. Twu, 1991 [267] has further improved this mixing rule, making it more complex.

G^E based mixing rules

Huron and Vidal (1979 [290]), as a consequence of Vidal's previous work (1978 [291]), observed that equations of state can be used to calculate the Gibbs excess energy as a function of pressure, if all pure components have a liquid phase root:

$$
g^E(P, T, \mathbf{N}) = RT\left(\sum x_i \ln\left(\frac{\varphi_i(P, T, \mathbf{N})}{\varphi_i^*(P, T)}\right)\right)
\tag{3.234}
$$

Knowing that the excess Gibbs energy models (see section 2.2.2.2.B, p. 61) provide generally good results for very strongly non-ideal mixtures, they propose to calculate the mixture a parameter of the cubic equation of state using this type of model, and assuming a linear mixing rule for the b parameter (3.224). Several approaches have been proposed on this principle, depending on the pressure chosen for solving the equation of state [284]. Table 3.25 shows the main representatives of the G^E-type mixing rules.

Table 3.25 Main representatives of the G^E-type mixing rules

Method name	Equation	
Huron-Vidal [290]	$$\frac{a_m}{b_m RT} = \sum_i x_i \frac{a_i}{b_i RT} - \frac{g^E}{q_0 RT}$$	(3.235)
MHV1 [292] [a]	$$\frac{a_m}{b_m RT} = \sum_i x_i \frac{a_i}{b_i RT} - \frac{1}{q_1}\left(\frac{g^E}{RT} + \sum_i x_i \ln \frac{b_i}{b}\right)$$	(3.236)
MHV2 [293] [a]	$$\frac{g^E}{RT} = \left(\sum_i x_i \ln \frac{b_i}{b}\right)$$ $$+ q_1\left(\frac{a_m}{b_m RT} - \sum_i x_i \frac{a_i}{b_i RT}\right) + q_2\left(\left(\frac{a_m}{b_m RT}\right)^2 - \sum_i x_i \left(\frac{a_i}{b_i RT}\right)^2\right)$$	(3.237)
LCVM [294] [b]	$$\frac{a_m}{b_m RT} = \left(\frac{\lambda}{q_0} + \frac{1-\lambda}{q_1}\right)\frac{g^E}{RT} + \frac{1-\lambda}{q_1}\sum_i x_i \ln \frac{b_i}{b} + \sum_i x_i \frac{a_i}{b_i RT}$$	(3.238)
Wong-Sandler [295]	$$\frac{a_m}{b_m RT} = \sum_i x_i \frac{a_i}{b_i RT} - \frac{g^E}{q_0 RT}$$	(3.239)

[a] MHV stands for Modified Huron Vidal; 1 and 2 mean first or second order.
[b] LCVM stands for Linear Combination Vidal Michelsen.

Parameters q_0, q_1 and q_2 are constants that depend on the equation of state, while parameter λ in the LCVM (Linear Combination Vidal Michelsen) mixing rule provides an additional degree of freedom (see Heidemann, 1996, for example of values of these parameters [284]).

All the rules in the table assume the linear mixing rule on the covolume (b_m) , equation (3.224), except the Wong and Sandler mixing rule, as they combine the theoretically more precise expression for the second virial coefficient with the Huron-Vidal expression:

$$B_m = \sum\sum x_i x_j B_{ij} = b_m - \frac{a_m}{RT} \tag{3.240}$$

as a consequence, they replace the conventional linear mixing rule on the b parameter (3.224) with:

$$b_m = \frac{\sum_i \sum_j x_i x_j \dfrac{\left(b_i - \dfrac{a_i}{RT}\right) + \left(b_j - \dfrac{a_j}{RT}\right)}{2}\left(1 - l_{ij}\right)}{1 - \left(\sum_i x_i \dfrac{a_i}{b_i RT} - \dfrac{g^E}{q_0 RT}\right)} \tag{3.241}$$

with another binary interaction parameter l_{ij} . Note that the use of this mixing rule on b makes this parameter temperature-dependent, resulting in possibly unphysical calculations (negative heat capacities) as pointed out by Satyro and Trebble (1996, 1998 [296, 297]). The volume of mixing for dense phases may also be wrongly calculated [38].

In all the above equations, the excess Gibbs energy g^E can still be chosen at will. Huron and Vidal [290] propose a slightly modified version of NRTL (see section 3.4.2.2.C, p. 177):

$$g^E = RT \sum_i x_i \sum_j \frac{b_j x_j \exp\left(-\alpha_{ji} \tau_{ji}\right)}{\sum_k b_k x_k \exp\left(-\alpha_{ki} \tau_{ki}\right)} \tau_{ji} \tag{3.242}$$

Compared with the original NRTL equation (3.119), the covolumes b_i are introduced in the equation. The major advantage of this last equation is that it is possible to recover the classical mixing rule (3.227) when appropriate parameters are chosen: taking $\alpha_{ij} = 0$, it is possible to find:

$$\tau_{ij} = \frac{q_0}{RT} \left(\frac{a_i}{b_i} - \frac{2\sqrt{a_i a_j}}{b_i + b_j} \left(1 - k_{ij}\right) \right) \tag{3.243}$$

Holderbaum and Gmehling [298] take advantage of the mathematical expressions of G^E-type mixing rules to incorporate UNIFAC in the equation of state. If SRK is used with the MHV1 mixing rule, the equation thus becomes PSRK (Predictive SRK). More recently, the same authors have also applied it to the Peng-Robinson EoS, correcting a number of inconsistencies thus improving predictions of asymmetric mixtures (light and heavy). This model is called the Volume Translated Peng Robinson equation (VTPR) [299, 300].

> The availability of G^E mixing rules considerably widens the scope of use of cubic equations of state: if their $\alpha(T)$ equation is adapted for each pure component, and if the adequate G^E model is used, virtually all complex systems can be described accurately. These equations require a large number of empirical parameters, however, which considerably reduces their predictive power.

3.4.3.5 The SAFT family equations of state

In the early 1990s, the publication of Chapman [218] on the SAFT equation of state changed the use of the equations based on the perturbation theory of statistical thermodynamics. Two main elements can explain this breakthrough. Firstly, the use of the Wertheim theory to describe association and, secondly, the physically meaningful expression of the chain formation.

A. The SAFT equation of state

Many authors have proposed improvements to the SAFT equation of state since it was initially proposed. As it is impossible to be exhaustive in referencing the various versions, the interested reader can consult one of following review papers [207, 301, 302]. The interest of using it in the petroleum industry has been discussed by de Hemptinne *et al.* [303] For practical applications, PC (Perturbed Chain)-SAFT [304, 305] is the most frequently used version because of its better ability to represent the vapour pressure of *n*-alkanes.

The SAFT EoS is rather complex in its derivation. However, it is good to recall on what assumptions it is constructed and what its molecular parameters are. It is written as a sum of terms as already discussed through equations (3.167) or (3.182):

$$A = A^{\#} + m\left(A_0^{hs} + A_0^{disp}\right) + A^{chain} + A^{ass} \tag{3.244}$$

using following assumptions:

- The reference term (A_0^{hs}) is the hard sphere of Carnahan Starling (see equation (3.171)). Hence, a hard sphere diameter appears, often called σ or d and generally expressed in ångström. It is often considered as temperature-dependent.

- The dispersion (A_0^{disp}) is expressed using a second order perturbation theory. This is one of the main distinctions between the various SAFT versions. Yet, they all use explicitly an energy well depth parameter, ε_{ij} (its meaning is shown in section 3.4.3.2.A, p. 193). It is often expressed as a ratio with the Boltzmann constant (ε/k), and the units are then kelvin. The SAFT-VR equation [306] uses a square well as description of the dispersive interactions among spheres. This is why it uses an additional, non-dimensional parameter λ to describe the well width.

- These spheres are connected tangentially to form a chain. The parameter m, explicitly present in (3.244), but also in the chain-forming term A^{chain}, is the number of segments. In the actual use of the equation, however, this number is used as an adjustable parameter and is therefore rarely an integer. It has no unit.

- In case of hydrogen bonding molecules, an additional association term, A^{ass} is often employed. Its use is discussed in section 3.4.3.2.B, p. 196. In this case, the equation must include two additional parameters: the volume of the hydrogen bond ($k^{A_iB_j}$) and its energy, often expressed as $(\varepsilon^{A_iB_j}/k)$ with units of temperature (kelvin).

For mixtures, the most traditional approach is to propose mixing rules that describe the fluid behaviour using the pure component parameters. Binary interaction parameters may be necessary, and are usually defined as:

$$\varepsilon_{ij} = \sqrt{\varepsilon_{ii}\varepsilon_{jj}}\left(1 - k_{ij}\right) \tag{3.245}$$

Developing the PC-SAFT equation, Gross and co-workers [304, 305] have also proposed parameters for many components. This database has been extended by Tihic *et al.* [307].

To extend the usability of the SAFT equations, several authors have developed group contribution methods for the parameters [72-76]. Among the most well-known (and applicable to PC-SAFT) are that of Tihic *et al.* [74], who use the concept of first and second-order groups, as already explained in section 3.1.2.2 (p. 124), but limited to non-associating components, and that of Passarello's group [72, 73], which has been extended to associating and polar components, as well as complex mixtures (called GC-PPC-SAFT).

Passarello and co-workers suggest [11, 12] the use of a predictive parameter k_{ij} for asymmetric mixtures, using the so-called "pseudo-ionisation energy" J_i, which is an additional pure component parameter:

$$1 - k_{ij} = \frac{2\sqrt{J_i J_j}}{J_i + J_j} \qquad (3.246)$$

This additional parameter can also be calculated using group contribution.

Lymperiadis *et al.* [76, 308] have proposed another group contribution method called VR-γ-SAFT. This more recent development using a variable range potential considers partly overlapping segments with an additional parameter γ. They use a heteronuclear approach with individually determined group-group interaction energies. Although less predictive (requires more data), this GC-SAFT version may provide more accurate results.

> The SAFT family of equations of state is one of the most mature molecular equations of state. Its physical basis makes it more suitable for extrapolating the results away from the fitting domain. This has been validated in particular using the group contribution methods.

B. Hybrid models

Hybrid equations of state have often been used in the engineering community. These equations combine energy interaction terms originating from different theories. We may mention the equation of Carnahan and Starling (1972) [210], for example, where the authors combine their own hard sphere repulsive equation with the attractive term of the cubic Redlich-Kwong equation.

a. Cubic Plus Association (CPA)

Along the same lines, Kontogeorgis *et al.* (1996) [309] suggested using the association term originating from SAFT [218] as an additional contribution to the classical cubic equation of state. They call it the Cubic Plus Association (CPA) EoS. Their equation is written as (using the Soave-Redlich-Kwong version of the attractive term):

$$A^{res} = A^{SRK} + A^{ass} \qquad (3.247)$$

The form of the association term is described in section 3.4.3.2.B, p. 196, with a simplified expression (3.175) of the interaction strength $\Delta^{A_j B_i}$:

$$\Delta^{A_j B_i} = b_{ij} g(v) \kappa^{A_j B_i} \left[\exp\left(\varepsilon^{A_j B_i} / kT \right) - 1 \right] \qquad (3.248)$$

where $g(v)$ is a very simple expression that depends on the cubic parameter b.

> The reason why this equation is of interest to petroleum engineers is that in the absence of associating compounds, the well-known cubic equation is recovered, and all the existing correlations can be used. The advantage of retaining the ability to work with complex hydrocarbon mixtures, including badly defined pseudo-components, is the main reason why, despite its dubious physical basis, this approach remains attractive in the petroleum engineering community [303].

The CPA EoS is now considered as the best approach for systems containing light hydrocarbons, water and alcohols [310-318]. Parameters have been published for these mixtures.

Note that these parameters do not originate from the corresponding states principle, but from a fit on vapour pressure and liquid molar volume data. As a result, they are not suitable for calculating a critical point. They may be combined with the classical cubic equation parameters in mixtures with hydrocarbons.

b. Group Contribution Association (GCA)

The Group Contribution-Association equation of state (originally from [319] and later combined with an association term by [80]) is a rather extreme example of hybridisation, which has in fact been shown to provide rather encouraging results on biomass applications. It is constructed from a sum of three terms, each originating from a totally different theory:

- A first term is a hard-sphere contribution which is essentially equivalent to the Carnahan-Starling expression (3.171), where the hard sphere diameter is made temperature dependent according to:

$$d = 1.06565 d_c \left(1 - 0.12 \exp\left(-\frac{2T_c}{3T} \right) \right) \tag{3.249}$$

thus using T_c (the critical temperature) and d_c, the hard sphere diameter at the critical temperature.
- The second term describes the dispersive interactions using a group contribution version of a density dependent NRTL equation. Parameters are group surface area and group interaction energy g_{ij}, that can be function of temperature. It also contains a group-group non-randomness parameter $\alpha_{ij} = \alpha_{ji}$.
- The third term is the association contribution, previously described in section 3.4.3.2.B, p. 196.

3.4.3.6 The lattice fluid equations of state

The lattice-fluid equations of state are constructed in a similar way as the excess Gibbs energy models [227]: a combinatorial contribution that takes into account volume and entropic effects and an enthalpic contribution accounting for molecular attractions. Hence the pressure (and therefore volume) dependence are explicitly taken into account. In the lattice models, each molecule is assumed to occupy a number of three-dimensional lattice cells, as already discussed in Flory's theory (see section 3.4.2.3.B, p. 180). In order to incorporate compressibility effects, two alternative approaches are possible:

- consider that a number of lattice sites are empty,
- allow variable-volume lattice sites.

The most well-known representative of this family is the Sanchez-Lacombe equation of state [320], which is constructed assuming empty sites. A reduced density is defined as the fraction of occupied lattice sites:

$$\tilde{\rho} = \frac{v^*}{v} = \frac{rN}{N_0 + rN} \tag{3.250}$$

where N is the number of molecules and r the number of sites one molecule occupies. N_0 is the number of vacant sites.

The energy of the lattice depends only on nearest-neighbour interactions. The lattice energy uses an energy parameter ε^* between two monomers. The final equation is written as:

$$P = \frac{RT}{v}\left(\frac{1}{r} - 1 - \frac{v}{v^*}\ln\left(1 - \frac{v^*}{v}\right) - \frac{v^*\varepsilon^*}{vRT}\right)$$ (3.251)

where, again, three parameters appear: the cell volume v^*, the number of segments in the molecule r and the interaction energy ε^*.

Panayiotou and Vera [233] have improved this equation introducing the quasi-chemical approximation for non-random mixing. The parameters are identical as in the original Sanchez-Lacombe version, but the equations are slightly more complex.

The so-called "GCLF" (Group Contribution Lattice Fluid) EoS has since been proposed by High and Danner [321], based on the Panayiotou-Vera equation. Group parameters have been determined using low molecular weight component densities and vapour pressures. This predictive approach is particularly worthwhile for polymer mixtures where it may be difficult to determine pure component parameters. The GCLF EoS has been applied to compute several properties (pure polymer density, VLE of both low and high molecular weight mixtures, weight fraction activity coefficients, liquid-liquid equilibria [78]).

In order to take into account hydrogen bonding, an extension to the lattice-fluid EoS has been proposed [322, 323] and called Non Random Hydrogen-Bonding model (NRHB). It adds onto a slightly modified Panayiotou-Vera EoS the possibility to consider the presence of proton acceptors and proton donors, not unlike the association approach considered in section 3.4.3.2.B, p. 196. The SAFT and NRHB approaches have been compared recently [324, 325]. It is shown that these two type of equations provide very similar, good quality predictions of both vapour-liquid and liquid-liquid equilibria of highly polar and hydrogen-bonding molecules.

3.4.3.7 Conclusions for equations of state

Equations of state are much more powerful than the activity coefficient models since they take density into account. As a result, they are able to describe the effect of pressure on phase equilibria. Only equations of state are able to produce full phase envelopes including critical points. They can deal both with gases and heavy components.

Table 3.26 summarises the use of equations of state. In the same way as for activity coefficient models, the predictive character of the equations is also indicated. "Correlative" means that the equation has no predictive character: experimental data are needed. "Corresponding states" means that the knowledge of the critical parameters and acentric factor are sufficient for using the model: this is essential for petroleum applications. Recently, a number of "group contribution" based models have been developed: the limitations are generally related to the number of groups considered by the authors.

Until 20 years ago, only virial and cubic equations of state were of any practical industrial use.

Virial equations of state, the most well-known of which is the Lee-Kesler [19] method, but that includes Benedict-Webb-Rubin (BWR) type equations, are very useful for calculating single phase properties (density, enthalpy, entropy, heat capacity, etc.), but too complex for phase equilibrium calculations.

The opposite is true for *cubic equations of state*: their main strength lies in rapid calculations of phase equilibria. They have been used with great success over many years and a large number of empirical correlations have been developed to extend their use to asymmetric mixtures and petroleum cuts. The use of cubic EoS requires attention to

(1) the choice of the $\alpha(T)$ (cohesion) function: very often, the Soave [326] function is used, but it also has drawbacks, and

(2) the choice of the mixing rule parameters. The group-contribution approach of Jaubert *et al.* (PPR78 [64]) is very useful from this point of view.

Nevertheless, both virial and cubic equations of state have the drawback of all empirical equations: they perform well when they can be tuned on data, but behave unexpectedly for extrapolations. Using Huron-Vidal (G^E-type) mixing rules [284], the SRK EoS has been combined with the predictive UNIFAC activity coefficient model to yield the so-called PSRK EoS, which is the most predictive of the "traditional" approaches: it is capable of calculating strongly non-ideal mixtures, including gas solubilities [56]. More recently, the VTPR EoS has been developed with higher accuracy [327].

More recently, however, using the strength of molecular simulation and statistical mechanics, *new equations of state* have emerged, whose physical background is much stronger. In the above discussion, focus has been mainly centered on the SAFT family of equations, but the lattice fluid equations (NRHB) have the same characteristics: the parameters refer to physical, molecular quantities, which can be determined by other means than simple regression on experimental data. These models have the great advantage to also take into account entropic effects that result from density differences between solvent and solute (resulting in high pressure liquid-liquid phase split). Even though accurate results can only be expected if experimental values can be used, an increasing number of authors propose

Table 3.26   Summary of important equations of state and their use

		Correlative	**Corresponding states**	**Group contribution**
Phase equilibria	Hydrocarbons	Cubic + $\alpha(T)$		
	Slightly non-ideal	Cubic + k_{ij} mixing rule		Cubic + Jaubert k_{ij} PC-SAFT (Tihic)
	Strongly non-ideal (enthalpic)	Cubic + G^E-type mixing rule CPA (water, alcohols, gases)		VR-γ-SAFT PSRK GCA (bio-applications)
	Enthalpic + entropic non-ideality	NRHB		GC-Lattice Fluid GC-PPC-SAFT
Phase properties		VR-SAFT MBWR	Lee Kesler Starling – BWR	

predictive ways of determining the equation parameters from the molecular structures. Group contributions represent the most widespread approach [328, 309, 310], but the use of quantum mechanical calculations for this purpose is quite promising [329-331].

As an intermediate solution, hybrid approaches have also been mentioned. They combine terms originating from different theories. Due to the nature of these equations, their use is system-specific: they can provide very good results on some systems, but be of no help for others. As examples, CPA [309] has proven to be of great help for natural gas applications where water, alcohols, acid gases and hydrocarbons coexist. GCA [80], on the other hand, is a predictive equation (based on group contributions), that has essentially been used on biofuel applications.

3.4.4 Phase-specific models

3.4.4.1 Pure solid phases

The calculation of a solid phase fugacity is provided in most textbooks [38, 148]. The fundamental hypothesis is based on the fact that the component crystallises as a pure component. The equilibrium between solid and liquid phases can therefore be written as:

$$f_i^{S*} = f_i^{L*} x_i \gamma_i^L = f_i^L \tag{3.252}$$

Using the Gibbs-Helmohltz equation, the identity between liquid and solid fugacities at the fusion temperature $(T_{F,i})$ can be transformed as follows to calculate the ratio of f_i^{S*} / f_i^{L*} at any pressure P and temperature T:

$$\ln\left(\frac{f_i^{S*}}{f_i^{L*}}\right) = \ln\left(x_i \gamma_i^L\right)$$

$$= \frac{\Delta h_{F,i}}{RT_{F,i}}\left(1 - \frac{T_{F,i}}{T}\right) + \frac{\Delta c_{P,F,i}}{R}\left[\frac{T_{F,i}}{T} - 1 - \ln\left(\frac{T_{F,i}}{T}\right)\right] + \frac{\Delta v_{F,i}\left(P - P_F\right)}{RT_{F,i}} \tag{3.253}$$

This expression is obtained by calculating the change in Gibbs energy between the pure solid and the pure component in a hypothetical supercooled liquid state $(T < T_{F,i})$. This is why $f_i^{L,*}$, the pure liquid fugacity, in the same pressure and temperature conditions as the system, is used. It is clear that this supercooled liquid is unstable in the given conditions, but the fugacity must be calculated as an intermediate step. It is extrapolated from an equation of state or vapour pressure correlation.

All parameters in (3.253) have a physical meaning and can be measured or found in databases: $\Delta h_{F,i} = h_i^L - h_i^S$ is the heat of fusion (positive value, the liquid has more energy than the solid).

We see from (3.253) that $\Delta v_{F,i} = v_i^L - v_i^S$, the volume difference upon melting (also generally positive, except for water) has an effect on the pressure dependence of the crystallisation. If this value is zero, it is considered that pressure has no effect on crystallisation. This approximation is often acceptable.

The difference in heat capacities between the liquid and the solid along the temperature path (from $T_{F,i}$ to T) $\Delta c_{P,F,i} = c_{P,i}^L - c_{P,i}^S$ is considered independent of temperature. This is not strictly correct, but may be considered as an acceptable approximation, as there is generally no experimental information for this value. Consequently, it may be considered as an adjustable parameter.

If the solid is not a pure component, equation (3.252) must be modified taking into account an activity coefficient for the fugacity of solid on the left side of the equality.

3.4.4.2 Hydrates

Natural gas hydrates are solid compounds consisting basically of water in the presence of light hydrocarbon or other gases. Their densities are lower than that of ice due to the presence of these other molecules in the internal structure of the solid and they are similar in appearance to ice. Hydrates are a very serious problem in gas production and surface facilities because they can cause plugging in lines and valves.

Different types of structure may be observed (Claussen, 1951 [332-334], von Stackelberg and Muller, 1951 [335] and Ripmeester, 1987 [336]) in the formation of hydrates where the light gases present are entrapped in cage-like cavities (the term clathrate used as a synonym describes this situation). Three structures of cages have been observed: two are cubic (sI, sII) and one hexagonal (sH).

A number of widely different methods based on equilibrium calculations have been developed, most based on the contributions of Van der Waals and Platteeuw in 1959 [337] and Parrish and Prausnitz in 1972 [338]. Equilibrium is based on the chemical potential or fugacity of water equalities between phases. Due to the specific structure of the hydrate, in which gas molecules are trapped inside cages of hydrogen bonded water molecules, the expression for the chemical potential of water is written as:

$$\Delta\mu_w^H = \mu_w^H - \mu_w^\beta = -RT\sum_m v_m \ln\left(1-\sum_j \theta_{mj}\right) \qquad (3.254)$$

where μ_w^β is the chemical potential of water in a hypothetical empty lattice while μ_w^H is the chemical potential of water in a filled cage. Similarly, v_m is the number of cavities of type m per molecule of water in the lattice and θ_{mj} is the fraction of type m cavities occupied by component j and given by the expression:

$$\theta_{mj} = \frac{C_{mj}f_j}{1+\sum_l C_{ml}f_l} \qquad (3.255)$$

where C_{mj} is the Langmuir constant and f_j is the fugacity of gas component j. Van der Waals and Patteeuw show that, using the Lennard-Jones-Devonshire cell theory, the Langmuir constant can be written by:

$$C_{mj}(T) = \frac{4\pi}{kT}\int_0^\infty \exp\left(\frac{-w_{mj}(r)}{kT}\right)r^2 dr \qquad (3.256)$$

where $w_{mj}(r)$ is the spherical cell potential. Some empirical correlations can also be used for $C_{mj}(T)$.

The reader is invited to consult the books by Caroll [339] and by Sloan [340] for further details.

3.4.4.3 Properties at infinite dilution

As mentioned earlier, the properties of a given component may depend to a large extent on its surrounding. The pure component properties were presented in section 3.2 (p. 142) and, in section 3.3 (p. 148), a number of theories allowing calculation of the properties in a mixture were discussed. Nevertheless, it is a well-known fact that these theories have difficulties to describe medium and high dilution properties simultaneously: at infinite dilution, when the solute molecules are entirely surrounded by a solvent their behaviour is quite different, which is why specific approaches are needed.

The infinite dilution properties may be expressed as Henry constant and are used with the asymmetric convention, as explained in section 2.2.3.1.A.c, p. 65 (this is generally the case for gases): it is the slope of the fugacity at infinite dilution in a solvent:

$$H_i = \lim_{x_i \to 0} \frac{f_i}{x_i} \tag{3.257}$$

Sometimes, however, the symmetric convention is used (the pure component properties must be known: for a liquid, $f_i^* \approx P_i^\sigma$), in which case we generally refer to the activity coefficient at infinite dilution. The relationship between these two properties is therefore:

$$H_i = \gamma_i^\infty f_i^* \tag{3.258}$$

In the same way as pure component properties are related using Maxwell's equation, the same can be said of infinite dilution properties. The most well-known one is:

$$H_i(P,T) = H_i(P_s^\sigma, T) \exp \frac{\overline{v_i}^\infty \left(P - P_s^\sigma \right)}{RT} \tag{3.259}$$

As a result, a full equation of state-type relationship $(v_i^\infty(P, T))$ may be constructed at infinite dilution [341]. This has been done for example for properties at infinite dilution in water. A review of the water solubility of many components is provided in section 4.2.4 (p. 293).

3.4.4.4 Gases in hydrocarbons: Grayson and Streed

Grayson and Streed [342] proposed the method below as an improvement on that of Chao and Seader [343]. It is very well known as such in the petroleum industry, even though a number of improvements have since been published [344]. The basic equation calculates the fugacities in both phases using a heterogeneous approach:

$$f_i^V = y_i \, P\varphi_i^V = x_i \, f_i^{L*}\gamma_i^\infty = f_i^L \tag{3.260}$$

The distribution coefficient can now be written as:

$$K_i = \frac{y_i}{x_i} = \frac{f_i^{L*}\gamma_i^\infty}{P\varphi_i^V} = \frac{\varphi_i^{L*}\gamma_i^\infty}{\varphi_i^V} \tag{3.261}$$

The fugacity coefficient of pure liquid is calculated with a Curl-Pitzer [345] corresponding states correlation:

$$\log \varphi_i^{L*} = \log \varphi_i^{(0)} + \omega_i \log \varphi_i^{(1)} \tag{3.262}$$

where $\log \varphi_i^{(0)}$ and $\log \varphi_i^{(1)}$ are a function of the reduced temperature and pressure and constants that may depend on the nature of the component.

A general expression (3.263) calculates the fugacity coefficient of the "pure" components in the liquid phase φ_i^{L*}. If i refers to a gas, this method is in fact a Henry constant method:

$$\varphi_i^{L*} = \frac{f_i^{L*}}{P} = \frac{H_i}{P\gamma_i^\infty} \tag{3.263}$$

If i refers to a liquid component (case "others" in the original paper), as described above,

this method in fact calculates a non-dimensional vapour pressure $\left(\varphi_i^{L*} \approx \frac{P_i^\sigma \wp_i}{P} \right)$.

The activity coefficient is calculated using the Scatchard-Hildebrandt theory for regular solutions (see section 3.4.2.2.B, p. 175) while the fugacity coefficient in the vapour phase is obtained using the Redlich Kwong [255] EoS.

3.5 WHAT ARE THE KEY COMPONENTS CONCENTRATION RANGE?

We know that fluid mixtures in the chemical and petrochemical industries generally contain a large number of components. So far, we have stressed the fact that each individual component should be characterised, and all binary parameters should be given suitable values before any calculation can be initiated. The number of binary sub-systems, however, may be very large in a $\mathcal{N}$-component system:

$$\mathcal{N}_{ij} = \binom{\mathcal{N}}{2} = \frac{\mathcal{N}!}{(\mathcal{N}-2)!2!} = \frac{1\cdot2\cdot3\cdot...\cdot(\mathcal{N}-2)\cdot(\mathcal{N}-1)\cdot\mathcal{N}}{1\cdot2\cdot3\cdot...\cdot(\mathcal{N}-2)\cdot1\cdot2} = \frac{(\mathcal{N}-1)\cdot\mathcal{N}}{2} \tag{3.264}$$

The question we want to investigate here is whether all of these subsystems are equally important. We may not be able to answer the question precisely in a general fashion, as it will be highly dependent on the design constraints of the process to be modelled. The process engineer is often in a better position than the thermodynamic specialist to provide a suitable answer. Thermodynamics may nevertheless provide an insight into the sensitivity of a given parameter to the final result.

The above question will be dealt with by using the concept of "key component(s)", a method of indicating that not all components have the same importance in the final result. It is the responsibility of the process engineer to distinguish between majority components (which make up the main part of the feed), and key components, which may or may not be majority components. The following component characteristics may be of help for this purpose:

- amount of component (often, the majority component may be considered key),
- reactivity (some components may react, and as result change the fluid properties),

- corrosivity (it may be essential to know the fate of components which may damage the equipment),
- catalyst poison (even in small concentrations, the presence of some components may deactivate catalysts),
- potential to freeze under process conditions (upon freezing, some components may deposit on the equipment walls, thus eventually blocking the flow: an example of this is discussed further in the case studies),
- unknown properties at design stage assumptions,
- etc.

For most single-phase property calculations (volume, enthalpy, viscosity, thermal conductivity, etc.) the mixture composition is somehow "averaged out" using mixing rules or other methods, so that the effect of each individual component is essentially proportional to its concentration. The binary interaction parameters are of less importance. The discussion will therefore be focused on the different types of phase equilibrium calculation.

3.5.1 Phase appearance or phase envelope calculations

When the limit of appearance of a new phase is required (condensation, vapourisation or crystallisation), the calculation is the same as for a phase envelope calculation: a phase diagram (or phase envelope) provides the boundaries between single-phase and two-phase regions, between two- and three-phase regions, etc.

A general observation is that the key component in these types of calculations is the component that will have most influence on the incipient phase properties (*i.e.* the phase whose possible appearance is investigated).

3.5.1.1 Vapour-Liquid Equilibrium (VLE) calculations

For VLE calculations, two possibilities are encountered: either a bubble point or a dew point has to be calculated.

At the **bubble point**, the incipient phase is a vapour phase. The vapour is rich in light components. Hence, the bubble conditions will be very sensitive to the concentration, but also to the parameterisation of the gases dissolved in the liquid.

For example, the effect of the parameter k_{ij} between methane (lightest component) and phenanthrene (heaviest component) is shown in figure 3.42. It is worth noting that the phenanthrene concentration in this mixture is very small. Here, the effect on the bubble pressure is obvious.

When calculating a **dew point**, the incipient phase is a liquid, which essentially contains heavy components. Sportisse [347] has shown that, even in very low concentrations, a heavy component has a very large effect on the dew curve. This is also illustrated in figure 3.43: the very small concentration in C_{11+} has a tremendous effect on the dew curve depending on whether it is described as $n\text{-}C_{11}$, $n\text{-}C_{16}$ or $n\text{-}C_{20}$ linear paraffins.

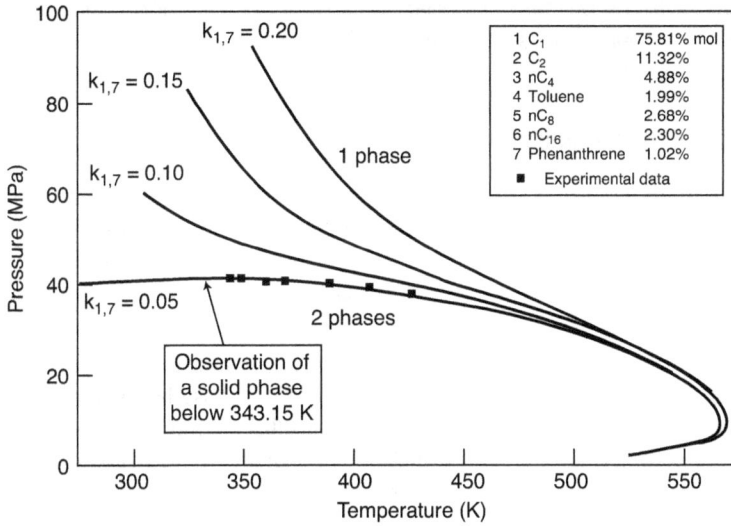

Figure 3.42

Effect of the binary interaction parameter ($k_{1,7}$) between the heaviest and the lightest component for the calculation of the VLE phase diagram of a synthetic hydrocarbon mixture [346] using the PR EOS.

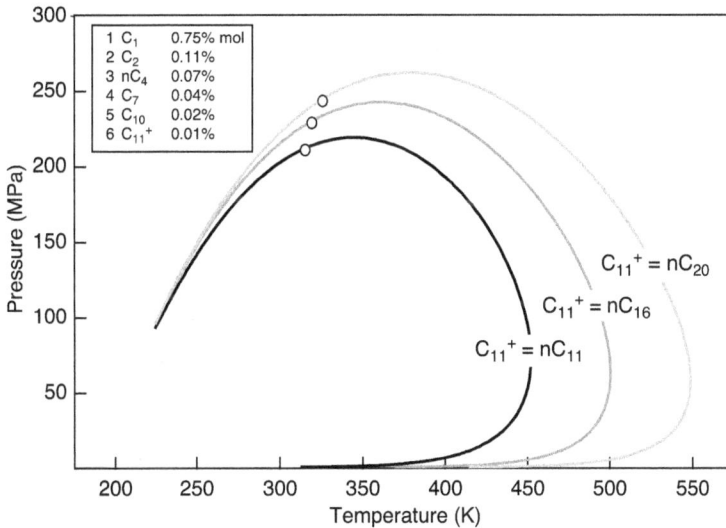

Figure 3.43

Effect of the description of the heavy end of a gas condensate on the phase envelope. The three curves are calculated in identical conditions, except that the C_{11} + pseudo-component is taken as $n\text{-}C_{11}$, $n\text{-}C_{16}$ or $n\text{-}C_{20}$, using the PR EoS.

> When calculating the phase boundary of vapour-liquid equilibrium, the most sensitive interaction parameters are between the lightest and the heaviest components in terms of volatility.

3.5.1.2 Liquid-Liquid Equilibrium (LLE) calculations

Volatility has no effect at all on liquid-liquid phase equilibrium. Only chemical affinity is at stake: molecules that interact strongly with themselves (as in auto-association: water, alcohols, or strongly polar molecules) will tend to phase split with molecules that do not have the same type of strong interactions (hydrocarbons, and in particular alkanes, are at the other end of that spectrum).

In section 3.3.1 of this chapter (p. 148), it has been shown how a slight change in the behaviour of the activity coefficient can dramatically change the phase behaviour. It is essential to investigate the interactions between the two components that are most unlike each other, but the presence of a co-solvent (molecule that is soluble in both of the other two components) can affect the phase equilibrium significantly. It is therefore recommended not only to rely on binary data, but to further investigate ternary equilibria. In figure 3.44, an example is shown for the glycerol + methanol + methyl-oleate (ester) mixture, important in the trans-esterification process where the purity of the ester and the glycerol phases are essential for the treatment of the products.

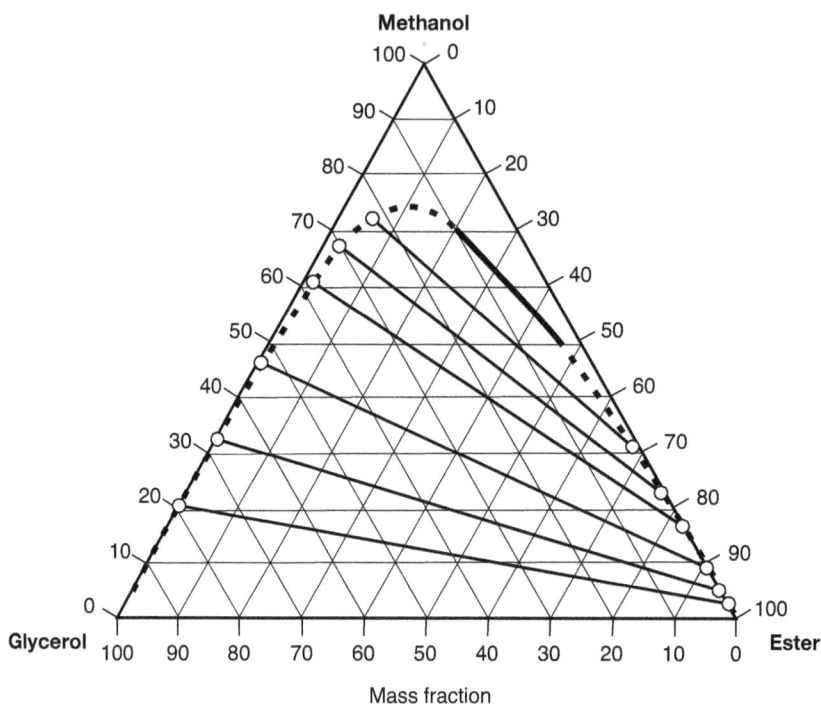

Figure 3.44

Ternary LLE diagram of the glycerol + methanol + methyl ester system.

In multicomponent mixtures, it may be useful to represent this ternary diagram, either by selecting a representative molecule for each group "solvent", "anti-solvent" and "co-solvent", or by grouping in one of the three sides the groups of components that behave similarly.

The difficulties related to the simultaneous description of LLE and VLE is further discussed by Cha and Prausnitz [348], who also concludes that the use of ternary data may be a great asset.

3.5.1.3 Liquid-Solid Equilibrium (LSE) calculations

When a solid phase can crystallise from a fluid system, it is essential to know what the solid composition should be. The factors that determine which component crystallises first are:

- the crystallisation temperature (discussed in section 3.1 of this chapter, p. 102), and
- the concentration of each component.

An example of such a case is discussed in the Liquefin application [349].

A good knowledge of the interactions between the crystallising component and the bulk fluid mixture is essential to develop a good model. As a first approximation, the bulk fluid composition will be reduced to its majority component, but it may also be worthwhile investigating the interactions with the other fluid components.

3.5.2 Distribution coefficients calculation

Distribution coefficients are used for all processes that use phase equilibrium for separation. In the case of stripping or extraction, the key component is generally well identified: always keep in mind that its properties strongly depend on its interactions with the bulk phase composition.

In case of distillation, components are separated based on their relative volatilities (see section 2.2.3.2, p. 81). Analysis and design of recent models used in distillation can be found in the recent book of Doherty and Malone [350]. Most distillation columns have two products: a heavy product (residue) and a light product (distillate). The purpose is often to split the feed at a boiling point range that is as narrow as possible (as shown in figure 3.45). This situation is called simple distillation. More complex situations may occur however, such as distillation of an azeotropic mixture or distillation with strict purity specifications. These cases are discussed below.

3.5.2.1 Simple distillation

When the purpose of the process is to split a feed according to a specific cut-point, the interactions between the components on either side of the cut-point will be vitally important when setting up the thermodynamic model.

Figure 3.45 illustrates how in fact the "cut point" is more generally a "cut zone". The feed is here considered as a continuum in terms of volatility (expressed using the vapour pressure P^{σ}). If the column had an infinite reflux ratio, or an infinite number of theoretical stages, it would be possible to attain a sharp, horizontal split between the two products. In reality, there will always be some of the lightest part of the heavy end that will end up in the distillate, and some of the heavy part of the light end that will be found in the residue. This "cut point zone" is where the attention should be focused for a good thermodynamic model: a good definition of their mutual interaction is essential.

Yet, it must be kept in mind, as always, that the components found in the cut point zone are in fact diluted in either the heavy end or the light end. As a result, following interactions should be taken into consideration:

- cut-point components among each other,
- cut-point components with the majority component(s) of the light end,
- cut-point components with the majority component(s) of the heavy end.

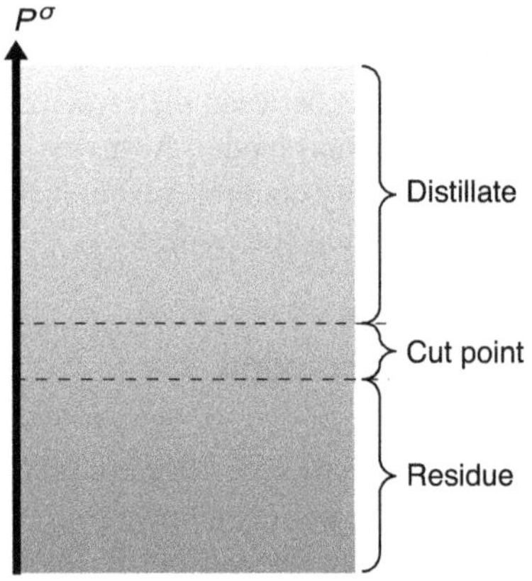

Figure 3.45

Illustration of the cut-point of a distillation unit. The components are listed according to their vapour pressure. The lightest components appear on top, and the heaviest on the bottom. The aim of distillation is to make the cut-point region as narrow as possible.

Examples of this situation are de-ethanisers, de-propanisers, etc. where the interactions between ethane and propane, or between propane and butane, respectively, are the most important to evaluate.

The importance of an accurate calculation of the relative volatility is highest when the volatility of the two keys are close, as shown in figure 3.46 (α is close to one). The error that may result on the number of trays to achieve a given separation can become very large.

3.5.2.2 Azeotropic distillation

The fact that in a mixture, the volatility of a given component may change significantly has been discussed extensively in section 3.4.1 of this chapter (p. 160). The result is that the

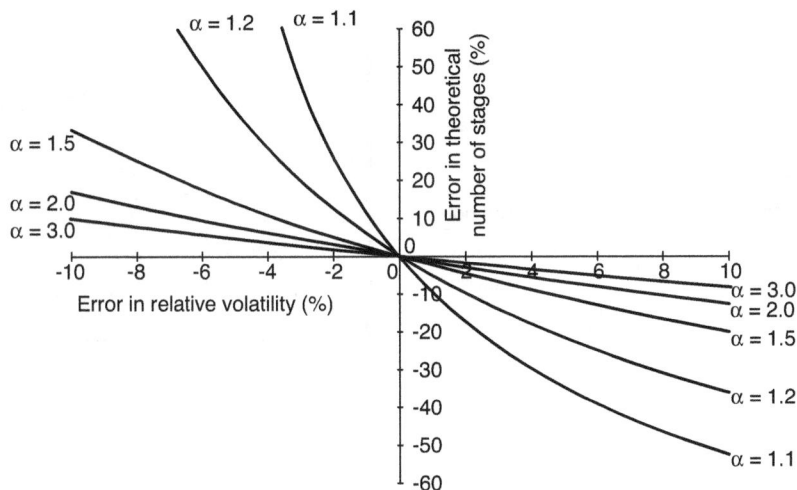

Figure 3.46

Evaluation of the effect of the error on the relative volatility $\left(\alpha_{ij} = \dfrac{K_i}{K_j} \right)$ on the error of the theoretical number of stages. This evaluation is made with the assumption of a constant relative volatility (taken from [351]).

components can no longer be listed according to their pure volatility, as in figure 3.45. Instead, they should be listed according to the product of their vapour pressure and their activity coefficient: the product $\gamma_i P_i^\sigma$. This is not possible as such, since the activity coefficient depends on the liquid composition, which itself varies along the distillation column.

Figure 3.47 attempts to help understanding how this change in volatility may affect the distillation. We only consider here positive deviations from ideality. Interactions between two components may be such that their relative volatilities are inverted within a certain concentration range (this is what happens in case they form an azeotrope). The figure shows the location of the binary azeotrope on the same scale as the volatility scale of the feed (see figure 3.47).

Two situations can occur: either the azeotrope lies entirely within the distillate or residue region (as shown in figure 3.47 a) or it will lie within the cut-point region (as shown in figure 3.47 b). In the former case, investigating the azeotropic binary will not improve the thermodynamic model; in the latter, it is essential for a good representation of the distillation column.

When the azeotropic pressure of a mixture lies above the cut-point, while the individual component volatilities lie in the residue region, a thorough thermodynamic investigation of the mixture is required: it indicates that in the mixture, the volatilities of these components are much higher than expressed by their vapour pressure, and that they will therefore most probably end up in the distillate.

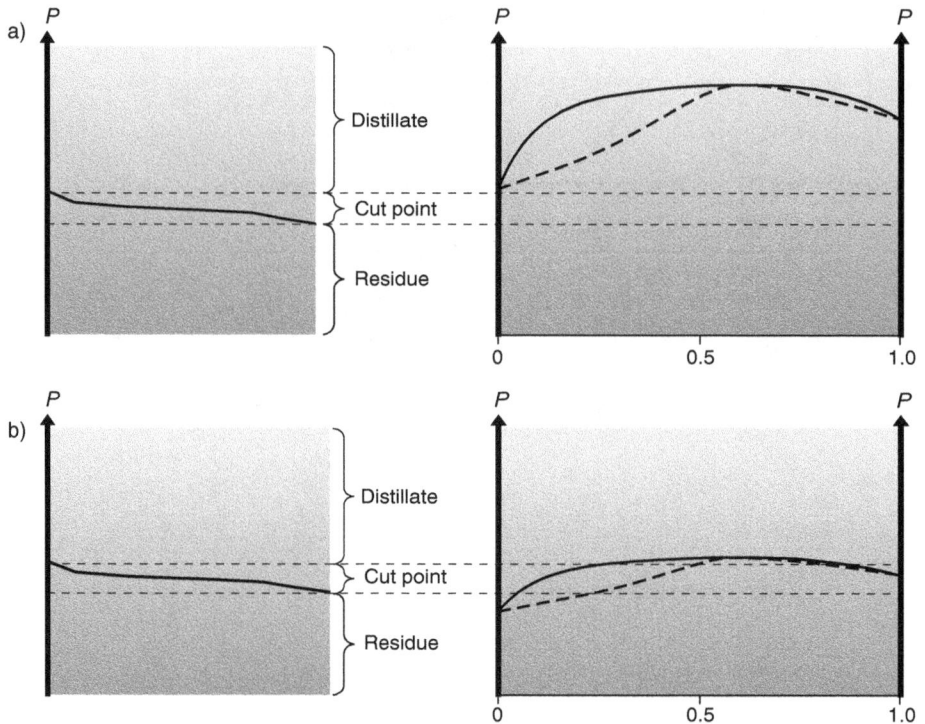

Figure 3.47

Position of the azeotrope with respect to the cut-point. a) the azeotrope lies within the distillate or residue region: no need to investigate further; b) the azeotrope lies in the cut-point region: the thermodynamic model must take into account the interactions between the azeotropic components.

3.5.2.3 Impurities

Knowing the fate of impurities in the process may be essential for various reasons, for example environmental, catalyst poisoning, or other possible risks. However, as already mentioned above and discussed in some detail in section 3.4.1 (p. 160), a component behaves very differently when in a mixture and when pure. The activity coefficient is used to express this deviation. It is often observed that the activity coefficient from the middle-composition is very different from the one at very high dilution (infinite dilution).

In order to illustrate this fact, figure 3.48 shows the methanol + *n*-hexane binary mixture with activity coefficient calculated from experimental data. It is clear that the NRTL model, used here to calculate the activity coefficient, although very good for medium concentrations, does not capture at all the infinite dilution activity coefficients. Therefore, if the high dilution region is of interest, a specific model must be used, or otherwise specific parameters must be fitted.

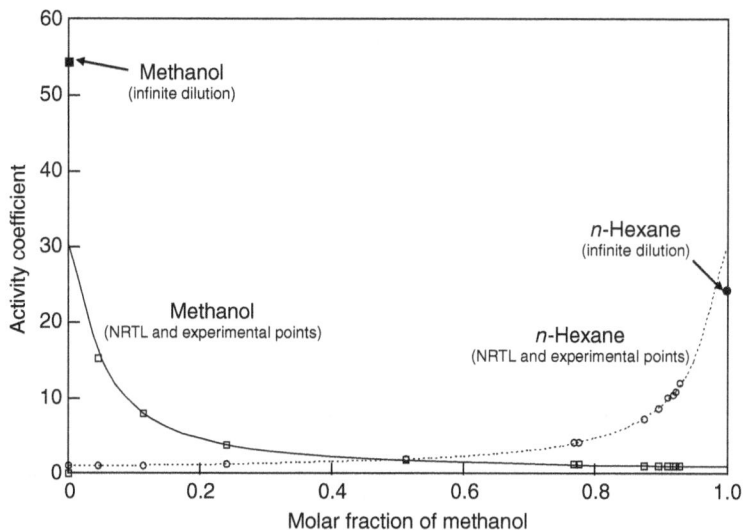

Figure 3.48

Comparison of activity coefficient at infinite dilution with extrapolation of activity coefficient from experimental data in the medium composition for an methanol + *n*-hexane mixture.

In conclusion, fitting a thermodynamic model on binary VLE data may not be sufficient to calculate infinite dilution or Henry constant values. Specific data at infinite dilution should be used. Predictive models (as UNIFAC) are often unable to describe infinite dilution behaviour.

This is why a close analysis of the process problem is essential.

REFERENCE LIST

[1] Poling, B. E., Prausnitz, J. M. and O'Connell, J. P. 'The Properties of Gases and Liquids", 5th Ed.; McGraw-Hill: New York, **2000**.

[2] Ewell, R. H., Harrison, J. M. and Berg, L. "Azeotropic Distillation" *Industrial & Engineering Chemistry* **1944**, **36**, 10, 871-875.

[3] Pitzer, K. S., Lippman, D. Z., Curl, R. F., Huggins, C. M. and Petersen, D. E. "The Volumetric and Thermodynamic Properties of Fluids. II. Compressibility Factor, Vapor Pressure and Entropy of Vapourisation" *Journal of the American Chemical Society* **1955**, **77**, 13, 3433-3440.

[4] Rowley, R.L., Wilding, W.V., Oscarson J.L., Yang Y., Zundel, N.A., Daubert, T.E. and Danner, R.P. "DIPPR® Data Compiltation of Pure Compound Properties", Design Institute for Physical Properties; AIChE, New York, **2003**.

[5] Laux, H., Butz, T., Meyer, G., Matthaï, M. and Hildebrand, G. "Thermodynamic Functions of the Solid-Liquid Transition of Hydrocarbon Mixtures" *Petroleum Science and Technology* **1999**, **17**, 9-10, 897-913.

[6] Briard, A-J., Bouroukba, M., Petitjean, D. and Dirand, M. "Models for Estimation of Pure *n*-Alkanes' Thermodynamic Properties as a Function of Carbon Chain Length" *Journal of Chemical & Engineering Data* **2003**, **48**, 6, 1508-1516.

[7] Calange, S. "Modélisation Thermodynamique Compositionelle de la Crystallisation des Bruts Paraffiniques", Université de Pau et des Pays de l'Adour, **1996**.

[8] Van Ness, K. and Van Westen, H. A. *"Aspects of the constitution of Mineral Oils"*; Elsevier, **1951**.

[9] Riazi, M. R. and Daubert, T. E. "Prediction of Molecular-type Analysis of Petroleum Fractions and Coal Liquids" *Industrial & Engineering Chemistry Process Design and Development* **1986**, **25**, 4, 1009-1015.

[10] Riazi, M. R. *"Characterization and Properties of Petroleum Fluids"*; American Society for Testing and Materials: Philadelphia, **2005**.

[11] Nguyen-Huynh, D., Passarello, J. P., Tobaly, P. and de Hemptinne, J. C. "Modeling Phase Equilibria of Asymmetric Mixtures Using a Group-Contribution SAFT (GC-SAFT) with a k(ij) Correlation Method Based on London's Theory. 1. Application to CO_2 + *n*-Alkane, Methane plus *n*-Alkane, and Ethane plus *n*-Alkane Systems" *Industrial & Engineering Chemistry Research* **2008**, **47**, 22, 8847-8858.

[12] Nguyen-Huynh, D., Tran, T. K. S., Tamouza, S., Passarello, J. P., Tobaly, P. and de Hemptinne, J. C. "Modeling Phase Equilibria of Asymmetric Mixtures Using a Group-Contribution SAFT (GC-SAFT) with a k(ij) Correlation Method Based on London's Theory. 2. Application to Binary Mixtures Containing Aromatic Hydrocarbons, *n*-Alkanes, CO_2, N_2, and H_2S" *Industrial & Engineering Chemistry Research* **2008**, **47**, 22, 8859-8868.

[13] Wilson, G. M. "A Modified Redlich-Kwong equation of state: Application to General Physical Data Calculations" Proceedings of the AIChE 65th National Meeting; Cleveland, Ohio, **1969**.

[14] Ledanois, J. M., Colina, C. M., Santos, J. W., Gonzalez-Mendizabal, D. and Olivera-Fuentes, C. "New Expressions for the Vapor Pressure of Pure Components Constructed on Characteristic Points" *Industrial & Engineering Chemistry Research* **1997**, **36**, 6, 2505-2508.

[15] Antoine, Ch. "Tension des vapeurs: nouvelle relation entre les tensions et les températures" *Comptes rendus hebdomadaires des séances de l'Académie des sciences* **1888**, 107, 681-684.

[16] Wagner, W. "New Vapour Pressure Measurements for Argon and Nitrogen and a New Method for Establishing Rational Vapour Pressure Equations" *Cryogenics* **1973**, **13**, 8, 470-482.

[17] Ambrose, D. and Walton, J. "Vapor-Pressures Up to Their Critical-Temperatures of Normal Alkanes and 1-Alkanols" *Pure and Applied Chemistry* **1989**, **61**, 8, 1395-1403.

[18] Daubert, T. E. and Danner, R. P. API Technical Data Book – Petroleum Refining, 6th ed.; American Petrolum Institute (API), **1997**.

[19] Lee, B.I. and Kesler, M.G. "A Generalized Thermodynamic Correlation Based on Three-Parameter Corresponding States" *AIChE Journal* **1975**, **21**, 3, 510.

[20] Rackett, H. G. "Equation of State for Saturated Liquids" *Journal of Chemical & Engineering Data* **1970**, **15**, 4, 514-517.

[21] Dymond, J. H. and Malhotre, R. "The Tait Equation, 100 Years on" *International Journal of Thermophysics* **1988**, **9**, 6, 941-951.

[22] Thomson, G. H., Brobst, K. R. and Hankinson, R. W. "An Improved Correlation for Densities of Compressed Liquids and Liquid-Mixtures" *AIChE Journal* **1982**, **28**, 4, 671-676.

[23] Aalto, M., Keskinen, K. I., Aittamaa, J. and Liukonen, S. "An Improved Correlation for Compressed Liquid Densities of Hydrocarbons. Part 1. Pure Compounds" *Fluid Phase Equilibria* **1996**, **114**, 1-2, 1-19.

[24] Aalto, M., Keskinen, K. I., Aittamaa, J. and Liukonen, S. "An Improved Correlation for Compressed Liquid Densities of Hydrocarbons. Part 2. Mixtures" *Fluid Phase Equilibria* **1996**, **114**, 1-2, 21-35.

[25] Pokki, J. P., Aalto, M. and Keskinen, K. I. "Remarks on Computing the Density of Dense Fluids by Aalto–Keskinen Model" *Fluid Phase Equilibria* **2002**, 194-197, 337-351.

[26] Younglove, B. A. "Thermophysical Properties of Fluids I. Argon, Ethylene, Parahydrogen, Nitrogen, Nitrogen Trifluoride and Oxygen" *Journal of Physical and Chemical Reference Data* **1982**, 11 (Supplement 1), 1-1-1-349.

[27] Younglove, B. A. "Thermophysical Properties of Fluids II. Methane, Ethane, propane, isobutane and Normal Butane" *Journal of Physical and Chemical Reference Data* **1987**, **16**, 4, 577-798.

[28] Frenkel, M., Kabo, G. J., Marsh, K. N., Roganov, G. N. and Wilhoit, R. C. Thermodynamics of Organic Compound in the Gas State (Volumes I and II); Thermodynamics Research Center (TRC): College Station, TX, **1994**.

[29] Edmonds, B. "PPDS-Physical Property Data Service" *CODATA Bulletin* **1985**, 58, 6-11.

[30] Shomate, C. H. "Heat Capacities at Low Temperatures of the Metatitanates of Iron, Calcium and Magnesium" *Journal of American Chemical Society* **1946**, **68**, 6, 964-966.

[31] Passut, C. A. and Danner, R. P. "Correlation of Ideal Gas Enthalpy, Heat Capacity, and Entropy" *Industrial & Engineering Chemistry Process Design and Development* **1972**, **11**, 4, 543-546.

[32] Aly, F. A. and Lee, L. L. "Self-consistent Equations for Calculating the Ideal Gas Heat Capacity, Enthalpy, and Entropy" *Fluid Phase Equilibria* **1981**, **3-4**, 169-179.

[33] Watson, K. M. "Thermodynamics of the Liquid State" *Industrial & Engineering Chemistry* **1943**, **35**, 4, 398-406.

[34] Anderko, A. "Equation of State for Pure Fluids and Mixtures Based on a Truncated Virial Expansion" *AIChE Journal* **1991**, **37**, 9, 1379-1391.

[35] Hayden, J. G. "Generalized Method for Predicting Second Virial Coefficients" *Ind. Eng. Chem. Proc. Des. Dev.* **1975**, **14**, 3, 209-216.

[36] Mathias, P. M. "The Second Virial Coefficient and the Redlich-Kwong Equation" *Industrial & Engineering Chemistry Research* **2003**, **42**, 26, 7037-7044.

[37] Mathias, P. M. "Comments on 'The Second Virial Coefficient and the Redlich-Kwong Equation' – Response" *Industrial & Engineering Chemistry Research* **2007**, **46**, 19, 6376-6378.

[38] Vidal, J. 'Thermodynamics: Applications in Chemical Engineering and the Petroleum Industry"; Editions Technip: Paris, **2003**.

[39] Tsonopoulos, C. "An Empirical Correlation of Second Virial Coefficients" *AIChE Journal* **1974**, 20, 263-272.

[40] Nielsen, T. L., Abildskov, J., Harper, P. M., Papaeconomou, I. and Gani, R. "The CAPEC Database" *Journal of Chemical and Engineering Data* **2001**, **46**, 5, 1041-1044.

[41] Ungerer, P., Tavitian, B. and Boutin, A. "Applications of Molecular Simulation in the Oil and Gas Industry" 1st Ed.; Editions Technip: Paris, **2005**.

[42] Lydersen, A. L. "Estimation of Critical Properties of Organic Compounds", Report, College of Engineering University of Wisconsin, **1955**.

[43] Benson, S. W. 'Thermochemical Kinetic"; Wiley, **1976**.

[44] Ambrose, D. "Correlation and Prediction of Vapour Liquid Critical Properties. I. Critical Temperatures of Organic Compounds" National Physical Laboratory, Teddington, Great Britain, NPL Rep. Chem. 92, **1978**.

[45] Ambrose, D. "Correlation and Prediction of Vapour Liquid Critical Properties. II. Critical Pressures and Volumes of Organic Compounds" National Physical Laboratory, Teddington, Great Britain, NPL Rep. Chem. 98, **1979**.

[46] Joback, K. G. "A Unified Approach to Physical Property Estimation using Multivariate Statistical Techniques", Massachussetts Institute of Technology, Cambridge (Ma), **1984**.

[47] Joback, K. G. and Reid, R. C. "Estimation of Pure Component Properties from Group Contributions" *Chemical Engineering Communications* **1987**, 57, 233-243.

[48] Marrero, J. and Gani, R. "Group-Contribution Based Estimation of Octanol/Water Partition Coefficient and Aqueous Solubility" *Industrial & Engineering Chemistry Research* **2002**, **41**, 25, 6623-6633.

[49] Marrero, J. and Gani, R. "Group-Contribution Based Estimation of Pure Component Properties" *Fluid Phase Equilibria* **2001**, 183, 183-208.

[50] Constantinou, L. and Gani, R. "New Group-Contribution Method for Estimating Properties of Pure Compounds" *AIChE Journal* **1994**, **40**, 10, 1697-1710.

[51] Constantinou, L., Gani, R. and Oconnell, J. P. "Estimation of the Acentric Factor and the Liquid Molar Volume at 298-K Using A New Group-Contribution Method" *Fluid Phase Equilibria* **1995**, **103**, 1, 11-22.

[52] Constantinou, L., Bagherpour, K., Gani, R., Klein, J. A. and Wu, D. T. "Computer Aided Product Design: Problem Formulations, Methodology and Applications" *Computers & Chemical Engineering* **1996**, **20**, 6-7, 685-702.

[53] Sandler, S. I. Quantum Mechanics: a New Tool for Engineering Thermodynamics, *Fluid Phase Equilibria* **2003**, 210, 147-160.

[54] Mavrovouniotis, M. L. Estimation of Properties form Conjugate Forms of Molecular-Structures – the Abc Approach, *Industrial & Engineering Chemistry Research* **1990**, 29, 1943-1953.

[55] Coniglio, L. and Daridon, J. L. "A Group Contribution Method for Estimating Ideal Gas Heat Capacities of Hydrocarbons" *Fluid Phase Equilibria* **1997**, **139**, 1-2, 15-35.

[56] Gmehling, J. "Present Status and Potential of Group Contribution Methods for Process Development" *The Journal of Chemical Thermodynamics* **2009**, **41**, 6, 731-747.

[57] Lydersen, A. L. "Estimation of Critical Properties of Organic Compounds" by the Method of Group Contributions, College of Engineering, University of Wisconsin, Eng. Exp., Sta. Rep., 3, Madison, WI, **1955**.

[58] Nannoolal, Y., Rarey, J. and Ramjugernath, D. "Estimation of Pure Component Properties. Part 2. Estimation of Critical Property Data by Group Contribution" *Fluid Phase Equilibria* **2007**, **252**, 1-2, 1-27.

[59] Vetere, A. "Methods to Predict the Critical Constants of Organic-Compounds" *Fluid Phase Equilibria* **1995**, **109**, 1, 17-27.

[60] Forman, J. C. and Thodos, G. "Critical Temperatures and Pressures of Hydrocarbons" *AIChE Journal* **2009**, **4**, 3, 356-361.

[61] Marrero-Morejon, J. and Pardillo-Fontdevila, E. "Estimation of Pure Compound Properties using Group-Interaction Contributions" *AIChE Journal* **1999**, **45**, 3, 615-621.

[62] Coniglio, L., Tracy, L. and Rauzy, E. "Estimation of Thermophysical Properties of Heavy Hydrocarbons through a Group Contribution Based Equation of State" *Industrial & Engineering Chemistry Research* **2000**, 39, 5037-5048.

[63] Abdoul, W., Rauzy, E. and Peneloux, A. "Group-Contribution Equation of State for Correlating and Predicting Thermodynamic Properties of Weakly Polar and Non-Associating Mixtures; Binary and Multicomponent Systems" *Fluid Phase Equilibria* **1991**, 68, 47-102.

[64] Jaubert, J. N. and Mutelet, F. "VLE Predictions with the Peng-Robinson Equation of State and Temperature Dependent kij Calculated through a Group Contribution Method" *Fluid Phase Equilibria* **2004**, 224, 285-304.

[65] Jaubert, J. N., Vitu, S., Mutelet, F. and Corriou, J. P. "Extension of the PPR78 Model (Predictive 1978, Peng-Robinson EOS with Temperature Dependent k(ij) Calculated through a Group Contribution Method) to Systems containing Aromatic Compounds" *Fluid Phase Equilibria* **2005**, **237**, 1-2, 193-211.

[66] Vitu, S., Jaubert, J. N. and Mutelet, F. "Extension of the PPR78 Model (Predictive 1978, Peng-Robinson EOS with Temperature Dependent kij Calculated through a Group Contribution Method) to Systems containing Naphtenic Compounds" *Fluid Phase Equilibria* **2006**, **243**, 1-2, 9-28.

[67] Privat, R., Mutelet, F. and Jaubert, J. N. "Addition of the Hydrogen Sulfide Group to the PPR78 Model (Predictive 1978, Peng-Robinson Equation of State with Temperature Dependent kij Calculated through a Group Contribution Method)" *Industrial & Engineering Chemistry Research* **2008**, **47**, 24, 10041-10052.

[68] Privat, R., Jaubert, J. N. and Mutelet, F. "Addition of the Nitrogen Group to the PPR78 Model (Predictive 1978, Peng Robinson EOS with Temperature-Dependent k(ij) Calculated through a Group Contribution Method)" *Industrial & Engineering Chemistry Research* **2008**, **47**, 6, 2033-2048.

[69] Privat, R., Jaubert, J. N. and Mutelet, F. "Addition of the Sulfhydryl Group (-SH) to the PPR78 Model (Predictive 1978, Peng-Robinson EOS with Temperature Dependent kij Calculated through a Group Contribution Method)" *Journal of Chemical Thermodynamics* **2008**, **40**, 9, 1331-1341.

[70] Vitu, S., Privat, R., Jaubert, J. N. and Mutelet, F. "Predicting the Phase Equilibria of CO_2 + Hydrocarbon Systems with the PPR78 Model (PR EOS and k(ij) Calculated through a Group Contribution Method)" *Journal of Supercritical Fluids* **2008**, **45**, 1, 1-26.

[71] Privat, R., Jaubert, J. N. and Mutelet, F. "Use of the PPR78 Model to Predict New Equilibrium Data of Binary Systems involving Hydrocarbons and Nitrogen. Comparison with other GCEOS" *Industrial & Engineering Chemistry Research* **2008**, **47**, 19, 7483-7489.

[72] Tamouza, S., Passarello, J. P., Tobaly, P. and de Hemptinne, J. C. "Group Contribution Method with SAFT EOS Applied to Vapor Liquid Equilibria of Various Hydrocarbon Series" *Fluid Phase Equilibria* **2004**, 222-223, 67-76.

[73] Tamouza, S., Passarello, J. P., Tobaly, P. and de Hemptinne, J. C. "Application to Binary Mixtures of a Group Contribution SAFT EOS" *Fluid Phase Equilibria* **2005**, 228-229, 409-419.

[74] Tihic, A., Kontogeorgis, G. M. V., von Solms, N., Michelsen, M. L. and Constantinou, L. "A Predictive Group-Contribution Simplified PC-SAFT Equation of State: Application to Polymer Systems" *Industrial & Engineering Chemistry Research* **2008**, **47**, 15, 5092-5101.

[75] Vijande, J., Pineiro, M. M., Bessieres, D., Saint-Guirons, H. and Legido, J. L. "Description of PVT Behaviour of Hydrofluoroethers using the PC-SAFT EOS" *Physical Chemistry Chemical Physics* **2004**, **6**, 4, 766-770.

[76] Lymperiadis, A., Adjiman, C. S., Galindo, A. and Jackson, G. "A Group Contribution Method for Associating Chain Molecules Based on the Statistical Associating Fluid Theory (SAFT-gamma)" *Journal of Chemical Physics* **2007**, **127**, 23.

[77] High, M. S. and Danner, R. P. "A Group Contribution Equation of State for Polymer Solutions" *Fluid Phase Equilibria* **1989**, 53, 323-330.

[78] Danner, R. P., Hamedi, M. and Lee, B. C. "Applications of the Group-Contribution, Lattice-Fluid Equation of State" *Fluid Phase Equilibria* **2002**, 194, 619-639.

[79] Skjold-Jorgensen, S. "Gas Solubility Calculations II Application of a New Group-Contribution Equation of State" *Fluid Phase Equilibria* **1984**, **16**, 3, 317-351.

[80] Gros, H. P., Bottini, S. and Brignole, E. A. "A Group Contribution Equation of State for Associating Mixtures" *Fluid Phase Equilibria* **1996**, **116**, 1-2, 537-544.

[81] Fredenslund, A., Jones, R. L. and Prausnitz, J. M. "Group Contribution Estimation of Activity Coefficients in Nonideal Liquid Mixtures" *AIChE Journal* **1975**, **21**, 6, 1089-1099.

[82] Fredenslund, A., Gmehling, J., Michelsen, M. L., Rasmussen, P. and Prausnitz, J. M. "Computerized Design of Multicomponent Distillation-Columns Using Unifac Group Contribution Method for Calculation of Activity-Coefficients" *Industrial & Engineering Chemistry Process Design and Development* **1977**, **16**, 4, 450-462.

[83] Kehiaian, H. V. "Thermodynamics of Binary-Liquid Organic Mixtures" *Pure and Applied Chemistry* **1985**, **57**, 1, 15-30.

[84] Kehiaian, H. V. "Group Contribution Methods for Liquid-Mixtures – A Critical-Review" *Fluid Phase Equilibria* **1983**, **13**, 10, 243-252.

[85] Derr, E. L. and Deal, C. H. "Analytical Solution of Groups: Correlation of Activity Coefficients Through Structural Group Parameters" *Institution of Chemical Engineers, Symposium Series* **1969**, 32, 44-51.

[86] Brennan, R. A., Nirmalakhandan, N. and Speece, R. E. "Comparison of Predictive Methods for Henrys Law Coefficients of Organic Chemicals" *Water Research* **1998**, **32**, 6, 1901-1911.

[87] Plyasunov, A. V. and Shock, E. L. "Group Contribution Values of the Infinite Dilution Thermo-dynamic Functions of Hydration for Aliphatic Noncyclic Hydrocarbons, Alcohols, and Ketones at 298.15 K and 0.1 MPa" *Journal of Chemical and Engineering Data* **2001**, **46**, 5, 1016-1019.

[88] Plyasunov, A. V., Plyasunova, N. V. and Shock, E. L. "Group Contribution Values for the Ther-modynamic Functions of Hydration at 298.15 K, 0.1 MPa. 4. Aliphatic Nitriles and Dinitriles" *Journal of Chemical and Engineering Data* **2006**, **51**, 5, 1481-1490.

[89] Plyasunov, A. V., Plyasunova, N. V. and Shock, E. L. "Group Contribution Values for the Ther-modynamic Functions of Hydration at 298.15 K, 0.1 MPa. 3. Aliphatic Monoethers, Diethers, and Polyethers" *Journal of Chemical and Engineering Data* **2006**, **51**, 1, 276-290.

[90] Lin, S. T. and Sandler, S. I. "Prediction of Octanol-Water Partition Coefficients using a Group Contribution Solvation Model" *Industrial & Engineering Chemistry Research* **1999**, **38**, 10, 4081-4091.

[91] Pedersen, K. S. and Christiansen, S. "Phase Behavior of Petroleum Reservoir Fluids"; CRC Press: Boca Raton, **2006**.

[92] Daubert, T. E. and Danner, R. P. "API Technical Data Book – Petroleum Refining, 6th ed. 6"; American Petrolum Institute: Washington D.C., **1997**.

[93] Daubert, T. E. "Petroleum Fraction Distillation Interconversions" *Hydrocarbon Processing* **1994**, **73**, 9, 75-78.

[94] Riazi, M. R. and Daubert T. E. "Analytical Correlations Interconvert Distillation Curve Types" *Oil & Gas Journal* **1986**, 84, 50-57.

[95] Wauquier, J. P. "Pétrole brut: Produits pétroliers; Schémas de fabrication", 1st Ed.; Editions Technip: Paris, **1994**.

[96] Wuithier, P. Le Pétrole: Raffinage et Génie chimique (Tome 1), Ed. Technip: Paris, **1972**.

[97] Winn, F. W. "Physical Properties by Nomogram" *Petroleum Refiner* **1957**, 36, 157-159.

[98] Cavett, R. H. "Physical Data for Distillation Calculations-Vapor-Liquid Equilibria", 27th Mid year Meeting, API Division of Refining, San Francisco, CA, May 15, **1965**.

[99] Kesler, M. G. and Lee, B. I. "Improve Prediction of Enthalpy of Fractions" *Hydrocarbon Processing* **1976**, **55**, 3, 153-158.

[100] Twu, C. H. "An Internally Consistent Correlation for Predicting the Critical Properties and Molecular Weights of Petroleum and Coal-tar Liquids" *Fluid Phase Equilibria* **1984**, **16**, 2, 137-150.

[101] API Technical Data Book – Petroleum Refining, 3rd Ed. 6; American Petroleum Institute: Washington, D.C., **1976**.

[102] Montel, F. and Gouel, P. L. "A New Lumping Scheme of Analytical Data for Compositional Studes" SPE: Houston, Tx, **1984**.

[103] Ruffier-Meray, V., Barreau, A. and Behar, E. "Optimal Reduction of the Analytical Data Necessary for the Thermodynamic Characterization of Natural Gases" **1990** SPE 20769; Houston, Tx.

[104] Leibovici, C. F. "A Consistent Procedure for the Estimation of Properties Associated to Lumped Systems" *Fluid Phase Equilibria* **1993**, **87**, 2, 189-197.

[105] Leibovici, C. F., Stenby, E. H. and Knudsen, K. "A Consistent Procedure for Pseudo-Component Delumping" *Fluid Phase Equilibria* **1996**, **117**, 1-2, 225-232.

[106] Leibovici, C. F., Barker, J. W. and Watche, D. "A Method for Delumping the Results of a Compositional Reservoir Simulation" Society of Petroleum Engineers **1998**, SPE 49068.

[107] Nichita, D. V., Broseta, D. and Leibovici, C. F. "Consistent Delumping of Multiphase Flash Results" *Computers & Chemical Engineering* **2006**, **30**, 6-7, 1026-1037.

[108] Nichita, D. V. and Leibovici, C. F. "An Analytical Consistent Pseudo-Component Delumping Procedure for Equations of State with Non-Zero Binary Interaction Parameters" *Fluid Phase Equilibria* **2006**, **245**, 1, 71-82.

[109] Nichita, D. V., Broseta, D. and Leibovici, C. F. "Reservoir Fluid Applications of a Pseudo-Component Delumping New Analytical Procedure" *Journal of Petroleum Science and Engineering* **2007**, **59**, 1-2, 59-72.

[110] Zbogar, A., Vidal da Silva Lopes, F. and Kontogeorgis, G. M. V. "Approach Suitable for Screening Estimation Methods for Critical Properties of Heavy Compounds" *Industrial & Engineering Chemistry Research* **2006**, **45**, 1, 476-480.

[111] Kontogeorgis, G. M. and Tassios, D. P. "Critical Constants and Acentric Factors for Long-Chain Alkanes Suitable for Corresponding States Applications. A Critical Review" *Chemical Engineering Journal* **1997**, **66**, 1, 35-49.

[112] Ungerer, P., Nieto-Draghi, C., Lachet, V., Wender, A., Di Lella, A., Boutin, A., Rousseau, B. and Fuchs, A. H. "Molecular Simulation Applied to Fluid Properties in the Oil and Gas Industry" *Molecular Simulation* **2007**, **33**, 4-5, 287-304.

[113] Ungerer, P., Nieto-Draghi, C., Rousseau, B., Ahunbay, G. and Lachet, V. "Molecular Simulation of the Thermophysical Properties of Fluids: From Understanding toward Quantitative Predictions" *Journal of Molecular Liquids* **2007**, **134**, 1-3, 71-89.

[114] Wilsak, R. A. and Thodos, G. "Critical-Assessment of 4 Vapor-Pressure Functions Over the Complete Vapor Liquid Coexistence Region" *Industrial & Engineering Chemistry Fundamentals* **1984**, **23**, 1, 75-82.

[115] Weir, R. D. and de Loos, T. "Measurement of the Thermodynamic Properties of Multiple Phases" 1st Ed.; Elsevier, **2005**.

[116] Frenkel, M., Chirico, R. D., Diky, V., Yan, X., Dong, Q. and Muzny, C. "ThermoData Engine (TDE): Software Implementation of the Dynamic Data Evaluation Concept" *Journal of Chemical Information and Modeling* **2005**, 45, 816-838.

[117] Clever, H. L., Young, C. L., Battino, R., Hayduck, W. and Wiesenburg, D. A. *Methane* 1st Ed.; Pergamon Press, **1987**.

[118] Hefter, G. T. and Tomkins, R. P. T. "The Experimental Determination of Solubilities" paperback; Wiley, **2003**.

[119] Prausnitz J. M., Lichtenthaler R. N., Gomes de Azevedo E.,"Molecular Thermodynamics of Fluid Phase Equilibria", 3rd Ed., Prentice Hall Int., **1999**.

[120] Van Ness, H. C., Byer, S. M. and Gibbs, R. E. "Vapor-Liquid Equilibrium: Part I. An Appraisal of Data Reduction Methods" *AIChE Journal* **1973**, **19**, 2, 238-244.

[121] Van Ness, H. C. "Thermodynamics in the Treatment of Vapor/Liquid Equilibrium (VLE) Data" *Pure and Applied Chemistry* **1995**, **67**, 6, 859-872.

[122] Fredenslund, A., Gmehling, J. and Rasmussen, P. "Vapor-Liquid Equilibria using UNIFAC-A Group Contribution Method"; Elsevier Scientific Publishing Company: New York, **1977**.

[123] Treybal, R. E. "Liquid Extraction"; McGraw-Hill Book Co: New York, **1963**.

[124] Othmer, D. F. and Tobias, P. E. "Liquid – Liquid Extraction Data – Toluene and Acetaldehyde Systems" *Industrial & Engineering Chemistry* **1942**, **34**, 6, 690-692.

[125] Hand, D. B. "Dineric Distribution" *The Journal of Physical Chemistry* **1930**, **34**, 9, 1961-2000.

[126] Englezos, P. and Kalogerakis, N. "Applied Parameter Estimation for Chemical Engineers" 1st Ed.; Marcel Dekker, Inc.: New York, Basel, **2001**.

[127] Ledanois, J. M., Lopez A.L., Pimentel J.A. and Pironti F. "Métodos Numéricos Aplicados en Ingeniería"; McGraw-Hill Interamericana: Caracas, **2000**.

[128] Chapra, S. C. and Canale, R. P. "Numerical Methods for Engineers" 5th. Ed.; McGraw-Hill International, **2006**.

[129] Clark, G. N. I., Haslam, A. J., Galindo, A. and Jackson, G. "Developing Optimal Wertheim-Like Models of Water for Use in Statistical Associating Fluid Theory (SAFT) and Related Approaches" *Molecular Physics* **2006**, **104**, 22-24, 3561-3581.

[130] Lopez, J. A., Trejos, V. M. and Cardona, C. A. "Objective Functions Analysis in the Minimization of Binary VLE Data for Asymmetric Mixtures at High Pressures" *Fluid Phase Equilibria* **2006**, **248**, 2, 147-157.

[131] Marquardt, D. W. "An Algorithm for Least-Squares Estimation of Nonlinear Parameters" *SIAM Journal of Applied Mathematics* **1963**, 11, 2, 431-441.

[132] Levenberg, K. "A Method for the Solution of Certain Non-Linear Problems in Least Squares" *Quarterly Journal of Applied Mathematics* **1944**, 11, 2, 164-168.

[133] Girard, A. "Calcul automatique en optique géométrique" *Revue d'Optique Théorique et Intrumentale* **1958**, 37, 225-241.

[134] Renon H., Asselineau L., Cohen G. "Calcul sur ordinateur des équilibres liquide-vapeur et liquide-liquide: application à la distillation des mélanges non idéaux et à l'extraction par solvants"; Editions Technip: Paris, **1971**.

[135] Touhara, H., Okazaki, S., Okino, F., Tanaka, H., Ikari, K. and Nakanishi, K. "Thermodynamic Properties of Aqueous Mixtures of Hydrophilic Compounds 2. Aminoethanol and Its Methyl-Derivatives" *Journal of Chemical Thermodynamics* **1982**, 14, 2, 145-156.

[136] Nath, A. and Bender, E. "Isothermal Vapor Liquid Equilibria of Binary and Ternary Mixtures Containing Alcohol, Alkanolamine, and Water with A New Static Device" *Journal of Chemical and Engineering Data* **1983**, **28**, 4, 370-375.

[137] Buslaeva, M. N., Tsetkov, V. B., Markova, V. B. and Kaimin, I. F. "Thermochemical Study of the Donator Capability of Monoethanolamine in Solutions" *Koordinatsionnaya Khimiya* **1983**, **9**, 6, 752-754.

[138] Dohnal, V., Roux, A. H. and Hynek, V. "Limiting Partial Molar Excess-Enthalpies by Flow Calorimetry – Some Organic-Solvents in Water" *Journal of Solution Chemistry* **1994**, **23**, 8, 889-900.

[139] Lenard, J. L., Rousseau, R. W. and Teja, A. S. "Vapor-Liquid Equilibria for Mixtures of 2-Aminoethanol + Water" *AIChE Symposium Serie* **1990**, **86**, 279, 1-5.

[140] Tochigi, K., Akimoto, K., Ochi, K., Liu, F. Y. and Kawase, Y. "Isothermal Vapor-Liquid Equilibria for Water + 2-Aminoethanol plus Dimethyl Sulfoxide and its Constituent Three Binary Systems" *Journal of Chemical and Engineering Data* **1999**, **44**, 3, 588-590.

[141] Dechema *Detherm* v 1.4.1.0.1, **2005**.

[142] Renon, H., Praunitz, J.M. "Local Composition in Thermodynamic Excess Functions for Liquid Mixtures" *AICHE.J.* **1968**, 14, 135-144.

[143] Schiller, M. "Measurement of Phase Equilibria and Their Prediction by a Modified UNIFAC Model", Universitaet Dortmund, **1988**.

[144] M. L. Michelsen "The Isothermal Flash Problem .1. Stability", *Fluid Phase Equilibria* **1982**, 9, 1, 1-19.

[145] M. L. Michelsen "The Isothermal Flash Problem .2. Phase-Split Calculation", *Fluid Phase Equilibria* **1982**, 9, 1, 21-40.

[146] M. L. Michelsen, J. Mollerup, Thermodynamic Models: "Fundamental and Computational Aspects", 1st Ed., Tie-Line Publications, **2004**.

[147] de Swaan Arons, J "The Systematic Study of Phase Behaviour and the Emerging Coherence of Phenomena" *Fluid Phase Equilibria* **1995**, 104, 97-118.

[148] Prausnitz, J. M., Lichtenthaler, R. N. and Gomes de Azevedo, E. "Molecular Thermodynamics of Fluid Phase Equilibria" 3rd Ed.; Prentice Hall Int., **1999**.

[149] Sandler, K. E. "Chemical and Engineering Thermodynamics" 3rd Ed.; J. Wiley & Sons: New York, **1999**.

[150] Kontogeorgis, G. M. and Folas, G. K. "Thermodynamic Models"; Elsevier, **2009**.

[151] Sum, A. K. and Sandler, S. I. "Use of ab initio Methods to Make Phase Equilibria Predictions using Activity Coefficient Models" *Fluid Phase Equilibria* **1999**, 160, 375-380.

[152] Garrison, S. L. and Sandler, S. I. "On the Use of ab initio Interaction Energies for the Accurate Calculation of Thermodynamic Properties" *Journal of Chemical Physics* **2002**, **117**, 23, 10571-10580.

[153] Klamt, A. "COSMO-RS From Quantum Chemistry to Fluid Phase Thermodynamics and Drug Design" 1st Ed.; Elsevier: Amsterdam, **2005**.

[154] Marsh, K. "COSMO-RS From Quantum Chemistry to Fluid Phase Thermodynamics and Drug Design" Book Review; *Journal of Chemical and Engineering Data* **2006**, **51**, 4, 1480.

[155] Scatchard, G. "Equilibria in Non-electrolyte Solutions in Relation to the Vapor Pressures and Densities of the Components" *Chemical Reviews* **1931**, **8**, 2, 321-333.

[156] Scatchard, G. "Interatomic Forces in Binary Alloys" *Journal of the American Chemical Society* **1931**, **53**, 8, 3186-3188.

[157] Hildebrand, J. H. "Solubility" *Journal of the American Chemical Society* **1916**, **38**, 8, 1452-1473.

[158] Hildebrand, J. H. and Wood, S. E. "The Derivation of Equations for Regular Solutions" *The Journal of Chemical Physics* **1933**, **1**, 12, 817-822.

[159] Scatchard, G. and Hildebrand, J. H. "Non-Electrolyte Solutions" *Journal of the American Chemical Society* **1934**, **56**, 4, 995-996.

[160] Alboudwarej, H., Akbarzadeh, K., Beck, J., Svrcek, W. Y. and Yarranton, H. W. "Regular Solution Model for Asphaltene Precipitation from Bitumens and Solvents" *AIChE Journal* **2003**, **49**, 11, 2948-2956.

[161] Akbarzadeh, K., Alboudwarej, H., Svrcek, W. Y. and Yarranton, H. W. "A Generalized Regular Solution Model for Asphaltene Precipitation from *n*-alkane Diluted Heavy Oils and Bitumens" *Fluid Phase Equilibria* **2005**, **232**, 1-2, 159-170.

[162] Yarranton, H. W., Fox, W. A. and Svrcek, W. Y. "Effect of Resins on Asphaltene Self-Association and Solubility" *Canadian Journal of Chemical Engineering* **2007**, **85**, 5, 635-642.

[163] Wilson, G. M. "Vapor-Liquid Equilibrium. XI. A New Expression for the Excess Free Energy of Mixing" *Journal of the American Chemical Society* **1964**, **86**, 2, 127-130.

[164] Elliott, J. R. and Lira, C. T. "Introductory Chemical Engineering Thermodynamics"; Prentice Hall PTR: Upper Saddle River, NJ, **1999**.

[165] Wertheim, M. S. "Fluids with Highly Directional Attractive Forces" *Journal of Statistical Physics* **1984**, 35, 35-47.

[166] Wertheim, M. S. "Fluids of Dimerizing Hard Spheres, and Fluid Mixtures of Hard Spheres and Dispheres" *Journal of Statistical Physics* **1986**, 85, 2929.

[167] Mengarelli, A. C., Brignole, E. A. and Bottini, S. B. "Activity Coefficients of Associating Mixtures by Group Contribution" *Fluid Phase Equilibria* **1999**, **163**, 2, 195-207.

[168] Ferreira, O., Macedo, E. A. and Bottini, S. B. "Extension of the A-UNIFAC Model to Mixtures of Cross- and Self-associating Compounds" *Fluid Phase Equilibria* **2005**, **227**, 2, 165-176.

[169] Flory, P. J. "Thermodynamics of Polymer-Solutions" *Abstracts of Papers of the American Chemical Society* **1976**, 12-12.

[170] Staverman, A. J. "The Entropy of Polymer Solutions" *Recueil des Travaux Chimiques des Pays-Bas et de la Belgique* **1950**, 69, 163-174.

[171] Guggenheim, E. A. "Mixtures"; Oxford University Press: Oxford, **1952**.

[172] Patterson, D. "Free Volume and Polymer Solubility. A Qualitative Review" *Macromolecules* **1969**, 2, 672.

[173] Elbro, H. S., Chen, F., Rasmussen, P. and Fredenslund, A. "Some Recent Developments in Group-Contribution Models for Polymer-Solutions" *Abstracts of Papers of the American Chemical Society* **1990**, 199, 22-IAEC.

[174] Elbro, H. S., Fredenslund, A. and Rasmussen, P. "Group Contribution Method for the Prediction of Liquid Densities As A Function of Temperature for Solvents, Oligomers, and Polymers" *Industrial & Engineering Chemistry Research* **1991**, **30**, 12, 2576-2582.

[175] Bogdanic, G. and Fredenslund, A. "Revision of the Group-Contribution Flory Equation of State for Phase-Equilibria Calculations in Mixtures with Polymers .1. Prediction of Vapor-Liquid-Equilibria for Polymer-Solutions" *Industrial & Engineering Chemistry Research* **1994**, **33**, 5, 1331-1340.

[176] Saraiva, A., Bogdanic, G. and Fredenslund, A. "Revision of the Group-Contribution Flory Equation of State for Phase-Equilibria Calculations in Mixtures with Polymers .2. Prediction of Liquid-Liquid Equilibria for Polymer-Solutions" *Industrial & Engineering Chemistry Research* **1995**, **34**, 5, 1835-1841.

[177] Redlich, O. and Kister, A. T. "Thermodynamics of Nonelectrolyte Solutions -x-y-t in a Binary System" *Industrial & Engineering Chemistry* **1948**, **40**, 2, 341-345.

[178] Huggins, M. L. "Solutions of Long Chains Compounds" *Journal of Chemical Physics* **1941**, **9**, 5, 440-440.

[179] Flory P. J. "Thermodynamics of High Polymer Solutions" *Journal of Chemical Physics* **1942**, **10**, 51-61.

[180] Flory P. J. "Thermodynamics of High Polymer Solutions" *Journal of Chemical Physics* **1941**, **9**, 8, 660-661.

[181] Huggins, M. L. "Theory of Solutions of High Polymers" *Journal of the American Chemical Society* **1942**, **64**, 7, 1712-1719.

[182] Abrams, D. S. and Prausnitz, J. M. "Statistical Thermodynamics of Mixtures: a New Expression for the Excess Gibbs Free Energy of Partly or Completely Miscible Systems" *AIChE Journal* **1975**, 21, 116-128.

[183] Fredenslund, A. and Rasmussen, P. "Correlation of Pure Component Gibbs Energy Using Unifac Group Contribution" *AIChE Journal* **1979**, **25**, 1, 203-205.

[184] Magnussen, T., Rasmussen, P. and Fredenslund, A. "Unifac Parameter Table for Prediction of Liquid-Liquid Equilibria" *Industrial & Engineering Chemistry Process Design and Development* **1981**, **20**, 2, 331-339.

[185] Gmehling, J., Rasmussen, P. and Fredenslund, A. "Vapor-Liquid-Equilibria by Unifac Group Contribution – Revision and Extension .2" *Industrial & Engineering Chemistry Process Design and Development* **1982**, **21**, 1, 118-127.

[186] Tiegs, D., Gmehling, J., Rasmussen, P. and Fredenslund, A. "Vapor-Liquid-Equilibria by Unifac Group Contribution .4. Revision and Extension" *Industrial & Engineering Chemistry Research* **1987**, **26**, 1, 159-161.

[187] Larsen, B. L., Rasmussen, P. and Fredenslund, A. "A Modified UNIFAC Group-Contribution Model for Prediction of Phase Equilibria and Heats of Mixing" *Industrial & Engineering Chemistry Research* **1987**, **26**, 11, 2274-2286.

[188] Hansen, H. K., Rasmussen, P., Fredenslund, A., Schiller, M. and Gmehling, J. "Vapor-Liquid-Equilibria by Unifac Group Contribution .5. Revision and Extension" *Industrial & Engineering Chemistry Research* **1991**, **30**, 10, 2352-2355.

[189] Kontogeorgis, G. M., Fredenslund, A. and Tassios, D. P. "Simple Activity-Coefficient Model for the Prediction of Solvent Activities in Polymer-Solutions" *Industrial & Engineering Chemistry Research* **1993**, **32**, 2, 362-372.

[190] Weidlich V., Gmehling J. "A Modified UNIFAC Model 1. Prediction of VLE, R^E and γ^{∞}" *Industrial & Engineering and Chemistry Research* **1987**, **26**, 7, 1372-1381.

[191] Gmehling J., Li J., Schiller M. "A Modified UNIFAC Model: 2 Present Parameter Matrix and Results for Different Thermodynamic Properties" *Ind. Eng. Chem. Res.* **1993**, **32**, 1, 178-193.

[192] Kikic, I., Fermeglia, M. and Rasmussen, P. "Unifac Prediction of Vapor-Liquid-Equilibria in Mixed-Solvent Salt Systems" *Chemical Engineering Science* **1991**, **46**, 11, 2775-2780.

[193] Zemaitis, J. F., Clark, D. M., Rafal, M. and Scrivner, N. C. "Handbook of Aqueous Electrolyte Thermodynamics"; Design Institute for Physical Properties (DIPPR), **1986**.

[194] Rafal, M., Bethold, J. W. and Scrivner, N. C. "Models for Electrolyte Solutions" in *Models for Thermodynamics and Phase Equilibria Calculations*; Ed. Sandler, S. I., Marcel Dekker, Inc, NY: New York, **1994**.

[195] Chen, C. C. and Evans, L. B. "A Local Composition Model for the Excess Gibbs Energy of Aqueous Electrolyte Systems" *AIChE Journal* **1986**, **32**, 3, 444-454.

[196] Sander, B., Fredenslund, A. and Rasmussen, P. "Calculation of Vapor-Liquid-Equilibria in Mixed-Solvent Salt Systems Using An Extended Uniquac Equation" *Chemical Engineering Science* **1986**, **41**, 5, 1171-1183.

[197] Li, J. D., Polka, H. M. and Gmehling, J. "A G(E) Model for Single and Mixed-Solvent Electrolyte Systems .1. Model and Results for Strong Electrolytes" *Fluid Phase Equilibria* **1994**, 94, 89-114.

[198] Polka, H. M., Li, J. D. and Gmehling, J. "A G(E) Model for Single and Mixed-Solvent Electrolyte Systems .2. Results and Comparison with Other Models" *Fluid Phase Equilibria* **1994**, 94, 115-127.

[199] Pitzer, K. S. "Thermodynamics of Electrolytes I: Theoretical Basis and General Equations" *Journal of Chemical Physics* **1973**, **77**, 2, 268-277.

[200] Pitzer, K. S. and Mayorga, G. "Thermodynamics of Electrolytes I. Activity and Osmotic Coefficients for Strong Electrolytes with One or Both Ions Univalent" *Journal of Chemical Physics.* **1973**, **77**, 19, 2300.

[201] Debye, P. and Hückel, E. "Zur Theorie der Elektrolyte I: Gefrierpunktserniedrigung und verwandte Erscheinungen" *Phys.Z.* **1923**, **24**, 9, 185-207.

[202] Debye, P. and Hückel, E. "Zur Theorie der Elektrolyte II: das Grensgesetz für die elektrische Leitfähigkeit" *Phys.Z.* **1923**, **24**, 15, 305-325.

[203] Pitzer, K. S. "Electrolytes. From Dilute Solutions to Fused Salts" *Journal of American Chemical Society* **1980**, **102**, 9, 2902-2906.

[204] van der Waals, J. D. "Over de Continuiteit van den Gas en Vloeistoftestand (On the Continuity of the Gas and Liquid State)", Hoogeschool te Leiden, **1873**.

[205] Wei, Y. S. and Sadus, R. J. "Equation of State for the Calculation of Fluid Phase Equilibria" *AIChE Journal* **2000**, **46**, 1, 169.

[206] Valderrama, J. O. "The State of the Cubic Equations of State" *Industrial & Engineering Chemistry Research* **2003**, **42**, 8, 1603-1618.

[207] Muller, E. A. and Gubbins, K. E. "Molecular-Based Equations of State for Associating Fluids: A Review of SAFT and Related Approaches" *Industrial & Engineering Chemistry Research* **2001**, 40, 2193-2211.

[208] Sengers, J. V., Kayser, R. F., Peters, C. J. and White, H. J. Jr "Equations of State for Fluids and Fluid Mixtures", First Ed.; Elsevier, **2000**.

[209] Nguyen-Huynh, D., Passarello, J. P., Tobaly, P. and de Hemptinne, J. C. "Application of GC-SAFT EOS to Polar Systems using a Segment Approach" *Fluid Phase Equilibria* **2008**, **264**, 1, 62-75.

[210] Carnahan, N. F. and Starling, K. E. "Intermolecular Repulsions and the Equation of State for Fluids" *AIChE Journal* **1972**, **18**, 6, 1184-1189.

[211] Mansoori, G. A., Carnahan, N. F., Starling, K. E. and Leland, T. W. Jr "Equilibrium Thermodynamic Properties of the Mixture of Hard Spheres" *Journal of Chemical Physics* **1971**, **54**, 4, 1523-1525.

[212] Alder, B. J., Young, D. A. and Mark, M. A. "Studies in Molecular Dynamics. X. Corrections to the Augmented van der Waals Theory for the Square Well Fluid" *Journal of Chemical Physics* **1972**, **56**, 6, 3013-3029.

[213] Barker, J. A. and Henderson, D. "Perturbation Theory and Equation of State for Fluids: the Square Well Potential" *Journal of Chemical Physics* **1967**, **47**, 8.

[214] Panayiotou, C. and Sanchez, I. C. "Hydrogen-Bonding in Fluids – An Equation-Of-State Approach" *Journal of Physical Chemistry* **1991**, **95**, 24, 10090-10097.

[215] Panayiotou, C. and Sanchez, I. C. "Statistical Thermodynamics of Associated Polymer-Solutions" *Macromolecules* **1991**, **24**, 23, 6231-6237.

[216] Heidemann, R. A. and Prausnitz, J. M. "Vanderwaals-Type Equation of State for Fluids with Associating Molecules" *Proceedings of the National Academy of Sciences of the United States of America* **1976**, **73**, 6, 1773-1776.

244   ***Chapter 3 • From Components to Models***

[217] Jackson, G., Chapman, W. G. and Gubbins, K. E. "Phase-Equilibria of Associating Fluids – Spherical Molecules with Multiple Bonding Sites" *Molecular Physics* **1988**, **65**, 1, 1-31.

[218] Chapman, W. G., Gubbins, K. E., Jackson, G. and Radosz, M. "New Reference Equation of State for Associating Liquids" *Industrial & Engineering Chemistry Research* **1990**, 29, 1709-1721.

[219] Huang, S. H. and Radosz, M. "Equation of State for Small, Large, Polydisperse and Associating Molecules: Extension to Fluid Mixtures" *Industrial Engineering and Chemistry Research* **1990**, 29, 2284-2294.

[220] Economou, I. G. and Donohue, M. D. "Chemical, Quasi-Chemical and Perturbation Theories for Associating Fluids" *AIChE Journal* **1991**, **37**, 12, 1875-1894.

[221] Gubbins, K. E. and Twu, C. H. "Thermodynamics of Polyatomic Fluid Mixtures – 1 Theory" *Chemical Engineering Science* **1978**, 33, 863-878.

[222] Kraska, T. and Gubbins, K. E. "Phase Equilibria Calculations with a Modified SAFT Equation of State: 1. Pure Alkanes, Alcohols and Water" *Fluid Phase Equilibria* **1996**, 35, 4727-4737.

[223] Kraska, T. and Gubbins, K. E. "Phase Equilibria calculations with a modified SAFT Equation of State: 2.Binary Mixtures of *n*-alkanes, 1-alcohols and Water" *Fluid Phase Equilibria* **1996**, 35, 4738-4746.

[224] Jog P. and Chapman, W. G. "Application of Wertheim's Thermodynamic Perturbation Theory to Dipolar Hard Sphere Chains" *Molecular Physics* **1999**, **97**, 3, 307-319.

[225] Karakatsani, E. K. and Economou, I. G. "Perturbed Chain-statistical Associating Fluid Theory Extended to Dipolar and Quadrupolar Molecular Fluids" *Journal of Physical Chemistry B* **2006**, **110**, 18, 9252-9261.

[226] Stell, G., Rasaiah, J. C. and Narang, H. "Thermodynamic Perturbation Theory for Simple Polar Fluids" *Molecular Physics* **1974**, **27**, 5, 1391-1414.

[227] Lambert, M. L., Song, Y. and Prausnitz, J. M. "Equations of State for Polymer Systems" in *Equations of state for fluids and fluid mixtures*; Ed. Sengers, J. V., Kayser, R. F., Peters, C. J, and White, H. J. Jr; Experimental Thermodynamics; Elsevier: Amsterdam, **2000**.

[228] Prigogine, I. "The Molecular Theory of Solutions"; North Holland: Amsterdam, **1957**.

[229] Beret, S. and Prausnitz, J. M. "Perturbed Hard-Chain Theory – Equation of State for Fluids Containing Small Or Large Molecules" *AIChE Journal* **1975**, **21**, 6, 1123-1132.

[230] Donohue, M. D. and Prausnitz, J. M. "Perturbed Hard Chain Theory for Fluid Mixtures: Thermodynamic Properties for Mixtures in Natural Gas and Petroleum Technology" *AIChE Journal* **1978**, 24, 849-860.

[231] Chapman, W. G., Jackson, G. and Gubbins, K. E. "Phase Equilibria of Associating Fluids: Chain Molecules with Multiple Bonding Sites" *Molecular Physics* **1988**, **65**, 1057-1079.

[232] Sanchez, I. C. and Panayiotou, C. G. "Equation of State Thermodynamics of Polymer and Related Solutions" in *Models for Thermodynamic and Phase Equilibria Calculations*; Ed. Sandler, S. I.; Chemical Industries; Marcel Dekker, Inc.: New York, **1994**.

[233] Panayiotou, C. and Vera, J. H. "An Improved Lattice-Fluid Equation of State for Pure Component Polymeric Fluids" *Polymer Engineering and Science* **1982**, **22**, 6, 345-348.

[234] Panayiotou, C. and Vera, J. H. "Statistical Thermodynamics of R-Mer Fluids and Their Mixtures" *Polymer Journal* **1982**, **14**, 9, 681-694.

[235] Privat, R., Privat, Y. and Jaubert, J. N. "Can Cubic Equations of State Be Recast in the Virial Form?" *Fluid Phase Equilibria* **2009**, **282**, 1, 38-50.

[236] Trusler, J. P. M. "The Virial Equation of State" in *Equations of State for Fluids and Fluid Mixtures*; Ed. Sengers, J. V., Kayser, R. F., Peters, C. J. and White, H. J. Jr; Experimental Thermodynamics series; Elsevier: Amsterdam, **2009**.

[237] Hayden, J. G. and O'Connell, J. P. "Generalized Method for Predicting Second Virial Coefficients" *Industrial & Engineering Chemistry Process Design and Development* **1975**, **14**, 3, 209-216.

[238] Jacobsen, R. T., Penoncello, S. G., Lemmon, E. and Span, R. "Multiparameter Equations of State" in *Equations of State for Fluids and Fluid Mixtures*; Ed. Sengers, J. V., Kayser, R. F., Peters, C. J. and White, H. J. Jr; Experimental Thermodynamics; Elsevier: Amsterdam, **2000**.

[239] Benedict, M., Webb, G. B. and Rubin, L. C "An Empirical Equation for Thermodynamic Properties of Light Hydrocarbons and Their Mixtures: I. Methane, Ethane, Propane, and *n*-Butane" *Journal of Chemical Physics* **1940**, **8**, 4, 334-345.

[240] Cooper, H. W. and Golfrank, J. C. *Hydrocarbon Processing* **1967**, **46**, 12, 141.

[241] Benedict, M., Webb, G. B. and Rubin, L. C "An Empirical Equation for Thermodynamic Properties of Light Hydrocatbons and Their Mixtures: Fugacities and Liquid-Vapor Equilibria (part 1)" *Chemical Engineering Progress* **1951**, **47**, 8, 419-454.

[242] Starling, K. E. "Fluid thermodynamic properties of light petroleum systems"; Gulf Publishing: Houston, TX, **1973**.

[243] Soave, G. "A Noncubic Equation of State for the treatment of Hydrocarbon Fluids at Reservoir Conditions" *Industrial & Engineering Chemistry Research* **1995**, **34**, 11, 3981-3994.

[244] Soave, G. "An Effective Modification of the Benedict Webb Rubin Equation of State" *Fluid Phase Equilibria* **1999**, **164**, 157-172.

[245] Plocker, U. "Calculation of High Pressure Vapor-Liquid Equilibria from a Corresponding States Correlation with Emphasis on Asymmetric Mixtures" *Industrial & Engineering Chemistry Process Design & Development* **1978**, **17**, 3.

[246] de Hemptinne, J. C. and Ungerer, P. "Accuracy of the Volumetric Predictions of Some Important Equations of State for Hydrocarbons, including a Modified Version of the Lee-Kesler Method" *Fluid Phase Equilibria* **1995**, **106**, 1-2, 81-109.

[247] de Hemptinne, J. C., Barreau, A., Ungerer, P. and Behar, E. "Evaluation of Equations of State at High Pressure for Light Hydrocarbons" in *Thermodynamic Modeling and Materials Data Engineering*; Ed. Caliste, J-P, Truyol, A. and Westbrooks, J. H.; Data and Knowledge in a Changing World; Springer, **1998**.

[248] Jacobsen, R. T. and Stewart, R. B. *J.Phys.Chem Ref Data* **1973**, 2, 757.

[249] Schmidt, R. and Wagner, W. "A New Form of the Equation of State for Pure Substances and Its Application to Oxygen" *Fluid Phase Equilibria* **1985**, **19**, 3, 175-200.

[250] Wagner, W. and Span, R. "Special Equations of State for Methane, Argon, and Nitrogen for the Temperature-Range from 270-K to 350-K at Pressures Up to 30-MPa" *International Journal of Thermophysics* **1993**, **14**, 4, 699-725.

[251] Van der Waals "The Equation of State for Gases and Liquids" *Nobel Lectures in Physics* **1873**, 1, 254-265.

[252] Clausius, R. "Ueber das Verhalten der Kohlensaure in Bezug auf Druck, Volumen und Temperatur" *Annalen der Physik und Chemie* **1880**, 3, 337-357.

[253] Clausius, R. "Sur une détermination générale de la tension et du volume des vapeurs saturées" *Comptes rendus hebdomadaires des séances de l'Académie des Sciences* **1881**, 97, 619-625.

[254] Berthelot, D. "Sur les thermomètres à gaz, et sur la réduction de leurs indications à l'échelle absolue des températures"; Gauthier-Villars: Paris, **1903**.

[255] Redlich, O. and Kwong, J. N. S. "On the Thermodynamics of Solutions: V. An Equation of State; Fugacities of Gaseous Solutions" *Chemical Reviews* **1949**, **44**, 1, 233-244.

[256] Soave, G. "Equilibrium Constants for a Modified Redlich-Kwong Equation of State" *Chemical Engineering Science* **1972**, 27, 1197-1203.

[257] Peng, D. Y. and Robinson, D. B. "A New Two-Constant Equation of State" *Industrial & Engineering Chemistry Fundamentals* **1976**, 15, 59-64.

[258] Robinson, D. B. "The Characterization of the Heptanes and Heavier Fractions for the GPA Peng-Robinson Programs" GPA, RR-28, **1978**.

[259] Abbott, M. M. "Cubic Equations of State: An Interpretative Review" *Advances in Chemistry* **1979**, 182, 47-70.

[260] Sandler, S. I. and Orbey, H. "Phase-Equilibrium of Complex-Mixtures with Equations of State – An Update" *Chinese Journal of Chemical Engineering* **1995**, **3**, 1, 39-50.

[261] Dupré, A. "Théorie mécanique de la chaleur"; Gauthier-Villars: Paris, **1869**.

[262] Bian, B. G., Wang, Y. R. and Shi, J. "Parameters for the PR and SRK Equations of State" *Fluid Phase Equilibria* **1992**, 78, 331-334.

[263] Graboski, M. S. and Daubert, T. E. "Modified Soave Equation of State for Phase-Equilibrium Calculations .1. Hydrocarbon Systems" *Industrial & Engineering Chemistry Process Design and Development* **1978**, **17**, 4, 443-448.

[264] Ledanois, J. M., Muller, E. A., Colina, C. M., Gonzalez-Mendizabal, D., Santos, J. W. and Olivera-Fuentes, C. "Correlations for Direct Calculation of Vapor Pressures from Cubic Equations of State" *Industrial & Engineering Chemistry Research* **1998**, **37**, 5, 1673-1678.

[265] Boston, J. F. and Mathias, P. M. "Phase Equilibria in a Third-Generation Process Simulator", Proceedings of the 2nd International Conference on Phase Equilibria and Fluid Properties in the Chemical Process Industries, *West Berlin*, 17-21 March 1980, 823-849.

[266] Mathias, P. M. and Copeman, T. W. "Extension of the Peng-Robinson Equation of State to Complex-Mixtures – Evaluation of the Various Forms of the Local Composition Concept" *Fluid Phase Equilibria* **1983**, **13**, 10, 91-108.

[267] Twu, C. H., Bluck, D., Cunningham, J. R. and Coon, J. E. "A Cubic Equation of State with a New Alpha Function and A New Mixing Rule" *Fluid Phase Equilibria* **1991**, 69, 33-50.

[268] Twu, C. H., Coon, J. E. and Cunningham, J. R. "A New Generalized Alpha-Function for A Cubic Equation of State .1. Peng-Robinson Equation" *Fluid Phase Equilibria* **1995**, **105**, 1, 49-59.

[269] Twu, C. H., Coon, J. E. and Cunningham, J. R. "A New Generalized Alpha-Function for A Cubic Equation of State .2. Redlich-Kwong Equation" *Fluid Phase Equilibria* **1995**, **105**, 1, 61-69.

[270] Neau, E., Hernandez-Garduza, O., Escandell, J., Nicolas, C. and Raspo, I. "The Soave, Twu and Boston-Mathias Alpha Functions in Cubic Equations of State: Part I. Theoretical Analysis of Their Variations According to Temperature" *Fluid Phase Equilibria* **2009**, **276**, 2, 87-93.

[271] Neau, E., Raspo, I., Escandell, J., Nicolas, C. and Hernandez-Garduza, O. "The Soave, Twu and Boston-Mathias Alpha Functions in Cubic Equations of State. Part II. Modeling of Thermodynamic Properties of Pure Compounds" *Fluid Phase Equilibria* **2009**, **276**, 2, 156-164.

[272] Valderrama, J. O. and Alfaro, M. "Liquid Volumes from Generalized Cubic Equations of State: Take it with Care" *Oil & Gas Science and Technology* **2000**, **55**, 5, 523-531.

[273] Peneloux, A., Rauzy, E. and Freze, R. "A Consistent Correction for Redlich-Kwong-Soave Volumes" *Fluid Phase Equilibria* **1982**, **8**, 1, 7-23.

[274] Mathias, P. M., Naheiri, T. and Oh, E. M. "A Density Correction for the Peng-Robinson Equation of State" *Fluid Phase Equilibria* **1989**, **47**, 1, 77-87.

[275] Tsai, J. C. and Chen, Y. P. "Application of a Volume-Translated Peng-Robinson Equation of State on Vapor-Liquid Equilibrium Calculations" *Fluid Phase Equilibria* **1998**, **145**, 2, 193-215.

[276] Fuller, G. G. "A Modified Redlich-Kwong-Soave Equation of State Capable of Representing the Liquid State" *Industrial & Engineering Chemistry Fundamentals* **1976**, **15**, 4, 254-257.

[277] Patel, N. C. and Teja, A. S. "A New Cubic Equation of State for Fluids and Fluid Mixtures" *Chemical Engineering Science* **1982**, **37**, 3, 463-473.

[278] Valderrama, J. O. "A Generalized Patel-Teja Equation of State for Polar and Non-Polar Fluids and Their Mixtures" *Journal of Chemical Engineering of Japan* **1990**, 23, 87.

[279] Martin, J. J. "Equations of State – Applied Thermodynamics Symposium" *Industrial & Engineering Chemistry* **1967**, **59**, 12, 34-52.

[280] Frey, K., Augustine, C., Ciccolini, R. P., Paap, S., Modell, M. and Tester, J. "Volume Translation in Equations of State as a Means of Accurate Property Estimation" *Fluid Phase Equilibria* **2007**, **260**, 2, 316-325.

[281] Twu, C. H. and Chan, H. S. "Rigorously Universal Methodology of Volume Translation for Cubic Equations of State" *Industrial & Engineering Chemistry Research* **2009**, **48**, 12, 5901-5906.

[282] Ungerer, P. and Batut, C. "Prediction of the Volumetric Properties of Hydrocarbons with an Improved Volume Translation Method" *Revue de l'Institut Français du Pétrole* **1997**, **52**, 6, 609-623.

[283] de Sant'Ana, H. B., Ungerer, P. and de Hemptinne, J. C. "Evaluation of an Improved Volume Translation for the Prediction of Hydrocarbon Volumetric Properties" *Fluid Phase Equilibria* **1999**, **154**, 2, 193-204.

[284] Heidemann, R. A. "Excess Free Energy Mixing Rules for Cubic Equations of State" *Fluid Phase Equilibria* **1996**, **116**, 1-2, 454-464.

[285] Hudson, G. H. and McCoubrey, J. C. "Intermolecular Forces between Unlike Molecules. A More Complete Form of the Comining Rules" *Trans.Faraday Soc.* **1960**, 56, 761.

[286] Coutinho, J. A. P., Vlamos, P. M. and Kontogeorgis, G. M. "General Form of the Cross-Energy Parameter of Equations of State" *Industrial & Engineering Chemistry Research* **2000**, **39**, 8, 3076-3082.

[287] Twu, C. H., Coon, J. E., Kusch, M. G. and Harvey, A. H. "Selection of Equation of Stat Models for Process Simulator" Simulations Sciences Inc. Workbook Meeting, **1994**.

[288] Panagiotopoulos, A. Z. and Reid, R. C. "New Mixing Rule for Cubic Equations of State for Highly Polar, Asymmetric Systems" *ACS Symposium Series* **1986**, 300, 571-582.

[289] Michelsen, M. L., Kistenmacher, H., "On Composition Dependent Interaction Cœfficients" *Fluid Phase Equilibria,* **1990**, **58**, 1-2, 229-230.

[290] Huron, M. J. and Vidal, J. "New Mixing Rules in Simple Equations of State for Representing Vapour-Liquid Equilibria of Strongly Non-ideal Mixtures" *Fluid Phase Equilibria* **1979**, **3**, 40, 255-271.

[291] Vidal, J. "Mixing Rules and Excess Properties in Cubic Equations of State" *Chemical Engineering Science* **1978**, **33**, 6, 787-791.

[292] Michelsen, M. L. "A Modified Huron-Vidal Mixing Rule for Cubic Equations of State" *Fluid Phase Equilibria* **1990**, 60, 213-219.

[293] Dahl, S., Fredenslund, A. and Rasmussen, P. "The MHV2 Model – A Unifac-Based Equation of State Model for Prediction of Gas Solubility and Vapor-Liquid-Equilibria at Low and High-Pressures" *Industrial & Engineering Chemistry Research* **1991**, **30**, 8, 1936-1945.

[294] Boukouvalas, C., Spiliotis, N., Coutsikos, P., Tzouvaras, N. and Tassios, D. "Prediction of Vapor-Liquid-Equilibrium with the LCVM Model – A Linear Combination of the Vidal and Michelsen Mixing Rules Coupled with the Original Unifac and the T-mPR Equation of State" *Fluid Phase Equilibria* **1994**, 92, 75-106.

[295] Wong, D. S. H., Orbey, H. and Sandler, S. I. "Equation of State Mixing Rule for Nonideal Mixtures using Available Activity Coefficient Model Parameters and that Allows Extrapolation over Large Ranges of Temperature and Pressure" *Industrial & Engineering Chemistry Research* **1992**, **31**, 8, 2033-2039.

[296] Satyro, M. A. and Trebble, M. A. "On the Applicability of the Sandler-Wong Mixing Rules for the Calculation of Thermodynamic Excess Properties – V^E, H^E, S^E, Cp^E" *Fluid Phase Equilibria* **1996**, **115**, 1-2, 135-164.

[297] Satyro, M. A. and Trebble, M. A. "A Correction to Sandler-Wong Mixing Rules" *Fluid Phase Equilibria* **1998**, **143**, 1-2, 89-98.

[298] Holderbaum, T. and Gmehling, J. "PSRK: A Group Contribution Equation of State Based on UNIFAC" *Fluid Phase Equilibria* **1991**, **70**, 2-3, 251-265.

[299] Ahlers, J., Gmehling, J. "Development of a Universal Group Contribution Equation of State.1. Prediction of Liquid Densities for Pure Compounds with a Volume Translated Peng-Robinson Equation of State" *Fluid Phase Equilibria* **2001**, **191**, 1-2, 177-188.

[300] Ahlers, J., Gmehling, J. "Development of a Universal Group Contribution Equation of State.2. Prediction of Vapor-Liquid Equilibria for Asymmetric Systems" *Industrial & Engineering Chemistry Research,* **2002**, **41**, 14, 3489-3498.

[301] Economou, I. G. "Statistical Associating Fluid Theory: A Succesful Model for the Calculation of Thermodynamic and Phase Equilibrium Properties of Complex Mixtures" *Industrial & Engineering Chemistry Research* **2002**, **41**, 5, 953-962.

[302] Tan, S. P., Adidharma, H. and Radosz, M. "Recent Advances and Applications of Statistical Associating Fluid Theory" *Industrial & Engineering Chemistry Research* **2008**, **47**, 21, 8063-8082.

[303] de Hemptinne, J. C., Mougin, P., Barreau, A., Ruffine, L., Tamouza, S. and Inchekel, R. "Application to Petroleum Engineering of Statistical Thermodynamics – Based Equations of State" *Oil & Gas Science and Technology-Revue de l'Institut Français du Pétrole* **2006**, **61**, 3, 363-386.

[304] Gross, J. and Sadowski, G. "Application of Perturbation Theory to a Hard-Chain Reference Fluid: an Equation of State for Square Well Chains" *Fluid Phase Equilibria* **2000**, **168**, 2, 183-199.

[305] Gross, J. and Sadowski, G. "Perturbed-Chain SAFT: An Equation of State based on a Perturbation Theory for Chain Molecules" *Industrial & Engineering Chemistry Research* **2001**, **40**, 4, 1244-1260.

[306] Gil-Villegas, A., Galindo, A., Whitehead, P. J, Mills, S. J., Jackson, G. and Burgess, A. N "Statistical Associating Fluid Theory for Chain Molecules with Attractive Potentials of Variable Range" *Journal of Chemical Physics* **1997**, **106**, 10, 4168-4186.

[307] Tihic, A., Kontogeorgis, G. M., von Solms, N. and Michelsen, M. L. "Applications of the Simplified Perturbed-Chain SAFT Equation of State using an Extended Parameter Table" *Fluid Phase Equilibria* **2006**, **248**, 1, 29-43.

[308] A. Lymperiadis, C. S. Adjiman, G. Jackson, A. Galindo, "A Generalisation of the SAFT-gamma Group Contribution Method for Groups Comprising Multiple Spherical Segments" *Fluid Phase Equilibria,* **2008**, **274**, 1-2, 85-104.

[309] Kontogeorgis, G. M., Voutsas, E. C., Yakoumis, I. V. and Tassios, D. P. "An Equation of State for Associating Fluids" *Industrial & Engineering Chemistry Research* **1996**, **35**, 11, 4310-4318.

[310] Derawi, S. O., Michelsen, M. L., Kontogeorgis, G. M. and Stenby, E. H. "Application of the CPA Equation of State to Glycol/Hydrocarbons Liquid-Liquid Equilibria" *Fluid Phase Equilibria* **2003**, **209**, 2, 163-184.

[311] Derawi, S. O., Kontogeorgis, G. M., Michelsen, M. L. and Stenby, E. H. "Extension of the Cubic-Plus-association Equation of State to Glycol-Water Cross-Associating Systems" *Industrial & Engineering Chemistry Research* **2003**, **42**, 7, 1470-1477.

[312] Folas, G. K., Berg, O. J., Solbraa, E., Fredheim, A. O., Kontogeorgis, G. M., Michelsen, M. L. and Stenby, E. H. "High-Pressure Vapor-Liquid Equilibria of Systems Containing Ethylene Glycol, Water and Methane – Experimental Measurements and Modeling" *Fluid Phase Equilibria* **2007**, **251**, 1, 52-58.

[313] Folas, G. K., Kontogeorgis, G. M., Michelsen, M. L., Stenby, E. H. and Solbraa, E. "Liquid-Liquid Equilibria for Binary and Ternary Systems containing Glycols, Aromatic Hydrocarbons, and Water: Experimental Measurements and Modeling with the CPA EoS" *Journal of Chemical and Engineering Data* **2006**, **51**, 3, 977-983.

[314] Garrido, N. M., Folas, G. K. and Kontogeorgis, G. M. "Modelling of Phase Equilibria of Glycol ethers Mixtures using an Association Model" *Fluid Phase Equilibria* **2008**, **273**, 1-2, 11-20.

[315] Kontogeorgis, G. M., Yakoumis, I. V., Meijer, H., Hendriks, E. and Moorwood, T. "Multicomponent Phase Equilibrium Calculations for Water-Methanol-Alkane Mixtures" *Fluid Phase Equilibria* **1999**, 160, 201-209.

[316] Lundstrom, C., Michelsen, M. L., Kontogeorgis, G. M., Pedersen, K. S. and Sorensen, H. "Comparison of the SRK and CPA Equations of State for Physical Properties of Water and Methanol" *Fluid Phase Equilibria* **2006**, **247**, 1-2, 149-157.

[317] Voutsas, E. C., Kontogeorgis, G. M., Yakoumis, I. V. and Tassios, D. P. "Correlation of Liquid-Liquid Equilibria for Alcohol/Hydrocarbon Mixtures using the CPA Equation of State" *Fluid Phase Equilibria* **1997**, **132**, 1-2, 61-75.

[318] Yakoumis, I. V., Kontogeorgis, G. M., Voutsas, E. C. and Tassios, D. P. "Vapor-Liquid Equilibria for Alcohol/Hydrocarbon Systems using the CPA Equation of State" *Fluid Phase Equilibria* **1997**, **130**, 1-2, 31-47.

[319] Skjold-Joergensen, S. "Group contribution equation of state (GC-EOS): a Predictive Method for Phase Equilibrium Computations over Wide Ranges of Temperature and Pressures up to 30 MPa" *Industrial & Engineering Chemistry Research* **1988**, **27**, 1, 110-118.

[320] Sanchez, I. C. and Lacombe, R. H. "Statistical Thermodynamics of Polymer-Solutions" *Macromolecules* **1978**, **11**, 6, 1145-1156.

[321] High, M. S. and Danner, R. P. "Application of the Group Contribution Lattice-Fluid EOS to Polymer Solutions" *AIChE Journal* **1990**, **36**, 11, 1625-1632.

[322] Panayiotou, C., Pantoula, M., Stefanis, E., Tsivintzelis, I. and Economou, I. G. "Nonrandom Hydrogen-Bonding Model of Fluids and their Mixtures. 1. Pure Fluids" *Industrial & Engineering Chemistry Research* **2004**, **43**, 20, 6592-6606.

[323] Panayiotou, C., Tsivintzelis, I. and Economou, I. G. "Nonrandom Hydrogen-Bonding Model of Fluids and their Mixtures. 2. Multicomponent Mixtures" *Industrial & Engineering Chemistry Research* **2007**, **46**, 8, 2628-2636.

[324] Grenner, A., Tsivintzelis, I., Economou, I. G., Panayiotou, C. and Kontogeorgis, G. M. "Evaluation of the Nonrandom Hydrogen Bonding (NRHB) Theory and the Simplified Perturbed-Chain-Statistical Associating Fluid Theory (SPC-SAFT). 1. Vapor-Liquid Equilibria" *Industrial & Engineering Chemistry Research* **2008**, **47**, 15, 5636-5650.

[325] Tsivintzelis, I., Grenner, A., Economou, I. G. and Kontogeorgis, G. M. "Evaluation of the Nonrandom Hydrogen Bonding (NRHB) Theory and the Simplified Perturbed-Chain-Statistical Associating Fluid Theory (sPC-SAFT). 2. Liquid-Liquid Equilibria and Prediction of Monomer Fraction in Hydrogen Bonding Systems" *Industrial & Engineering Chemistry Research* **2008**, **47**, 15, 5651-5659.

[326] Soave G. "Direct Calculation of Pure-Compound Vapor-Pressures Through Cubic Equations of State" *Fluid Phase Equilibria,* **1986**, **31**, 2, 203-207.

[327] Schmid, B, Gmehling, J. "Revised Parameters and Typical Results of the VTPR Group Contribution Equation of State" *Fluid Phase Equilibria* **2012**, **317**, 110-126.

[328] Emami, F. S., Vahid, A., Elliott, J. R. and Feyzi, F. "Group Contribution Prediction of Vapor Pressure with Statistical Associating Fluid Theory, Perturbed-Chain Statistical Associating Fluid Theory, and Elliott-Suresh-Donohue Equations of State" *Industrial & Engineering Chemistry Research* **2008**, **47**, 21, 8401-8411.

[329] Singh, M., Leonhard, K. and Lucas, K. "Making Equation of State Models Predictive – Part 1: Quantum Chemical Computation of Molecular Properties" *Fluid Phase Equilibria* **2007**, **258**, 1, 16-28.

[330] Leonhard, K., Van Nhu, N. and Lucas, K. "Making Equation of State Models Predictive – Part 3: Improved Treatment of Multipolar Interactions in a PC-SAFT Based Equation of State" *Journal of Physical Chemistry* **2007**, **111**, 43, 15533-15543.

[331] Leonhard, K., Van Nhu, N. and Lucas, K. "Making Equation of State Models Predictive – Part 2: An improved PCP-SAFT Equation of State" *Fluid Phase Equilibria* **2007**, **258**, 1, 41-50.

[332] Claussen, W. F. "Suggested Structures of Water in Inert Gas Hydrates" *Journal of Chemical Physics* **1951**, **19**, 2, 259-260.

[333] Claussen, W. F. "Suggested Structures of Water in Inert Gas Hydrates – Erratum" *Journal of Chemical Physics* **1951**, **19**, 5, 662-662.

[334] Claussen, W. F. "A Second Water Structure for Inert Gas Hydrates" *Journal of Chemical Physics* **1951**, **19**, 22, 1425-1426.

[335] von Stackelberg, M. and Muller, H. R. "On the Structure of Gas Hydrates" *Journal of Chemical Physics* **1951**, **19**, 10, 1319-1320.

[336] Ripmeester, J. A., Tse, J. S., Ratcliffe, C. I. and Powell, B. M. "A New Clathrate Hydrate Structure" *Nature* **1987**, **325**, 6100, 135-136.

[337] Van der Waals, J. H. and Platteeuw, J. C. "Clathrate solutions" *Advance in Chemical Physics* **1959**, **2**, 1, 1-57.

[338] Parrish, W. R. and Prausnitz, J. M. "Dissociation Pressure of Gas Hydrates Formed by Gas Mixtures" *Industrial & Engineering Chemistry Process Design and Development* **1972**, **11**, 1, 26-35.

[339] Carroll, J. J. "Natural Gas Hydrates: A Guide for Engineers"; Gulf Professional Publishing, **2009**.

[340] Sloan, E. D. and Koh, C. A. "Clathrate Hydrates of Natural Gases" 3rd Ed.; CRC/ Taylor & Francis, **2008**.

[341] Harvey, A. H., Sengers, J. M. H. L. and Tanger, J. C. "Unified Description of Infinite-Dilution Thermodynamic Properties for Aqueous Solutes" *Journal of Physical Chemistry* **1991**, **95**, 2, 932-937.

[342] Grayson, H. G. and Streed, C. W. "Vapor-Liquid Equilibria for High Temperature, High PresSure Hydrogen-Hydrocarbon Systems" 6th World Congress for Petroleum; Frankfurt, **1963**.

[343] Chao, K. C. and Seader, J. D. "A General Correlation of Vapor-Liquid Equilibria in Hydrocarbon Mixtures" *AIChE Journal* **1961**, **7**, 4, 598-605.

[344] Jin, Z. L., Greenkorn, R. A. and Chao, K. C. "Correlation of Vapor-Liquid-Equilibrium Ratio of Hydrogen" *AIChE Journal* **1995**, **41**, 6, 1602-1604.

[345] Curl, R. F. and Pitzer, Kenneth "Volumetric and Thermodynamic Properties of Fluids Enthalpy, Free Energy, and Entropy" *Industrial & Engineering Chemistry* **1958**, **50**, 2, 265-274.

[346] de Hemptinne, J. C. and Ungerer, P. "HP/HT Characterisation of Reservoir Fluids" IFP, Rep. 40 601, **1993**.

[347] Sportisse, M., Barreau, A. and Ungerer, P. "Modeling of Gas Condensates Properties using Continuous Distribution Functions for the Characterisation of the Heavy Fraction" *Fluid Phase Equilibria* **1997**, **139**, 1-2, 255-276.

[348] Cha, T. H. and Prausnitz, J. M. "Thermodynamic Method for Simultaneous Representation of Ternary Vapor Liquid and Liquid Liquid Equilibria" *Industrial & Engineering Chemistry Process Design and Development* **1985**, **24**, 3, 551-555.

[349] de Hemptinne, J. C. "Benzene Crystallization Risks in the LIQUEFIN Natural Gas Process" *Process Safety Progress* **2005**, **24**, 3, 203-212.

[350] Doherty, M. F. and Malone, M. F. "Conceptual Design of Distillation Systems"; McGraw Hill: New York, **2001**.

[351] Kister, A. T. "Distillation Design"; McGraw Hill: New York, **1992**.

4

From Phases to Method (Models) Selection

It is impossible to choose correctly a model if the phase behaviour of the system is unknown. The aim of this fourth chapter is therefore to provide a phenomenological understanding of the various phases that may exist in mixtures of industrial interest, and the behaviour of the properties as a function of pressure and temperature. Obviously, the reader should not expect an exhaustive review, since it would extend far beyond the scope of this book. The interested reader will be invited to consult other documents for further study [1-3].

Generalities about phase behaviour are discussed in section 2.1.3 (p. 27). The phases are often labelled as vapour, liquid and solid. Solid phases are generally easy to differentiate from the fluid phases because of their specific crystalline character. Amorphous (*i.e.* highly viscous, but non-crystalline) solids may be treated as liquids, and will not be considered here. It may not always be obvious to distinguish vapour and liquid phases. The best criterion considers the density of the phase and its compressibility with respect to pressure. The critical vapour-liquid zone may require special attention since the mathematical expressions used for property calculations must display continuous behaviour.

In this chapter, we intend to present the recommended thermodynamic methods, in the sense that was introduced in the first chapter: a method is a combination of models that provide a final numerical value for the requested property(-ies), possibly through the use of an algorithmic computation.

The recommendations depend on the phases at hand. This is why these various phases will be first defined in somewhat more details. The fluid property behaviour within these phases will be discussed. The recommended calculation method for single phase properties will be provided, both for pure components as for mixtures. On this subject, it is important to mention the well-known book of Poling *et al.* 2001 [4] and the previous versions of Reid *et al.* [5, 6], who investigate and recommend a large number of methods in very specific conditions.

In the second section of this chapter, the type of phase equilibrium behaviour will be reviewed considering the main components in the mixture. For each subsection, the conditions of occurrence of the various phase equilibria will be mentioned; the order of magnitude of the solubilities will be given and some recommended models will be proposed.

In a concluding section, some general guidelines are provided, including a decision tree that covers the most frequently encountered situations.

4.1 SINGLE PHASE PROPERTIES

Single phase properties can be plotted as a function of temperature and pressure only if the composition is fixed.

> Whether dealing with a pure component or a mixture, the general shape of these plots is almost identical.

The behaviour of carbon dioxide is chosen as an example in this section. For the fluid phase, the corresponding states principle (presented in section 2.2.2.1.C, p. 57) is often used in order to calculate single phase properties, based on the sole knowledge of critical temperature, critical pressure and acentric factor. This principle is very powerful for non-polar systems, but can only be used as a guideline for polar components. High quality plots for a large number of often-used components and some mixtures can be produced using the "Refprop" package available at NIST (www.nist.gov/srd/nist23.htm).

4.1.1 PVT plot

The easiest property to discuss is the volume. Figure 4.1 illustrates the behaviour in a well-known fashion. It shows the volume behaviour of carbon dioxide (CO_2) as a function of pressure and temperature in the fluid region.

The various types of phase behaviour can be identified: on the right, the isotherms follow the ideal-gas type behaviour: this is the vapour phase; on the left, the isotherms are almost vertical lines, indicating that the compressibility is very small: this is the liquid phase behaviour. In between the two zones, a dome appears, defining the two-phase region. The isotherms are here strictly horizontal, since they link the saturated liquid volume with the saturated vapour volume at the component vapour pressure (equilibrium is isobaric and isothermal). If the system molar volume is within this region, the lever rule can be used to determine the amount of vapour and liquid phase (in this case θ is the mole fraction of vapour):

$$\theta = \frac{v - v^{L,\sigma}}{v^{v,\sigma} - v^{L,\sigma}} \tag{4.1}$$

Figure 4.1 shows that upon increasing temperature, the liquid saturated volume (the dotted line on the left) increases, while the vapour saturated volume (the dashed line on the right) decreases. When the critical temperature is reached, the two volumes coincide. All phase properties coincide at this point in fact: this is the fluid critical point (see also section 2.1.3.1, p. 28).

> The critical point is defined as the point where all properties of two distinct fluid phases coincide.

Figure 4.1 also helps understand the difference between "vapour behaviour" and "liquid behaviour" in the supercritical domain (*i.e.* above the critical temperature). On the right hand side of the diagram, in the "vapour region", the fluid continues to behave as a vapour (high compressibility), even at very high temperatures. This is the case, for example, with

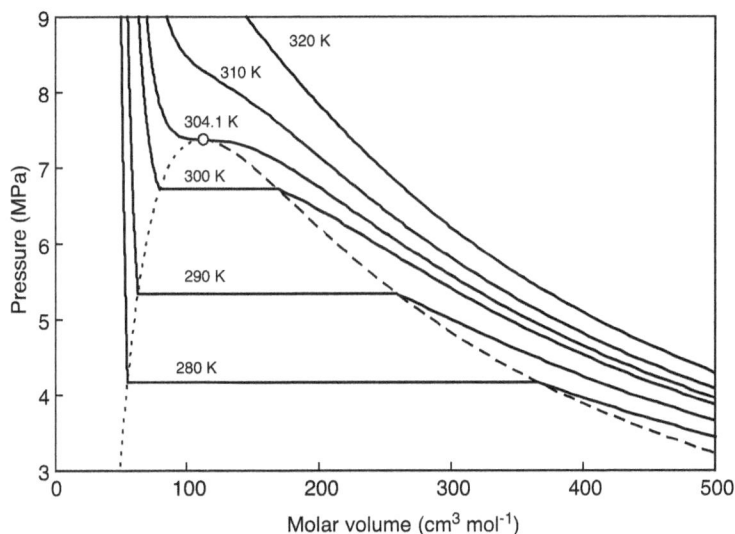

Figure 4.1

Sketch of carbon dioxide PvT behaviour.

carbon dioxide, methane, and to an even larger extent hydrogen, at ambient temperature and pressure: they are supercritical vapours.

If carbon dioxide is compressed further at the supercritical temperature its compressibility will decrease steadily and, without passing through a vapour-liquid condensation stage, it will become almost incompressible at high pressure. In other words, its behaviour passes from a vapour-type behaviour to a liquid-type behaviour in a continuous fashion.

The limiting conditions where the "fluid" can no longer be called vapour is up to the operator to decide: there is no absolute definition. It is sometimes convenient to use the critical density as a limiting criterion, but this is a purely semantic choice.

As an illustration of the effect of pressure on density for two immiscible fluids with very different behaviour, the case of water and supercritical carbon dioxide is shown in figure 4.2. What was thought to be the vapour phase (CO_2 rich) will become denser than the aqueous phase at high pressure. This density inversion depends on temperature. Specific models capable of describing the fluid changing compressibility are required. "Pseudo-experimental" equations of state are designed for that purpose (see section 3.4.3.3.C, p. 200).

In the critical region, the volume (or density) becomes very sensitive to pressure. It has in fact been shown that this extreme sensitivity cannot be described by an analytical equation [7] and several methods are therefore used to approach this limiting behaviour [8]. The so-called "cross-over" equations are nothing but a rescaling of the equation of state in the vicinity of the critical point, in order to find the correct asymptotic critical behaviour. The renormalisation scheme used for this purpose is very computer-time consuming, however.

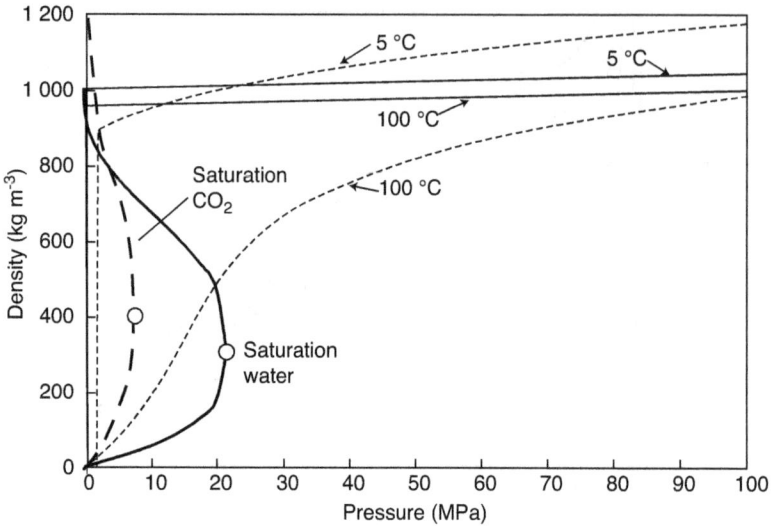

Figure 4.2

Pressure dependence of density for pure carbon dioxide (CO_2) and pure water (H_2O).

4.1.2 Enthalpy and entropy plots

Various types of presentation showing the behaviour of energy or entropy can be used. One of great interest is the enthalpy *vs.* temperature *HT* diagram. It was traditionally used for manual calculations and diagrams for almost all light hydrocarbons have been published in basic handbooks such as Maxwell (1950 [9]), API (1976 [10]) and Wuithier (1972 [11]).

Note that the origin of the enthalpy may be taken at will, as mentioned in section 2.1.1.5, p. 26.

The *HT* diagram is very easy to use (figure 4.3). One of the most interesting piece of information contained in this representation is the evolution of the heat capacity at constant pressure. This property is the slope of any isobaric curve shown on the figure 4.3.

$$c_p = \frac{\partial h}{\partial T}\bigg|_P \qquad (4.2)$$

The upper curve, at low pressure, shows the ideal gas behaviour for which models have been discussed in section 3.1.1.2.E, p. 116. The vapour-liquid equilibrium zone is clearly visible, with the vapour enthalpy on top and the liquid enthalpy below. The difference between these two values is the enthalpy of vapourisation (Δh^σ). In the liquid phase, pressure has almost no effect on the enthalpy. The diagram shows that when pressure increases, the temperature of phase change (bubble temperature) moves upwards and the vapourisation enthalpy decreases. At the critical point, the vapourisation enthalpy becomes zero, and the slope of the enthalpy with temperature (the heat capacity, according to (4.2)), becomes infinite.

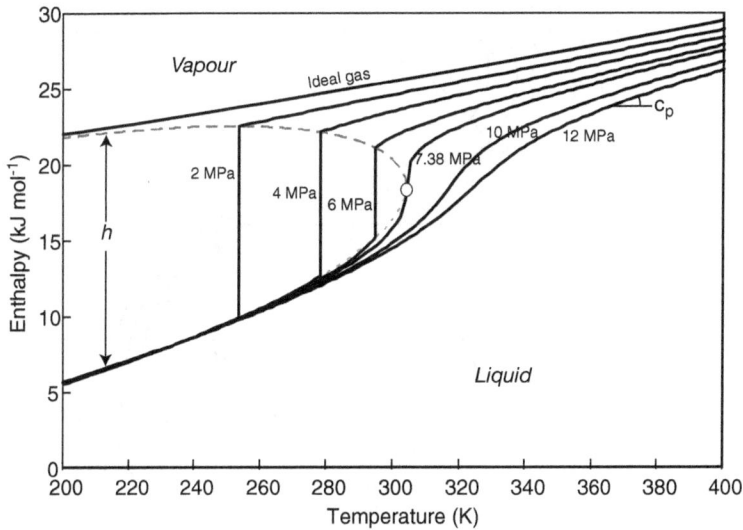

Figure 4.3

Sketch of carbon dioxide isobars on a enthalpy *vs.* temperature plot *HT*.

Another important diagram for enthalpy is the pressure *vs.* enthalpy *PH* plot (figure 4.4). It is widely used for manual calculations [12], as it shows the effect of isenthalpic changes that may occur in valves, as well as that of isentropic changes as in ideal pumps, compressors and turbines design (a dimensionless version is given by Danner and Daubert [13] to be used with the corresponding states principle). The main advantage of this diagram compared to figure 4.3 is the wider separation between the isotherms near the saturation region. An even more expanded plot can be obtained using a logarithmic scale for pressure as shown on figure 4.4. Enthalpy is larger on the right hand side of the diagram: this is the vapour region. Here, the isotherms are almost vertical lines: in the limit of the ideal gas behaviour, pressure has no effect on enthalpy. The vapour-liquid coexistence region is easy to recognise, where the isotherms are horizontal lines because the liquid and the vapour enthalpy have identical pressure and temperature. The critical point is on top of the vapour-liquid region. On the left, the almost vertical lines show that in the liquid phase, pressure has little effect on the enthalpy. This is no longer true at very high pressures where a change in slope is observed. This is the Joule-Thomson inversion phenomenon that is discussed below.

Other kinds of representation may include the entropy as a state variable. The two basic charts are the temperature *vs.* entropy *TS* diagram and the enthalpy *vs.* entropy *HS* chart, known as Mollier's chart [14]. These diagrams are particularly used for turbines (isentropic expansion) and pumps (isentropic compressions). They are often used with pure water in power generation due to the capacity of "expanding" information near the vapour saturation line and allow easy reading in this area. The two diagrams are shown on figures 4.5 and 4.6 respectively. Note that on the *HS* chart the pure component critical point is not on the top of the vapour-liquid coexistence region.

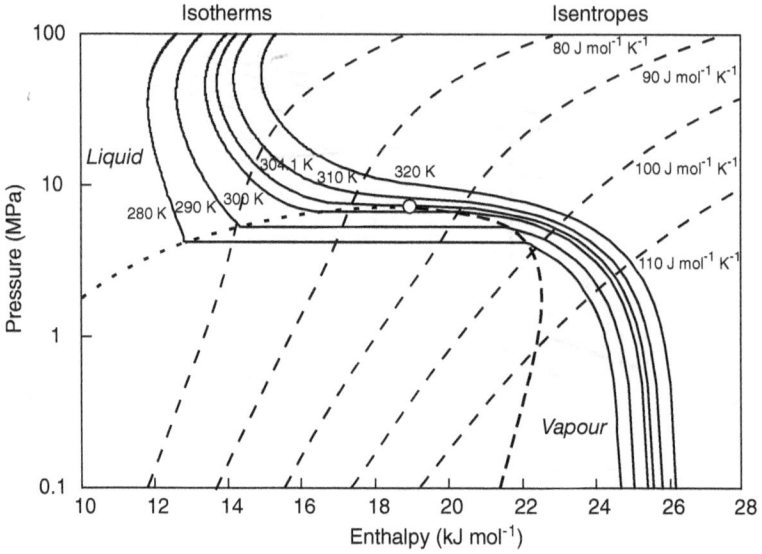

Figure 4.4

Sketch of carbon dioxide isotherms and isentropes on a pressure *vs.* enthalpy plot *PH*.

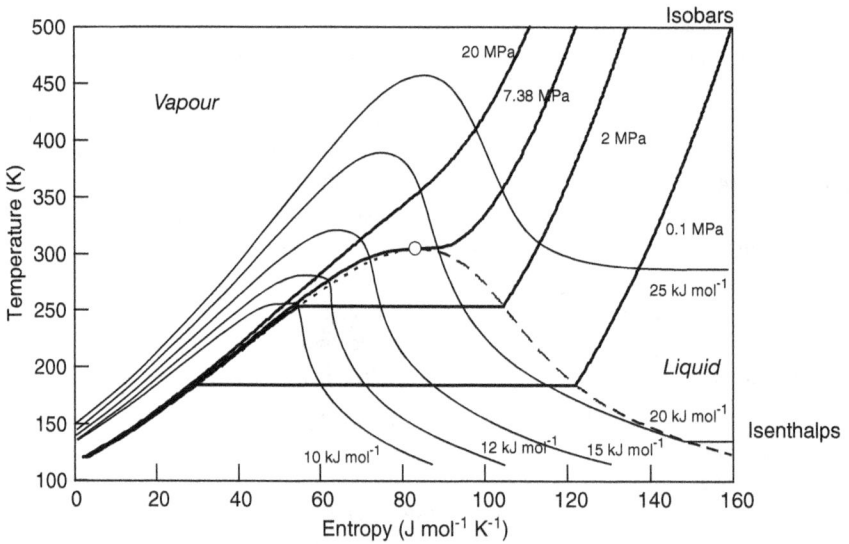

Figure 4.5

Sketch of carbon dioxide isobars (full lines) and isenthalps (dashed lines) on a temperature *vs.* entropy plot *TS*.

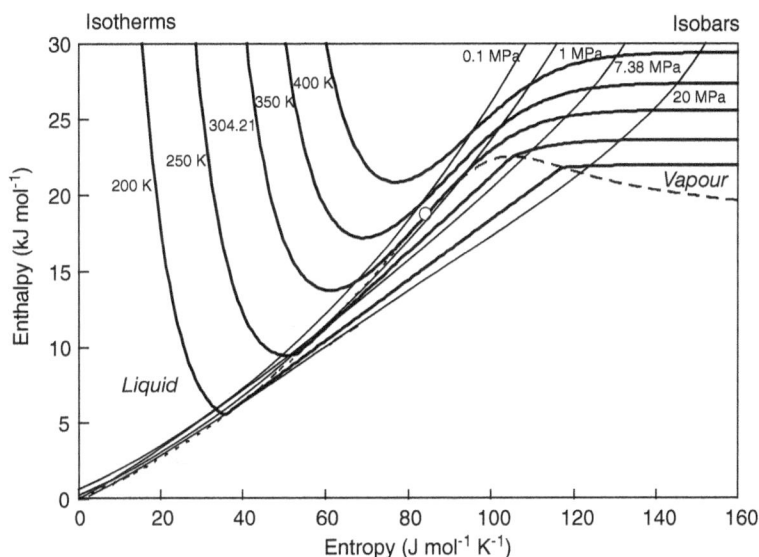

Figure 4.6

Sketch of carbon dioxide isotherms (full lines) and isobars (dashed lines) on an enthalpy *vs.* entropy plot *HS*.

4.1.3 Derived properties

The so-called derived properties based on the volume or enthalpic behaviour can be obtained from the thermodynamic derivatives. The isobaric heat capacity, already mentioned while describing the *HT* diagram, the Joule-Thomson coefficient and the speed of sound are discussed here because of their potentially important industrial applications.

4.1.3.1 Heat capacities

The heat capacity is defined from equation (4.2). It is the slope of the isobars plotted in figure 4.3. The figure shows that the critical isotherm has a vertical inflection point at the critical conditions, meaning that at this point, the heat capacity has an infinite value. This is made visible in figure 4.7. This behaviour can only be reproduced using an equation of state. Away from the critical point, different equations, with their own parameter sets, can be used for each phase. Yaws (2005 [15]), for example, offers three complete tables for vapour, liquid and solid phases.

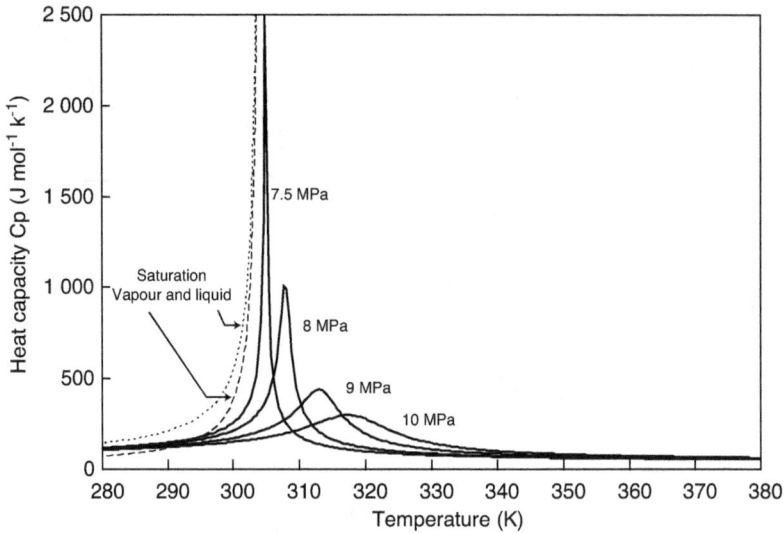

Figure 4.7

Behaviour of carbon dioxide heat capacity as a function of temperature at different pressures near the critical point (7.38 MPa).

4.1.3.2 Joule-Thomson coefficient

The Joule-Thomson effect is observed when a high pressure fluid is expanded through a valve. It has already been discussed using the example 2.5 of chapter 2 page 54, for an ideal gas. The heat exchange is neglected (isenthalpic change), and the pressure drop is known. The fluid transformation can then be depicted as a vertical line on the *PH* diagram, figure 4.4. The temperature behaviour is expressed using the Joule-Thomson coefficient, defined as:

$$\mu = \left.\frac{\partial T}{\partial P}\right|_H \tag{4.3}$$

The Joule-Thomson effect generally gives rise to strong cooling ($\mu > 0$). Yet, when a fluid is dense (like a liquid) it can also heat up ($\mu < 0$) upon pressure drop. Hence, a line in the *TP* phase diagram can be defined where the Joule-Thomson coefficient is zero. It is called the Joule-Thomson inversion curve, and is shown in figure 4.8 for CO_2: the inversion curve is found where the isenthalpic lines on a *TP* plot reach a maximum, or in the *PH* diagram figure 4.4 where the isotherms have a vertical slope.

It is observed that the Joule Thomson coefficient is negative for low temperature liquids (reduced temperature, $T_r < 0.6$), as well as for high temperature gases ($T_r > 3$). The inversion curve is continuous from liquid to gas but goes to very high pressures. The maximum pressure above which the Joule Thomson coefficient is always negative is $P_r > 13.5$.

Recent papers on this topics can be found in Colina *et al.* (1997-2008 [16-21]), Kortekass (1997-1998 [22, 23]), Segura (2003 [24]), Nichita *et al.* (2006-2008 [25, 26]).

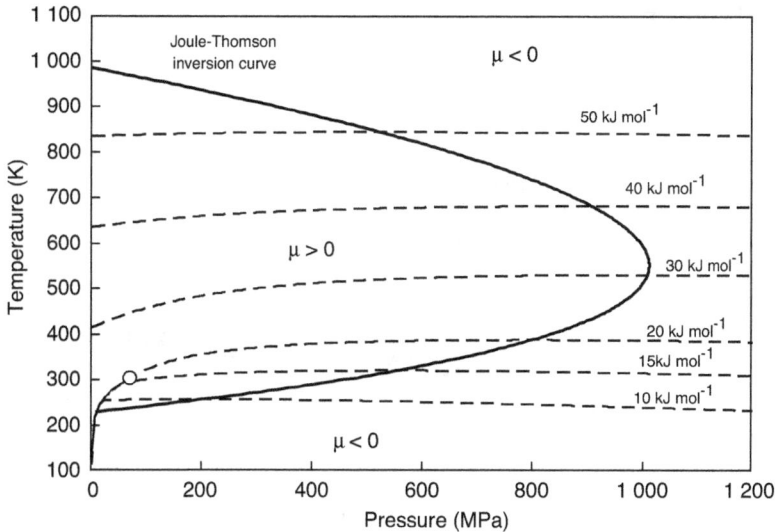

Figure 4.8

Position of the Joule-Thomson inversion curve with respect to the isenthalpic (dashed) lines in a *TP* diagram for carbon dioxide. The vapour pressure, ending with the critical point, is shown as a bold line.

4.1.3.3 Speed of sound

The speed of sound, u, is another thermodynamic property that may be of use in fluid mechanics applications. It is defined from the fluid PvT behaviour using:

$$\frac{1}{u^2} = -\frac{1}{v^2}\frac{\partial v}{\partial P}\bigg|_S \qquad (4.4)$$

The speed of sound behaviour as a function of pressure and temperature is depicted in figure 4.9. The speed of sound is large in the liquid or dense phase. It decreases with temperature. The opposite behaviour is observed in the vapour phase: here, the speed of sound, much lower, slightly increases with temperature and decreases with pressure. The result of this behaviour is the crossing of isobars: the same speed of sound may be found, at the same temperature, for a high pressure liquid and a low pressure vapour.

4.1.4 Model recommendations

In the above diagrams, the behaviour of the pure component CO_2 is illustrated. The plots have been generated using a BWR-type equation of state, which makes it possible to have a reasonable approximation of properties including the critical point. However, the fluid may also be a mixture. In that case, two main families of methods may be recommended, depending on the degree of non-ideality of the fluid mixture.

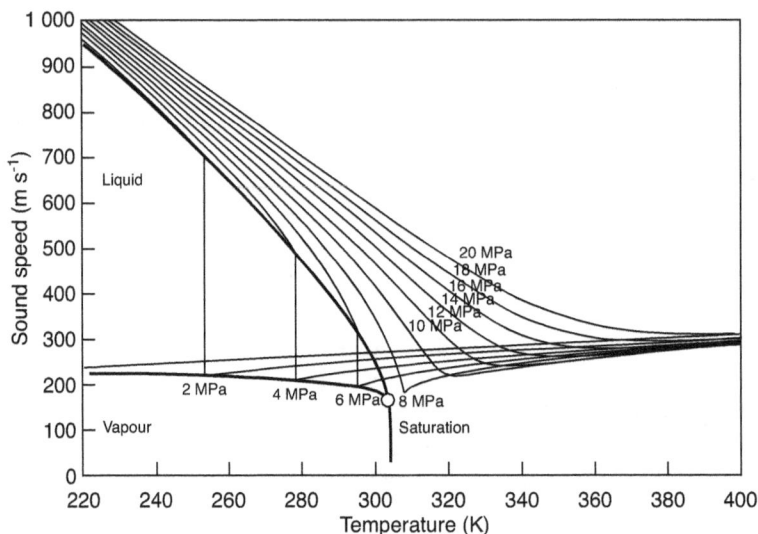

Figure 4.9

Speed of sound for carbon dioxide.

4.1.4.1 Use of the one-fluid approximation

For pure components, or for mixtures that behave in a manner that is close to ideal, the one-fluid approximation can be used. This means that the mixture is considered as a pseudo-pure component. The parameters of the equation of state are calculated using some mixing rule from the pure component parameters, in the same way as shown in section 3.4.3.3.C.d (p. 202).

Model Recommendations

Extended virial equations of state

The best type of equations are those developed using the virial approach, as discussed in section 3.4.3.3.C, p. 200 (as for example Lee-Kesler, recommended for hydrocarbons [28]). Span and Wagner at NIST have proposed very accurate equations of this type [29-33], often called "Modified BWR" (MBWR) equations of state. For mixtures, they can only be used for natural gases or other mixtures without strong intermolecular interactions. The *Refprop* package, available at NIST, uses this kind of equation and generates high quality plots of thermodynamic properties.

Cubic EoS

It may be acceptable to use cubic EoS to calculate enthalpy or entropy of complex (*i.e.* polar) mixtures on the condition that the mixing rule (see section 3.4.3.4.D, p. 211) has been validated for that purpose.

One of the major advantages of the equation of states is that they are able to describe at least approximately the critical point region (as described above, the true critical exponents cannot be calculated correctly using an equation of state, but require more complex approaches, as the cross-over method [27]).

4.1.4.2 Use of mixing rules and excess properties

The mixing rules for fluid properties have been discussed in section 2.2.2.2 (p. 60): knowing the pure component properties at given pressure and temperature (these can be found either with an equation of state as discussed in section 3.4.3 (p. 189) or, more often, from a liquid correlation presented in section 3.1.1, p. 102), the properties of the mixture can be readily calculated for a strictly ideal mixture. In case the mixture is not ideal, an excess property (activity coefficient) model must be used. These are further discussed in section 3.4.2 (p. 171).

> **Model Recommendations**
>
> **Mixing rules**
>
> The calculation method is detailed using table 2.15 (page 63).
>
> 1. Pure component properties:
> Pure component properties can be calculated using either correlations (for saturated liquid properties, see the correlations proposed in section 3.1 (p. 102): they can be used up to approximately 1 MPa above the saturation pressure), or "pseudo-experimental" equations of state (see section 3.4.3.3.C, p. 200).
>
> 2. Excess properties:
> Excess properties are calculated from activity coefficient models. Generally, only NRTL or UNIQUAC are used for calculating excess properties in this context: the model must have a sufficient physical foundation and allow flexibility in terms of temperature dependence of the parameters: the thermodynamic derivatives used are most often temperature derivatives (see section 2.2.2.2.B, p. 61).

Note the limitations of this approach (mixing rules + excess properties):

- It is impossible to calculate properties near the critical point.
- Considering that the models are independent of pressure, it is impossible to use them at high pressure. Generally, the use of this approach is limited to less than 1 MPa above the bubble pressure.
- Activity coefficient models are unable to calculated excess volume. Fortunately, excess volume is generally very small.
- Because thermodynamic derivatives are used for calculating the excess properties, it is essential to validate the results on experimental data: default parameters, obtained from a fit on phase equilibrium data, are generally unable to provide good values for other excess properties (see for example the VLE fit in section 3.3.2.2, p. 156).

Example 4.1 Calculation of the vapourisation enthalpy of the acetone + water mixture with NRTL at a given pressure (1 bar)

Analysis:

The vapourisation enthalpy is required. This is the difference between the vapour and liquid enthalpy of a mixture:

$$\Delta h^\sigma = h^V - h^L \tag{4.5}$$

Since vapour enthalpy is always larger than liquid enthalpy, this number is always positive. The condensation enthalpy has the opposite sign and is therefore always negative.

The enthalpy of a phase is calculated using the ideal gas and the residual contribution.

$$h = h^\# + h^{res} \tag{4.6}$$

The ideal gas contribution is calculated using the ideal gas heat capacity of the individual components and the knowledge that the ideal gas is an ideal mixture:

$$h^\# = \sum_i x_i \int_{T_0}^{T} c_{pi}^\# dT \tag{4.7}$$

In this equation, the reference enthalpy is taken to be zero in the thermodynamic state defined as the pure component ideal gas at temperature T_0. The pressure of the reference state need not be specified since pressure has no effect on the ideal gas enthalpy.

The residual contribution to equation (4.6) can be neglected in the vapour phase (pressure less than 0.5 MPa), and calculated in the liquid phase using either an equation of state (see section 2.2.2.1.B, p. 55) or liquid property model (section 2.2.2.2.B, p. 61). In the latter case, the pure component residual contributions must be known through another means. In this example, the mixture (acetone + water) behaves in a strongly non-ideal manner, and it is therefore recommended to use an excess Gibbs energy model. It is possible to calculate the excess enthalpy from the excess Gibbs energy using the Gibbs-Helmholtz relationship (section 2.2.2.2.B (p. 62), eq (2.101)) The NRTL excess energy model is proposed (introduced in section 3.4.2.2.C, p. 177). The liquid phase residual enthalpy is then calculated as:

$$h^{res,L} = \sum_i x_i h_i^{L,*} + h^E(T,x) \tag{4.8}$$

where the pure component liquid enthalpies should be computed using the same reference state as for the vapour state (as in equation (4.6)):

$$h_i^{L,*} = h_i^\# + h_i^{res,L} \approx h_i^\# - \Delta h_i^\sigma \tag{4.9}$$

where the pure component residual enthalpy $h_i^{res,L}$ is assumed to be equal to the opposite of the vapourisation enthalpy. This assumption is acceptable provided that the saturated vapour behaves as an ideal gas at the temperature of interest.

Note that for a mixture, the temperature of the dew and the bubble points are different. Therefore, the temperature at which h^V and h^L should be calculated are different. These temperatures are also calculated using the NRTL method, through the activity coefficients:

for the bubble point, the temperature must be found solving the following equation, at given composition (x):

$$P = \sum_i P_i^\sigma(T) x_i \gamma_i(T,x) = 0.1 \text{MPa} \tag{4.10}$$

for the dew point, the temperature must be found using the same kind of equation, at given composition (y):

$$\sum_i \frac{y_i}{P_i^\sigma(T)\gamma_i(T,x)} = \frac{1}{P} = 10 \text{ MPa}^{-1} \tag{4.11}$$

In this last equation, x must be computed at each iteration from the phase equilibrium equations. The algorithm is given in more details in section 2.2.3.1.C, p. 70.

Solution:

From the above equations, it is clear that many different models come into play before the actual condensation enthalpy can be calculated:

- ideal gas heat capacity for each component ($c_{pi}^\#$),
- vapourisation enthalpy for each component (Δh_i^σ),
- vapour pressure for each component (P_i^σ),
- the NRTL equation that is used both for calculating activity coefficients (γ_i) and excess enthalpy (h^E).

Each of these models should be validated against experimental data.

Table 4.1 provides the results in tabular form.

To calculate the condensation enthalpy of a mixture, we can write:

1. Calculate dew temperature using equation (4.11);
2. Calculate ideal gas enthalpies at the dew temperature, using equation (4.7). Considering the low pressure condition, this can be regarded as the vapour phase enthalpy;
3. Calculate the boiling temperature using equation (4.10);
4. Calculate the ideal gas enthalpies at the bubble temperature, using equation (4.7);
5. Calculate the vapourisation enthalpy at the bubble temperature, for each component;
6. Calculate the excess enthalpy (function of bubble temperature and composition) of the mixture;
7. Use equations (4.8) and (4.6) for calculating the actual liquid phase enthalpy.

The difference between the calculations in point 2 and in point 7 provides the vapourisation enthalpy.

Table 4.1 Intermediate calculations required to calculate the enthalpy of vaporisation of the acetone + water mixture

Acetone (1) mole fraction	T_{dew}	$h^\#$ (T_{dew})	T_{bubble}	$h^\#$ (T_{bubble})	Δh^σ (acetone)	Δh^σ (water)	h^E	h^L	$h^L - h^\#$ $= -\Delta h^\sigma$
	K	J mol^{-1}	K	J mol^{-1}	J mol^{-1}	J mol^{-1}	J mol^{-1}	J mol^{-1}	J mol^{-1}
0.00	372.80	2521.68	372.80	2521.68	-27186.33	-40813.94	0.00	-38292	40813.94
0.10	369.97	2762.76	341.67	1659.99	-28905.84	-42113.94	388.86	-38745	41507.28
0.20	366.87	2961.99	337.24	1661.81	-29142.63	-42298.11	535.28	-37470	40432.04
0.30	363.44	3112.44	335.60	1755.51	-29230.28	-42366.54	577.22	-36093	39205.41
0.40	359.58	3204.63	334.44	1859.42	-29291.87	-42414.71	572.62	-34734	37938.09
0.50	355.15	3224.98	333.39	1958.90	-29347.46	-42458.24	545.73	-33398	36623.01
0.60	349.95	3152.87	332.39	2051.36	-29400.66	-42499.95	503.78	-32085	35237.80
0.70	343.62	2955.31	331.41	2136.55	-29452.22	-42540.42	443.61	-30799	33753.40
0.80	335.84	2597.40	330.47	2214.73	-29502.27	-42579.74	354.06	-29549	32145.94
0.90	330.44	2353.72	329.58	2288.47	-29549.32	-42616.74	215.96	-28352	30705.26
1.00	328.90	2371.31	328.90	2371.31	-29584.95	-42644.79	0.00	-27214	29584.95

It is worth observing in table 4.1 that the excess enthalpy is very small compared to the vapourisation enthalpy. Neglecting the excess enthalpy results in over-evaluating the vapourisation of 1 to 2%.

This example is discussed on the website:
http://books.ifpenergiesnouvelles.fr/ebooks/thermodynamics

4.2 PHASE EQUILIBRIUM BEHAVIOUR OF INDUSTRIALLY SIGNIFICANT MIXTURES

All thermodynamic calculations first require an evaluation of the number of phases present. Only after the different phases have been identified can a property calculation be performed. Most often, vapour-liquid equilibrium is evaluated by default. However, if there is any risk that a second liquid phase or a solid phase could appear, a specific calculation must be performed. Again, it is the responsibility of the user to request this type of additional check.

A more detailed description of the phase diagrams and how they should be interpreted is given by de Swaan Arons and de Loos [1]. The authors propose a classification of the phase diagrams according to Van Konynenburg and Scott [34], now widely recognised for fluid phase behaviour. We must stress the fact that this classification implies continuity between vapour and liquid phases, more generally called the "fluid" phases. Solid phases are not described within this classification, although they are always present at sufficiently low temperature. It is quite possible that some low temperature phenomena are masked by the appearance of a solid phase. A general description of all kinds of phase transitions can be found in Stanley (1971 [35]) or Papon *et al.* (2002 [36]). Sadus, 1992, also describes the phase behaviour of industrially important fluids [3].

The discussion will be separated into two items: fluid phase equilibrium, and fluid-solid equilibrium. It is assumed at this stage that the reader is familiar with phase envelopes. Some information to help reading them is provided in chapter 2. We will simply recall some of the fundamentals.

PT plots

The multi-dimensional pressure-temperature-composition diagram is often presented in a simple *PT* plot. If the feed composition is known, these plots can be very convenient in order to visualise the conditions under which one, two or more phases may coexist. They can be used for any number of components.

The same type of plots can be used to visualise synthetically all the possible types of phase behaviour that can occur for a given system, *i.e.* for any composition. These are projections of the most remarkable features on the *PT* plane. The plots therefore become much more complex. In principle, they can also be conceived for any number of components, but they are generally drawn for binary systems only. It contains various kinds of points:

- pure component critical points (two phases merge and become identical),
- pure component triple points (three phases coexist),
- mixture quadruple points (four phases coexist),

and a number of lines:

- pure component vapour pressure, sublimation and crystallisation lines,
- loci of vapour-liquid or liquid-liquid critical points,
- three-phase coexistence lines,
- azeotropic lines.

Txy or Pxy plots

PT plots provide very synthetic information, but it is sometimes very difficult to analyse them quantitatively. Experience shows that complex phase behaviour is better understood on fixed pressure or fixed temperature cuts. Limited to binary systems, they can be read in two different ways: either by reading the bubble or dew pressure (or temperature) as a function of the composition (vertical reading in figures 2.4 or 2.5 for example), or by reading the composition of the phases at equilibrium as a function of pressure or temperature (horizontal reading in figures 2.4 or 2.5 for example).

Note that because of the low compressibility of liquid or solid phases, the pressure generally has a limited influence on liquid-liquid, liquid-solid or solid-solid equilibria. More generally therefore, these types of equilibrium are represented on fixed pressure plots (*Txy* diagrams). Phase equilibria in the presence of a vapour phase are equally sensitive to pressure or temperature.

4.2.1 Phase equilibrium classification

4.2.1.1 Fluid phases equilibrium

Fluid phase equilibria include vapour-liquid equilibrium, vapour-liquid-liquid equilibrium and liquid-liquid equilibrium.

A simplified classification proposed by O'Connell and Haile, 2005 [37] for binary mixtures is shown (slightly modified to be compatible with Scott's nomenclature [38]), in figure 4.10.

UCEP and LCEP are abbreviations for Upper and Lower Critical End Point. These are the intersections (and end points) of the three-phase vapour-liquid-liquid equilibrium lines and the critical locus (either vapour-liquid or liquid-liquid). Scott and Van Konynenburg [39, 40] suggest in a more simplified form the classification presented in figure 4.11. These figures represent *PT* projections of binary mixture phase diagrams. Three dimensional representations (figure 4.12) of different phase types can help understand the real phase behaviour of such complex shapes. The other way consists in studying the isothermal *Pxy* (figure 4.13 and figure 4.14) or isobaric *Txy* cuts. The latter represent a better visualisation mode in case of liquid-liquid equilibrium, as in figure 4.15.

The type I diagram is the simplest and most encountered. The corresponding *TPxy* diagram is shown in figure 2.3 (also on 4.11 at the top left and 4.12, at the left) for hydrocarbon mixtures of similar size. The two lines ending in the empty dot correspond to the vapour pressure of pure components, *i.e.* the transition line between the liquid phase (in the upper part) and the vapour phase (in the lower part). The third dimension (composition of the

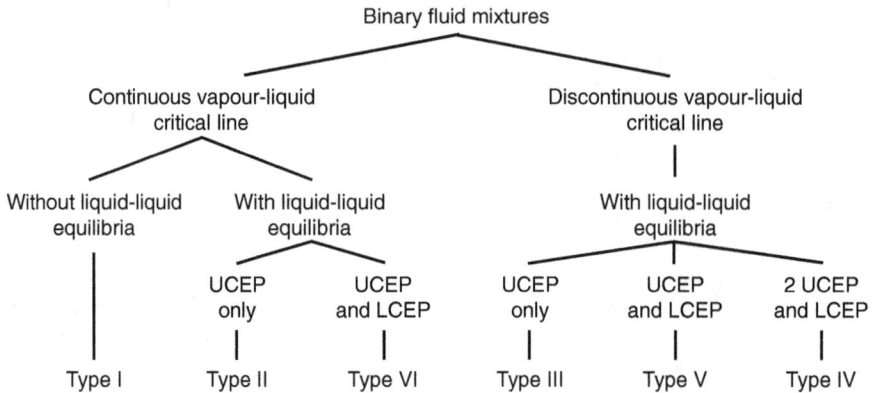

Figure 4.10

Simplified classification of fluid phase behaviour (modified from O'Connell and Haile, 2005 [37]).

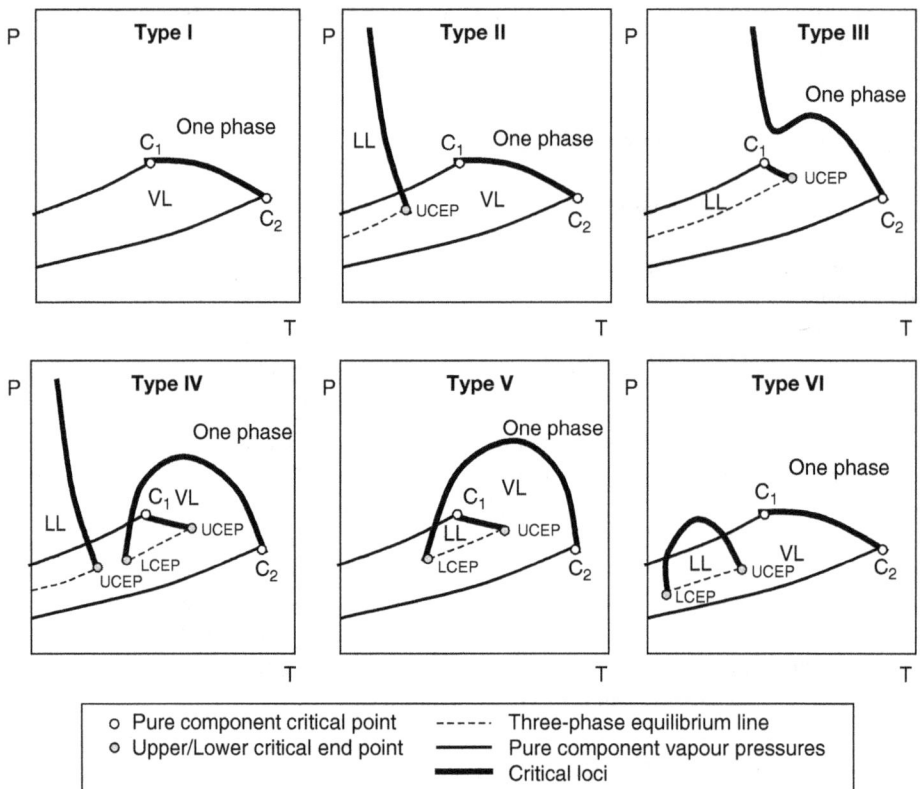

Figure 4.11

Simplified classification of fluid phase behaviour (from Peters 1993 [46]).

binary mixture) contains the envelope limiting the space where the mixture exhibits a vapour-liquid equilibrium. The envelope is limited by the dew surface (on the bottom – low pressure) and the bubble surface (on the top – high pressure); these two surfaces merge at the critical locus. As an example, mixtures of hydrocarbons of similar molar mass or mixtures containing CO_2 with light alkanes up to $n\text{-}C_5$ belong to this type (Schneider, 2004 [41]).

Type II has also a continuous critical locus, but a vapour-liquid-liquid equilibrium line appears at low temperature (*PT* projection in figure 4.11 and 3D plot in figure 4.12). This kind of behaviour is observed in many systems where the molecules have a strong preference for remaining with their own kind, such as for example when auto-association occurs (alcohol C_5 to C_{18} + water (Schneider, 2004 [41])). Water + diethyl ether (Clark *et al.* 2008 [42]) or CO_2 + alkanes between C_6 and C_{12} are other examples that belong to this type.

In Type III diagrams, two distinct critical loci are found, each initiating in one of the pure component critical points (*PT* projection in figure 4.11 and 3D plot in figure 4.12). The dotted line located inside the two vapour pressure lines on figure 4.11 corresponds to the vapour-liquid-liquid three-phase equilibrium. This line can also be on top of both vapour pressure lines when a hetero azeotrope exists. It is typically observed in a wide variety of water + hydrocarbon mixtures or water + CF_4 (Smits *et al.*, 1997, 1998 [43, 44]) or CF_4 + $n\text{-}C_4H_{10}$ to $n\text{-}C_{30}H_{62}$ (Schneider, 2002 [45]). CO_2 + alkanes beyond C_{14+} or N_2 + alkanes belong to this type.

Types I, II and III are the most common and a three dimensional *PTxy* sketch of figure 4.12 give a better idea of the different phase zones.

Type IV shows two liquid-liquid equilibrium regions: one at low temperature that ends with a UCEP and another at high temperature that starts with a LCEP. The liquid-liquid phase split at high temperature is a result of density differences between the phases (Wang and Sadus, 2003 [48, 49]). This is typically found when small non-polar molecules are mixed with large non-polar molecules as in solvent + polymer mixtures. Alcane mixtures

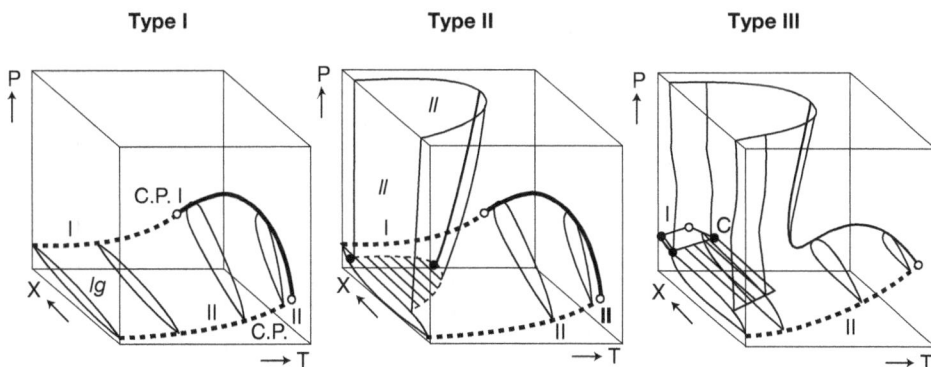

Figure 4.12

Three dimensional phase representation of some different types (from Schneider 1983 [47]).

Reprinted from [47], with permission from Elsevier.

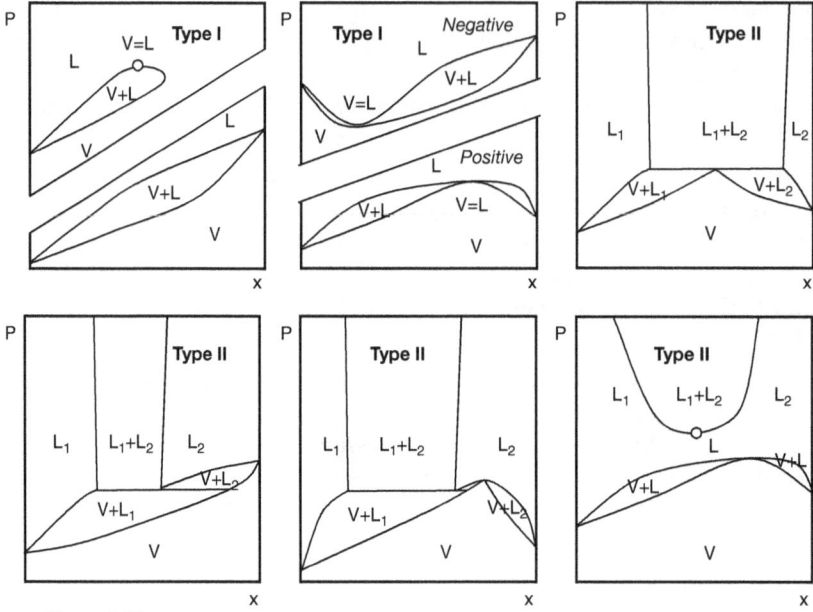

Figure 4.13

Schematic isothermal *Pxy* behaviour of types I and II (inspired from [55] and [56]).

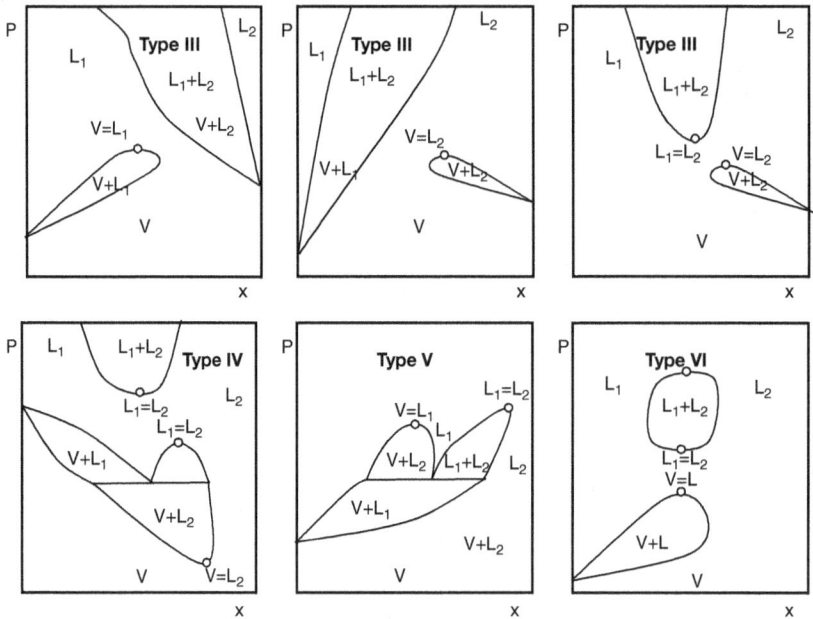

Figure 4.14

Schematic isothermal *Pxy* behaviour of types III to VI (inspired from [55], and [56]).
Reprinted from [55, 56], with permission from Elsevier.

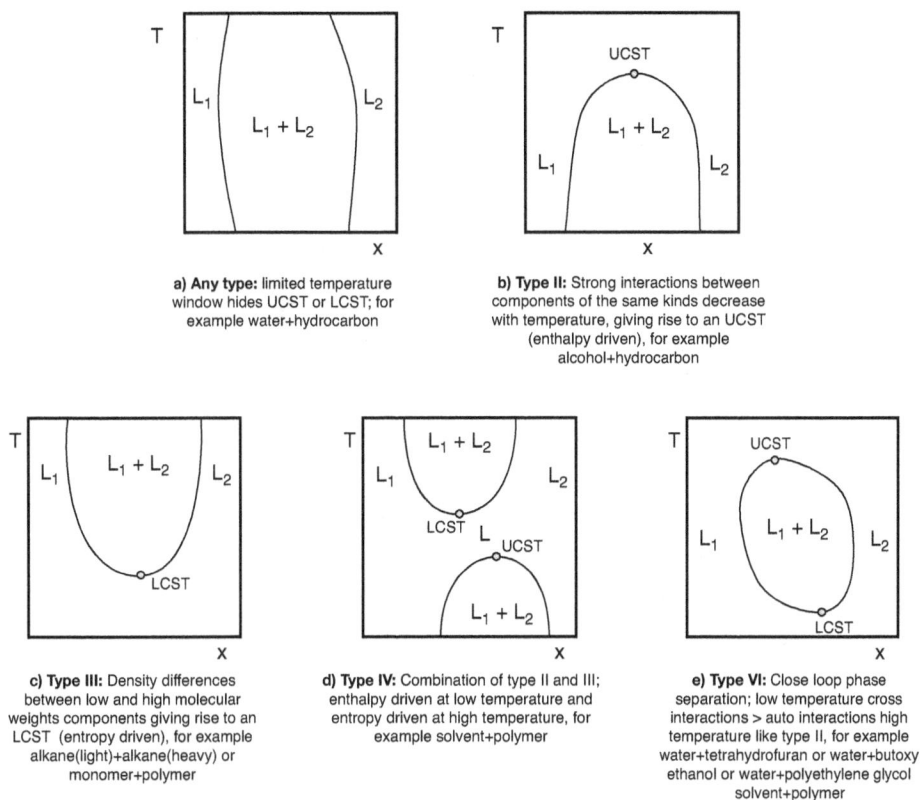

a) **Any type:** limited temperature window hides UCST or LCST; for example water+hydrocarbon

b) **Type II:** Strong interactions between components of the same kinds decrease with temperature, giving rise to an UCST (enthalpy driven), for example alcohol+hydrocarbon

c) **Type III:** Density differences between low and high molecular weights components giving rise to an LCST (entropy driven), for example alkane(light)+alkane(heavy) or monomer+polymer

d) **Type IV:** Combination of type II and III; enthalpy driven at low temperature and entropy driven at high temperature, for example solvent+polymer

e) **Type VI:** Close loop phase separation; low temperature cross interactions > auto interactions high temperature like type II, for example water+tetrahydrofuran or water+butoxy ethanol or water+polyethylene glycol solvent+polymer

Figure 4.15

Various types of liquid-liquid phase split behaviour [57, 58]. LCST stands for Lower Critical Solution Temperature, and UCST for Upper Critical Solution Temperature.

Reprinted from [57, 58], with permission from Elsevier.

may exhibit this behaviour for very asymmetric systems at low temperature, as discussed in section 4.2.2.2 below (p. 280). $CO_2 + n$-tridecane is the only binary of this family to show this type of phase behaviour (Aparicio-Martinez and Hall, 2007 [50]).

Type V is identical to type IV except that no liquid-liquid phase split occurs at low temperature. It is also observed in monomer + polymer mixtures and in the $NH_3 + KI$ mixture (Brandt *et al.*, 2001 [51]).

Type VI leads to "closed loop" liquid-liquid phase splits. It is observed for mixtures that tend to display type II phase behaviour, but because of their highly oriented specific interactions (e.g. cross-association) that become stronger at low temperature, they mix again below a certain temperature. This is the case, for example, of water + 2-butoxyethanol [52]. Other examples are discussed in section 4.2.6.3 (p. 319).

Shaw and Behar [53] indicate that using a continuous path it is possible to pass in sequence from type III to type IV to type II and finally to type I. These transitions have already been described by Van Konynenburg and Scott. Three other extensions of the phase classification (type Vm, VII and VIII) have been defined (Wang and Sadus, 2003 [48, 49]). Various studies provide an exhaustive analysis of the geometric global phase behaviour using dimensionless parameters for different kinds of EoS (cubic and more complex) and have clearly positioned each type and the transitions between them. The papers of Scott (1999 [38]) and Yelash and Kraska (1999 [54]) as well as the book of Sadus (1992 [3]) give a good coverage of this topic.

The phase envelopes for specific mixtures of industrial interest will be investigated below in more detail. The most useable diagram from this point of view will be the *Txy* or *Pxy* plots since they explicitly show the solubility behaviour as a function of pressure or temperature. In some cases, however, the equilibrium coefficient can be discussed directly. It provides exactly the same type of information.

4.2.1.2   Fluid-solid equilibrium

From a practical point of view, a solid phase can be seen as a material with a very high viscosity. However, we will here only look at the "thermodynamic" solid phase, which is crystalline. Highly viscous phases as gels, glasses and amorphous polymer phases would require an entire book to discuss [59]. For phase equilibrium calculations, all of these are generally treated as liquids [60].

> The solid phase is found at temperatures below the crystallisation (or melting or fusion) temperature: a phase transition is always observed.

When solid phases are taken into account, the behaviour can become quite complex. Shaw and Behar [61] have discussed vapour-liquid-liquid-solid (VLLS) phase diagrams for mixtures of petroleum interest.

Several types of solid phases may be encountered. A distinction is made between the so-called "stoichiometric" solid phases (immiscible solid phases) which have a fixed composition (generally defined by their crystal structure), and the solid solutions, whose composition can vary [1].

The extreme case of a stoichiometric solid is a **pure solid**, as in the case of ice, carbon dioxide or other pure compounds. Crystallisation processes are of great importance for purification of specific components (e.g. para-xylene). The model used to calculate fugacities (*i.e.* phase equilibrium) in this case is discussed in section 3.4.4.1 (p. 222).

Scale (solid salt precipitation from an aqueous solution) is another example of a stoichiometric solid. Yet, because of the electro-neutrality constraint, the number of anions and cations are related, and the liquid-solid equilibrium is often considered as if it were a pure solid. In fact, even though this is a phase equilibrium calculation, when the salt is completely dissociated, it may be treated as a chemical equilibrium calculation. This is extensively discussed by Zemaitis *et al.* [62] and by Rafal *et al.* [63].

In many cases, the solid composition can vary within limits. A practical example of this situation is found with **paraffin** deposits (also called **wax deposits**). Hence, not only pure component properties are required but in addition the solid mixture may have non-ideal behaviour.

Clathrate hydrates [64] are another example of a specific non-stoichiometric crystalline water structure. They will be discussed in some more details below (section 4.2.4.3, p. 304).

> The fugacity of solid phases must be calculated using a different model from that used for fluid phases. Equations of state are not suitable for solid phases; activity coefficient models can be used but must be regressed specifically on solid phase data.

4.2.2 Phase equilibrium in organic (hydrocarbon) mixtures

According to Peters [65], the two types of fluid phase behaviour that can be observed in alkane mixtures are the type I and type V in the classification of Van Konynenburg and Scott. Type I is more usual: only vapour-liquid is observed in the temperature region of interest. Type IV (occurrence of liquid-liquid demixing at low temperature) appears when the asymmetry between the components increases (methane + n-hexane and heavier; ethane + n-octadecane (C_{18}) and heavier; propane + n-triacontane (C_{30}) and heavier). In fact, solidification of the heavier component often makes it impossible to visualise the full fluid phase diagram, in which case the diagram is type V.

4.2.2.1 Vapour-liquid equilibrium in organic mixtures

As long as the molar mass difference between the components of the mixture is not too large, phase equilibrium diagrams with regular and closed shapes are found. The critical loci of these binary mixtures have been described by Polishuk *et al.* [66, 67] up to octane using different cubic EoS. Typical examples of type I are shown in figure 4.16. Only the critical loci are shown on this figure.

When the size difference between the molecules increases, the critical locus moves to higher pressures. In extreme cases, for example methane + n-tetracosane, it is possible to find critical points (and therefore vapour-liquid equilibrium conditions) up to 110 MPa (Arnaud *et al.* 1996 [68]).

The phase diagrams can also be presented as a series of isothermal cuts (for example methane + n-nonane in figure 4.17).

Note that if the mixture had been ideal, the bubble pressure curve would have been a straight line up to close to the critical point (see Henry Slope in figure 4.17). In this case it is not ideal because the size of the two molecules is noticeably different. The deviation from ideality is negative, as testified by the fact that the line curves upwards. The higher the temperature, the straighter is the bubble pressure curve.

It is worth noticing that at low temperature, the vapour composition is almost pure methane. This behaviour is typical for light + heavy mixtures: both components mix well in the liquid phase, but the heavy component is almost absent from the vapour phase. The higher the temperature, the larger the heavy component concentration in the vapour phase.

Note also that the critical pressure is close to that of pure methane at low temperature. It increases to a maximum when temperature increases and then decreases again to reach the heavy component critical pressure. This behaviour is the same as that observed in figure 4.16.

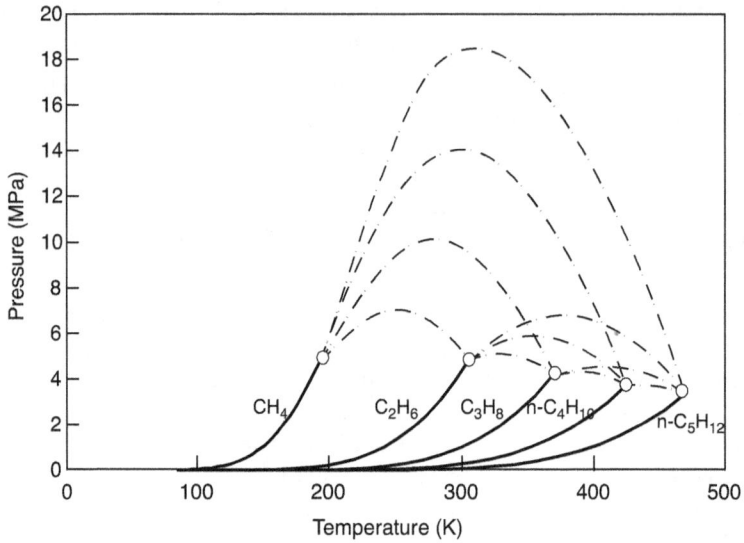

Figure 4.16

PT projection of vapour pressure and critical loci of methane + light alkanes.

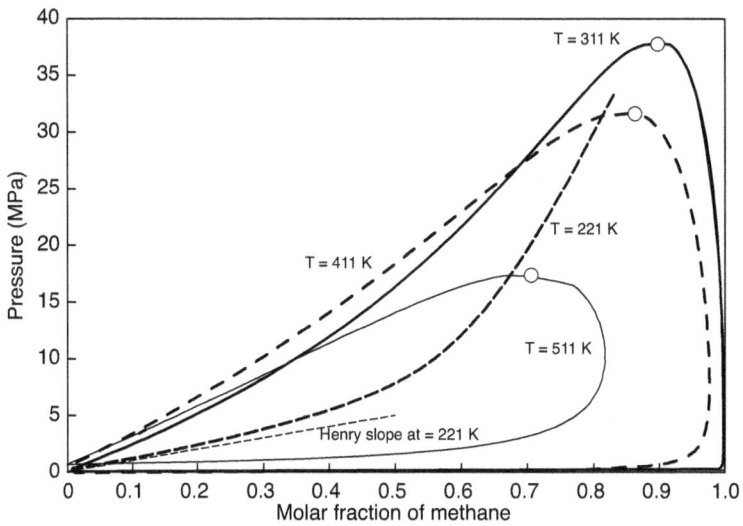

Figure 4.17

Pxy isothermal projections of the phase diagram of methane + *n*-nonane.

A. Gas solubility

The *Pxy* cuts also provide information concerning the solubility of the supercritical component in the liquid phase. The inverse Henry constant can be used as a measure of this solubility for the liquid phase at low concentration or pressure (see section 2.2.3.1.A, p. 63 for the use of the Henry constant). As a first approximation, at moderate pressure (as shown for the 221 K isotherm in figure 4.17):

$$x_i = \frac{Py_i}{H_i} \qquad (4.12)$$

A good evaluation of the solubility as a function of temperature is found in Fogg and Gerrard [69]. They note that the molar solubility of methane rises very slightly with increasing chain length of the solvent (figure 4.18). When expressed in mass fraction, the solubility decreases strongly with molar mass of the solvent.

Fogg and Gerrard [68] also note that a minimum is observed in the plot of solubility as a function of temperature, as with 1-methylnaphtalene and diphenylmethane (figure 4.19). This minimum is located at rather high temperature (around 420 K), and as a result it can not be made visible with low molecular weight hydrocarbon solvents whose critical temperature is smaller than this value. This is why very often, the methane solubility is considered as a purely decreasing function of temperature. Some values of Henry's constants for hydrocarbons other than methane can be found in Zuliani, 1993 [70].

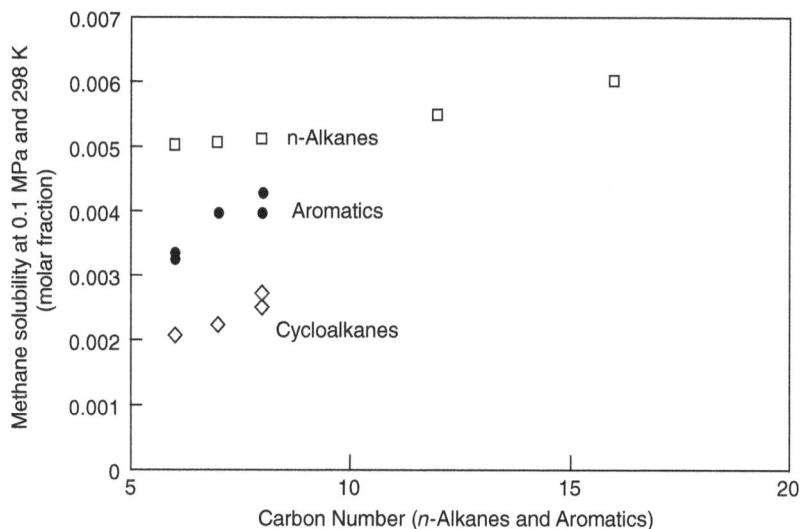

Figure 4.18

Solubility of methane in various hydrocarbon families at 298.15 K and atmospheric pressure (Fogg and Gerrard [69]).

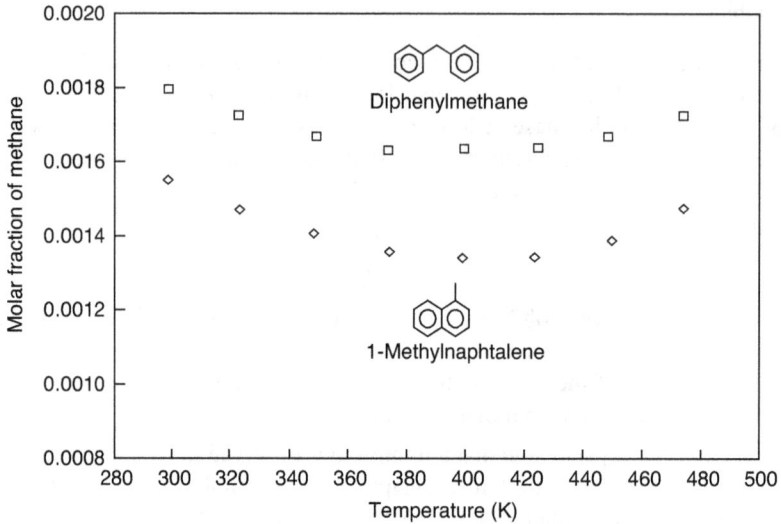

Figure 4.19

Solubility of methane as a function of temperature (Fogg and Gerrard [69]).

B. Mixtures of hydrocarbon families

Hydrocarbons contain only hydrogen and carbon atoms. Depending on their organisation (chemical families), a small polarity may be observed. The resulting deviation from the ideal mixture behaviour can be quantified looking at the infinite dilution activity coefficient (as discussed in section 3.4.1, p. 160). As an example, for C_6 hydrocarbons, table 4.2 shows some numerical values at 298.15 K:

Table 4.2   Activity coefficient at infinite dilution of some C_6 hydrocarbons at 298.15 K from Dechema [71]

Solute \ Solvent	n-hexane	iso-hexane	1-hexene	cyclo-hexane	benzene
n-hexane	1		1.1	1.18	2.21
iso-hexane	0.936	1	1.07	1.2	2.2
1-hexene			1		
cyclo-hexane	1.14		1.21	1	1.75
benzene	1.8				1

Considering n-hexane as non-polar (iso-hexane shows a negative deviation from ideality indicating entropic effect: size and shape difference), increasing polarity is found in the following order:

$$n\text{-alkane} < \text{olefin} < \text{cycloalkane} < \text{aromatic}$$

As a result of these slight differences, mixtures of these components of almost similar volatility may result in azeotropic behaviour. As an example, figure 4.20 presents the benzene + cyclohexane mixture.

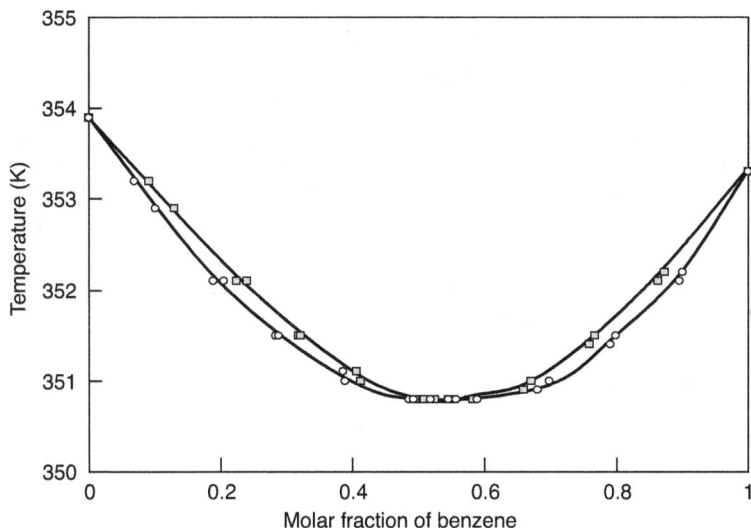

Figure 4.20

Narrow azeotrope of benzene + cyclohexane mixture at atmospheric pressure
(data from Boldyrev *et al.*, 1973 [72]).

Model Recommendations

The ideal mixture

Mixtures of hydrocarbons provide a good example of an ideal mixture, with certain constraints:

• If the components belong to different chemical families, non-idealities may be observed, especially when the vapour pressures are similar (examples are cyclohexane or *n*-hexane with benzene as in figure 4.20). In this case, the regular solution or Scatchard-Hildebrand method (section 3.4.2.2.B, p. 175) provides a simple predictive (*i.e.* not very accurate) method for calculating non-idealities.

• If the number of carbon atoms of the components differ by more than four, a negative deviation from ideality is observed. In this case, a Flory correction (predictive; section 3.4.2.3.B, p. 180) helps to improve the predictions.

The main advantage of using the ideal mixture model is that relatively simple plots can be used to evaluate the pressure or temperature effect of the equilibrium coefficients: the temperature effect can be estimated from the low pressure approximation of equation (2.128):

$$K_i = \frac{P_i^\sigma(T,i)\wp_i(T,P)}{P\varphi_i^v(T,P)} \qquad (4.13)$$

It is clear that the equilibrium coefficient increases with temperature in the same way as the vapour pressure (ln $K_i \sim 1/T$). As a first approximation, the equilibrium coefficient is inversely proportional to pressure. These pressure behaviour is further analysed in example 4.2. below.

Cubic EoS – Classical mixing rules

For thermodynamic simulation of hydrocarbon mixtures, cubic EoS, like traditional SRK [73] and PR [74, 75] are widely used in industry. Many variants for these equations are proposed in simulators, from modifications of the "alpha" function to different mixing rules (see section 3.4.3.4, page 204). Most of them have been tested thoroughly against experimental data. Some recent publications of Nasrifar *et al.* [76, 77] have compared the use of different EoS for LNG dew point calculations. The same kind of analysis was proposed by Pfohl and Dohrn in 1998 [78].

As an example, the work of Kordas *et al.* [79] who propose a correlation for the Peng-Robinson k_{ij} between methane and other hydrocarbons must be mentioned. They observe that this parameter shows discontinuous behaviour, increasing up to C_6, and then decreasing. They suggest the following correlation with the heavy compound acentric factor:

$$\text{for } C_n \leq 20 \quad k_{ij} = -0.13409\,\omega + 2.28543\,\omega^2 - 7.61455\,\omega^3 + 10.46565\,\omega^4 - 5.2351\omega^5 \quad (4.14)$$

$$\text{for } C_n > 20 \qquad\qquad\qquad k_{ij} = -0.04633 - 0.4367\,\omega \qquad\qquad\qquad (4.15)$$

It is very important to use this kind of correlation (like the Kordas correlation) exclusively with the model and the alpha function used for their development. Misuse can lead to results worse than using $k_{ij} = 0$.

Cubic EoS – Group Contribution mixing rules (PPR78)

Jaubert's group [80] proposed a predictive, group contribution method for the binary interaction parameters, as discussed in section 3.4.3.4.E (p. 214).

UNIFAC

The UNIFAC group contribution method incorporates both a combinatorial (entropic) and residual (enthalpic) contributions to the activity coefficient calculation. It can therefore be used successfully for VLE calculations of hydrocarbon mixtures at moderate pressures (UNIFAC, like all heterogeneous methods, are unable to calculate a full phase envelope).

PSRK

In order to use UNIFAC for high pressures, mixtures of gas and liquid components, or critical point calculations, it must be incorporated in an equation of state. This is the principle of the PSRK EoS, whose theory is presented in section 3.4.3.4.E (p. 216).

More recently, the Volume Translated Peng-Robinson EoS (VTPR) has been developed using the same philosophy. Its calculations are more precise, but the number of group parameters that are available is smaller.

SAFT

It is also worth bearing in mind that the molecular equations of state, and in particular SAFT, are particularly well-suited for describing mixtures of small and long hydrocarbons. The group contribution developed by several groups (Passarello [81, 82], Kontogeorgis [83-85], Jackson [86-87]) has proven to be of great interest in this case. For mixtures of small and large molecules, it is important to include a binary interaction parameter (k_{ij}) that can be predicted using the pseudo-ionisation parameter, see section 3.4.3.5.A (p. 216).

Example 4.2 Distribution coefficients in an ideal mixture (propane + *n*-pentane)

If equilibrium is considered at high pressure, fugacity coefficients and Poynting corrections must be taken into account as shown in equation (4.13). Assuming an ideal mixture (*i.e.* the equilibrium coefficient is independent of composition), it can be shown that, as a first approximation, the equilibrium coefficients change with pressure as follows:

$$\ln(K_i P) = \alpha_i(T) + \beta_i(T)P \qquad (4.16)$$

where

$$\alpha_i(T) = \ln(P_i^\sigma) - \frac{P_i^\sigma(v_i^L - B_i)}{RT} \qquad (4.17)$$

and

$$\beta_i(T) = \frac{(v_i^L - B_i)}{RT} \qquad (4.18)$$

Both α_i and β_i depend only on pure component properties (liquid molar volume v_i^L, vapour pressure P_i^σ and second virial coefficient B_i). Equation (4.16) shows that the product of equilibrium constant and pressure should be considered. Raoult's law suggests that this product is only a function of temperature. In practice, a linear relationship with pressure is found above 0.5 MPa. The resulting distribution coefficients for propane and *n*-pentane at 343.15 K are shown in figure 4.21. We can see that their respective slopes are such that the equilibrium coefficients converge at higher pressures. The practical meaning of this observation is that the volatility of the two components becomes closer when pressure is increased, and therefore more equilibrium stages will be needed to reach the same separation.

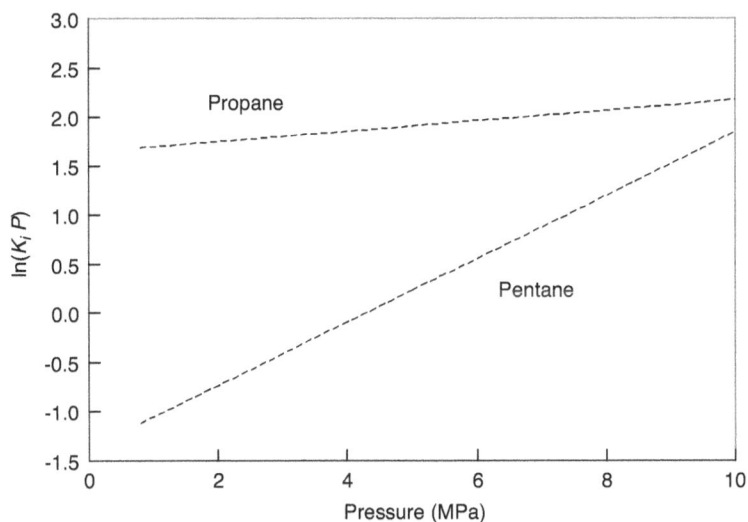

Figure 4.21

Change of propane and pentane equilibrium coefficients at 343.15 K with pressure.

This example is discussed on the website:
http://books.ifpenergiesnouvelles.fr/ebooks/thermodynamics

A general message to remember from the above exercise is that:

> The higher the pressure (and therefore the temperature), the more difficult a given separation becomes.

Example 4.3 Comparison of phase envelope prediction for ethane + *n*-pentane mixture.

Various models can be used to predict a phase envelope. For the ethane + *n*-pentane mixture, the Reamer *et al.* data of 1960 is available [88]. Bubble and dew curves are represented on figure 4.22 with solid dots and empty dots respectively.

Given that ethane is supercritical at this temperature, and a critical point is observed on the diagram, only an equation of state (homogeneous approach) can be used.

The cubic Peng-Robinson EoS with binary interaction parameters (BIP) equal to zero gives a very good approximation of the complete envelope. There is a small under-prediction of the critical point but the general shape is correct. A small improvement can be made using a small BIP value ($k_{ij} = 0.014$).

Finally, a modern equation of state such as SAFT, in this case using the group contribution version (GC-SAFT), allows accurate representation of the whole envelope based only on predictive behaviour of molecules as described by Nguyen-Huynh *et al.* (2009 [89]).

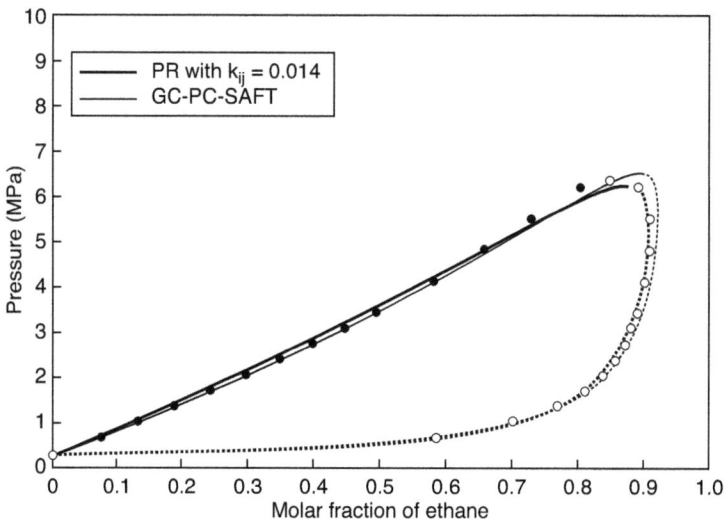

Figure 4.22

Phase envelope of ethane + pentane equilibrium (experimental and models) at 344.26 K (data from [88]).

This example is discussed on the website:
http://books.ifpenergiesnouvelles.fr/ebooks/thermodynamics

Example 4.4 Behaviour of a methane + *n*-decane mixture and its models

Experimental values of vapour-liquid equilibrium of methane + *n*-decane mixture are available from Lin *et al.* (1979 [90]). The large difference in molecular size results, as explained before, in an asymmetric mixture with entropic deviation from ideality. In addition, since methane is largely supercritical, a critical point appears at intermediate concentration. As shown on figure 4.23, the bubble points (solid dots) and the dew points (empty dots) show a large gap between both curves, indicating the experimental difficulty in locating the critical point. The Peng-Robinson model with BIP equal to zero can adequately predict the dew curve, although it excessively under-predicts the bubble curve. With the BIP predicted by the Kordas *et al.* [79] method, the bubble line is better fitted with a small loss of precision for the dew line. The Jaubert *et al.* [80] predictive method PPR78 is even better for the bubble line but is worse for the dew line representation.

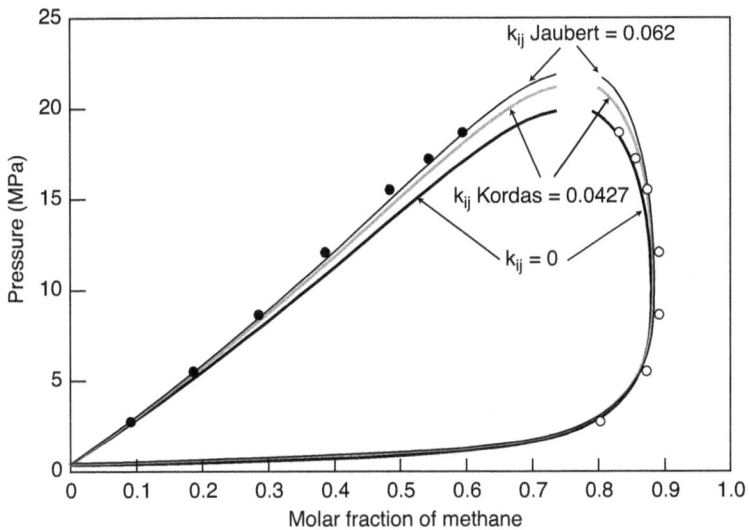

Figure 4.23

Phase envelope of methane + decane equilibrium (experimental and models) at 510.95 K (data from [90]).

This example is discussed on the website:
http://books.ifpenergiesnouvelles.fr/ebooks/thermodynamics

Example 4.5 Behaviour of the benzene + *n*-hexane mixture and its models

Benzene and hexane have very similar volatility but different polarity. As a result, a positive deviation from ideality is expected (as discussed in table 4.1 above). In figure 4.24, the Peng-Robinson equation of state is used and compared with the ideal mixture behaviour and the experimental data of Susarev and Chen (1963 [91]). The behaviour is nearly azeotropic, which the ideal mixture is unable to represent. The equation of state with zero interaction parameter catches the positive deviation, but using a small value for the binary interaction parameter ($k_{ij} = 0.0074$) improves the plot. For the purpose of comparison, the BIP has been calculated using the PPR78 predictive method of Jaubert *et al.* [80], giving a very similar result ($k_{ij} = 0.00865$).

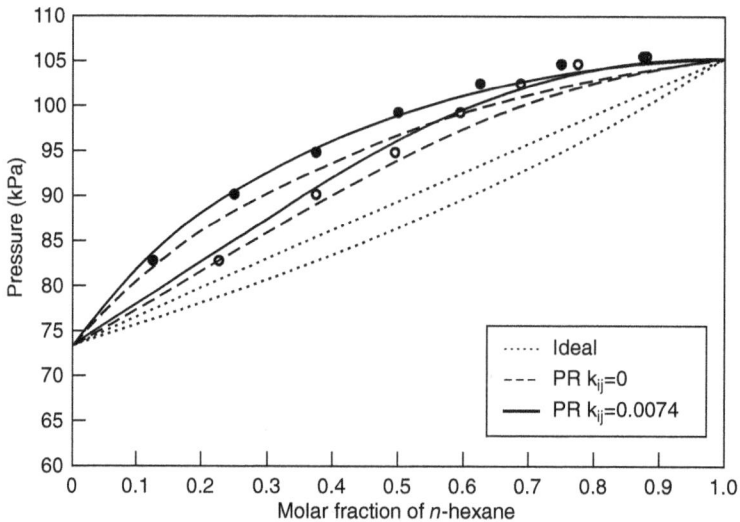

Figure 4.24

Comparison of experimental VLE and different models for a mixture of hexane + benzene at 343.15 K (data from [91]).

This example is discussed on the website:
http://books.ifpenergiesnouvelles.fr/ebooks/thermodynamics

4.2.2.2 Liquid-liquid in organic mixtures

At low temperature it has been shown that mixtures of small and heavy hydrocarbons may split in two liquid phases [1, 92, 93]. The reason for this phase split is based on the large size difference between the mixture components, resulting in the same kind of phase behaviour as is seen in the case of polymer + monomer mixtures. The phase diagram of these mixtures is type IV, as shown on figures 4.6 and 4.7.

The positions of the lower critical end point (LCEP) and the upper critical end point (UCEP) obviously depend on the type of mixture. Figure 4.25 shows how these temperatures depend on the mixture of interest. For example the mixtures of methane + C_6 to C_8 are on the left corner of the figure at almost 200 K. Longer chain alkanes freeze at this low temperature. Ethane + C_{18}-C_{28} are in the middle of the figure at near 300 K. Finally the mixtures of propane + C_{30}-C_{70} are on the right hand side with a liquid-liquid zone at 350-370 K. A three-phase vapour-liquid-liquid equilibrium exists between the LCEP and the UCEP.

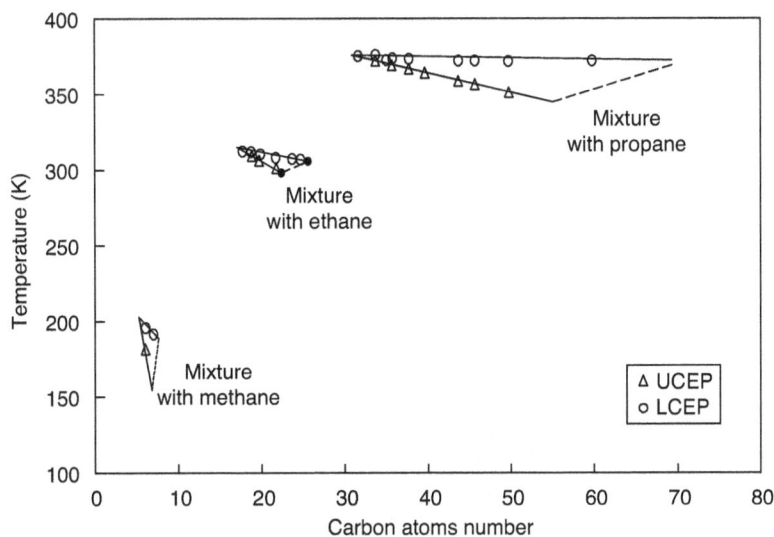

Figure 4.25

Position of the liquid-liquid equilibrium in asymmetric alkanes mixtures [46]. © Kluwer Academic publishers. Reprinted with permission from Applied Sciences, vol. 273, 1993.

Model Recommendations

SAFT, Lattice-Fluid, Free Volume EoS

The liquid-liquid phase split is of entropic origin as indicated in chapter 3: it is driven by a difference in the densities of the various phases. As a result, the model to be used must be an equation of state, such as those adapted for polymer mixtures. Both the SAFT or the lattice-fluid EoS [94] should be well-adapted. The free-volume equations of state (entropic free volume, UNIFAC free volume [95]) are also constructed for that purpose.

4.2.2.3 Fluid-solid equilibrium in organic mixtures

The onset of crystallisation of a component out of a fluid phase depends on the melting temperatures of the components and on their concentrations. Crystallisation of hydrocarbons may interfere with the process conditions on several occasions.

A. Single component crystallisation

Often, it may be considered that the solid crystallises as a single component. This allows very good separation, unfortunately at a high energy cost. Crystallisation may also result in an unwanted solid deposit. Examples are the formation of scale out of aqueous solutions or low temperature benzene crystallisation in the natural gas liquefaction process. This last example is described more extensively by de Hemptinne [96].

Crystallisation only occurs at temperatures below the crystallisation temperature of the pure component. Table 3.1 provides a number of such values. It is important to note that the order of crystallisation temperature is generally unrelated to the component volatilities as discussed in section 3.1.1.1.F, p. 105.

The second factor for crystallisation is the relative abundance of the component that may potentially crystallise. The lower the concentration, the lower the temperature can be before it will crystallise. In case of ideal solutions, this relationship is easy to identify, otherwise, a fluid model capable of describing the component activities in the liquid must be used (an activity coefficient model or an equation of state).

Model Recommendations

As discussed in section 3.4.4.1, p. 222 both a fluid phase fugacity and a solid phase fugacity are needed. The solid state fugacity can be computed using the pure crystalline equation (3.253), considering however that if the process temperature is well below the crystallisation temperature, the extrapolation must be validated. The fluid phase fugacity of the crystallising component can be calculated with any method (typically a cubic equation of state), but great care must be taken with the concentration range: it may be close to infinite dilution, which means that the mixing rule must be properly validated. It is best to use a G^E type mixing rule.

Example 4.6 Calculation of the eutectic of ortho- and para-xylenes

Para-xylene is an important chemical stock for the synthesis of terephthalic acid, which is further polymerised to form polyester resins and fibres. Industrial production of pure p-xylene is costly, largely due to the difficulties associated with the separation of xylene isomers: they have similar molecular structures and close boiling points, making them difficult to separate by distillation. Nevertheless, their crystallisation behaviour are quite different, as a result of their different molecular structures: para-xylene can pack nicely in a crystalline lattice while ortho-xylene shows steric hindrance. As a result, the melting temperatures and fusion enthalpies have significantly different values (table 4.3):

Table 4.3 Crystallisation parameters of ortho- and para-xylenes

	o-xylene	*p*-xylene
Melting point (K)	247.98	286.41
Heat of fusion at melting point (kJ/mol)	13.6	17.1

An additional advantage offered by separation of para-xylene from the liquid through crystallisation is that the solid phase is almost pure. The thermodynamic conditions for this solid-liquid separation must be determined however.

Analysis:

The property to be investigated is the liquid-solid equilibrium, which again means fugacity calculations.

Since the mixture investigated here contains aromatics of equivalent molecular weight, the mixture can be considered ideal.

In addition, pressure is moderate (atmospheric), so that the liquid phase fugacity can be written as:

$$f_i^L = x_i f_i^{L,*} \tag{4.19}$$

The solid phase fugacity is calculated from equation (3.253), assuming that the heat capacity upon melting, $\Delta c_{p,F,i}$, and the pressure effect are negligible:

$$\ln\left(\frac{f_i^S}{f_i^{L,*}}\right) = \frac{\Delta h_{F,i}}{RT_{F,i}}\left(1-\frac{T_{F,i}}{T}\right) \tag{4.20}$$

Solution:

The pure liquid fugacity is approximated with the vapour pressure, which is extrapolated below the triple point:

$$f_i^{L,*} \approx P_i^\sigma = \exp\left(A_i + \frac{B_i}{T} + C_i \ln T + D_i T^{E_i}\right) \tag{4.21}$$

At equilibrium, fugacities are equal, meaning:

$$f_i^L = x_i f_i^{L,*} = f_i^S \tag{4.22}$$

when para-xylene is crystallising, the liquid phase composition is:

$$x_{p-xyl} = \frac{f_{p-xyl}^S}{f_{p-xyl}^{L,*}} \tag{4.23}$$

while the o-xylene composition is the complement:

$$x_{o-xyl} = 1 - x_{p-xyl} \tag{4.24}$$

If crystallising from a melt containing large amounts of o-xylene, the solid phase that first appears is pure o-xylene, and the subscripts in the two previous equations must be switched.

The result is shown in figure 4.26 where the eutectic temperature corresponds to the intersection of the two LSE curves.

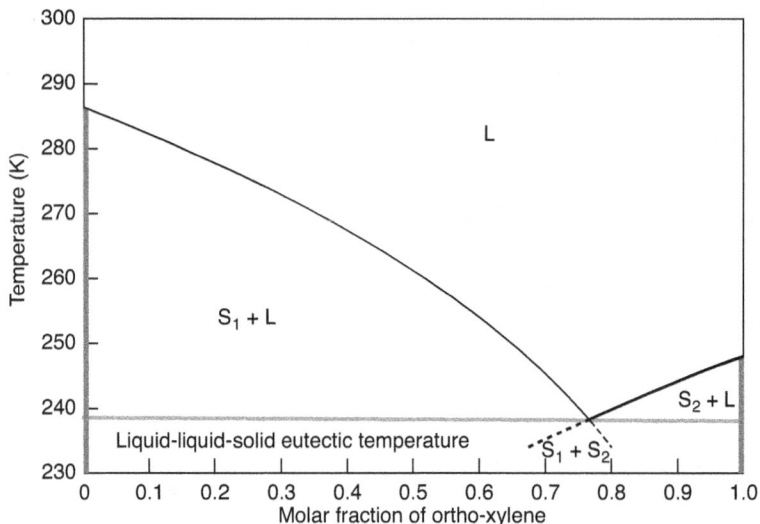

Figure 4.26

Eutectic diagram calculation for the binary *p*-xylene + *o*-xylene mixture.

This example is discussed on the website:
http://books.ifpenergiesnouvelles.fr/ebooks/thermodynamics

B. Crystallisation of solid solutions (e.g. wax deposit in paraffinic crudes)

Wax (*n*-paraffins) deposit in paraffinic crudes occurs at the so-called *Wax Appearance Temperature*, WAT. This phenomenon is essentially identical to that described above, except for the fact that waxes may crystallise as mixtures. This phenomenon makes the description of the phase equilibrium somewhat more complex as possible non-ideal behaviour in the solid phase should be taken into consideration.

Long chain alkanes may crystallise at moderate temperature, resulting in serious industrial problems. Won has published different papers on this subject, starting in 1986 [97]. Vidal [98] describes different types of liquid-solid phase equilibrium that can be encountered. A very clear example of high pressure wax crystallisation is provided by Ungerer *et al.* [99] (figure 4.27). Two different mixtures have been investigated over a wide range of temperatures and pressures. The WAT in both cases varies very little with pressure.

Gas-Liquid-Solid of different methane-rich oils

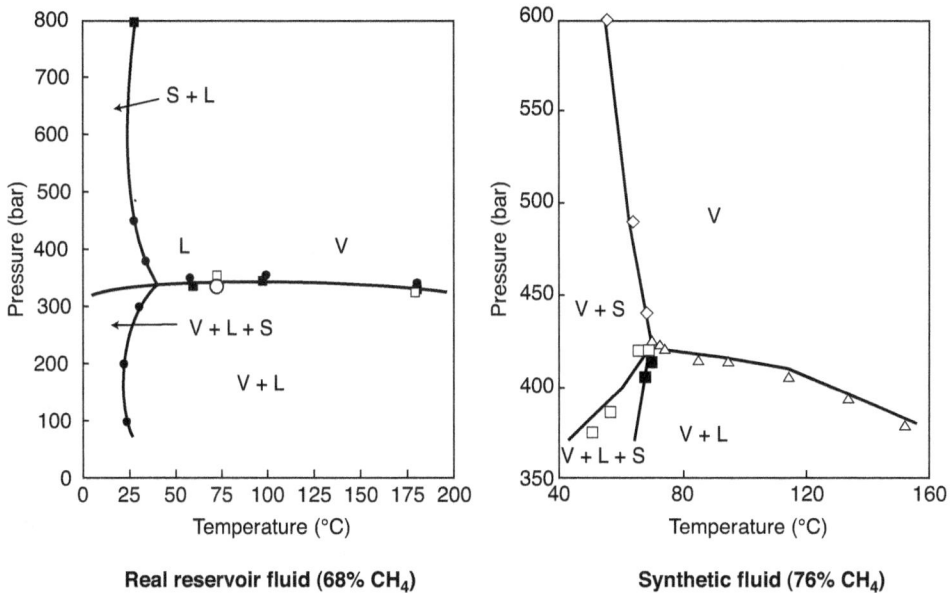

Real reservoir fluid (68% CH₄)

Synthetic fluid (76% CH₄)

Figure 4.27

Phase diagram of two mixtures: on the left, a reservoir fluid containing nearly 5% of C_{11}^{+} (up to C_{40}) and 68% methane; on the right a synthetic mixture consisting of 2.3% $n\text{-}C_{16}$, 1% phenanthrene and 76% methane. A solid phase is observed at temperatures lower than 293-303 K, depending on pressure (from [99]) – Symbols are experimental points and lines calculations.

Reprinted from [99], with permission from Elsevier.

Model Recommendations

The model for the vapour-liquid equilibrium is well fitted with a cubic Peng-Robinson equation of state, coupled with a good mixing rule (see above).

Extrapolation methods for the fusion properties of the pure component long-chain paraffins exist (section 3.1.1.1.F, p. 105). For solid state mixtures, non-ideal behaviour may also exist. The activity coefficient in the solid phase is generally described with activity models, similar to those used in the liquid phase (see section 3.4.2, p. 171, e.g. UNIQUAC, whose parameters must be re-fitted). This approach has been used by Coutinho *et al.* [100, 101] and Esmaeilzadeh *et al.* [102]. Ji *et al.* [103] propose a good summary of the existing models. The G^E models have also been applied to wax deposits by Pauly *et al.* [104].

Note that in wax deposits several solid phases may co-exist and phase transitions may occur in the solid phase. Vidal [98] and Jensen *et al.* [105] propose to improve the WAT model by taking this phenomenon into consideration.

4.2.2.4 Asphaltene precipitation

Petroleum fluids often include a very heavy residue that contains large, complex molecules of relatively unknown molecular structure [106, 225]. They are generally referred to as resins and asphaltenes and are identified by various standards. Resins for example are soluble in *n*-heptane (ASTM D3279, NFT 60115 – IP 143); asphaltenes are not, but are soluble in toluene. Other standards define solubility using propane or *n*-pentane as solvent. Asphaltenes are in fact heavy hydrocarbons of molar mass in the range 400-1500 g mol^{-1}. containing heteroatoms such as oxygen, nitrogen and sulphur. They have a H/C ratio of about 1.2.

The presence of asphaltene in a petroleum fluid may result in phase separation and deposition of a heavy, very viscous material, which in some cases is identified as a solid, in other cases as a liquid. Deposition occurs either as a result of mixing, for example in refinery operations, or upon pressure change in reservoir draining and extraction (see Pina *et al.,* 2006 [107]).

The first type of flocculation phenomenon generally occurs at atmospheric pressure, when crude oils from different origins are mixed. This can be understood by visualising the solubility limits on a triangular plot [108-110], representing the asphaltene concentration, the concentration of asphaltene solvent (resins, aromatics), and the concentration of asphaltene anti-solvent (alkanes, saturated components) as in the insert of figure 4.28. For a given fluid, the same solubility limit can also be shown on a plot presenting the solvent (often toluene) concentration on one axis and the anti-solvent (often *n*-heptane) concentration on the other axis, as shown in figure 4.28. A line is drawn in the phase split region of the triangular phase diagram. It illustrates how a mixture of two stable phases can yield an unstable blend. If the "crude oil" corner of the diagram is magnified, the departure of the stability limit line is shown in the large diagram. In the upper part of the plot, the asphaltene is unstable, or forms a second phase, while below the solubility limit the asphaltene remains in solution. The ordinate at $x = 0$ of this line gives the onset of flocculation of the fluid. If the onset is greater than zero, the fluid is naturally stable. If however the onset is negative, an anti-flocculent must be used in order to prevent precipitation.

The second type of asphaltene precipitation is observed under pressure reducing conditions. This is illustrated in figure 4.29. Asphaltene solubility is high when the pressure is high, decreasing as the pressure is reduced down to the bubble point of the oil. Below the bubble point, the compositional effect becomes larger than the pressure effect: when the gases (very poor solvents) disappear, the asphaltene solubility in the remaining oil increases.

There is no method or systematic approach to predict the onset of flocculation, whatever the cause (pressure effect or mixing). Experimental data are always necessary to discuss industrial cases. Buckley *et al.* (1998 [112] [113]) recently proposed an interesting relation between the onset of flocculation and the refractive index (RI). This parameter seems better correlated than the solubility parameter used in the past.

However, the major difficulty with this type of phase behaviour is that the asphaltene structure is very difficult to reproduce, and may even vary depending on its environment or history as mentioned by Maham *et al.* 2005 [114].

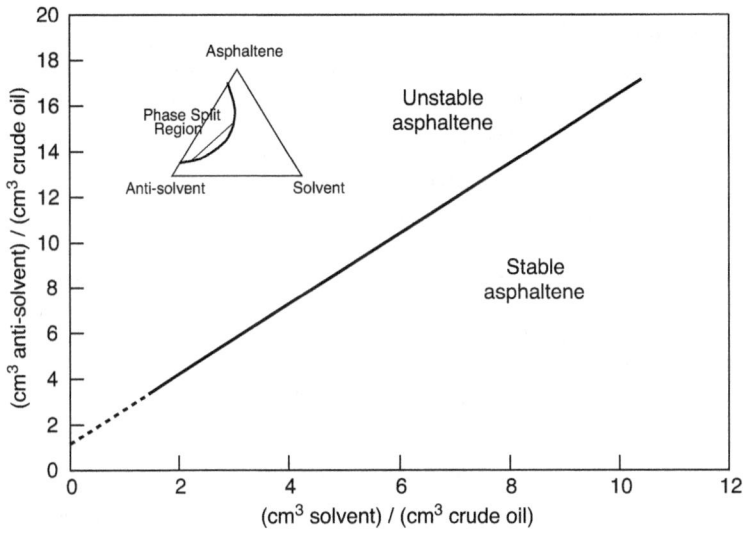

Figure 4.28

Qualitative representation of asphaltene flocculation in a ternary diagram. Example of a stable fluid.

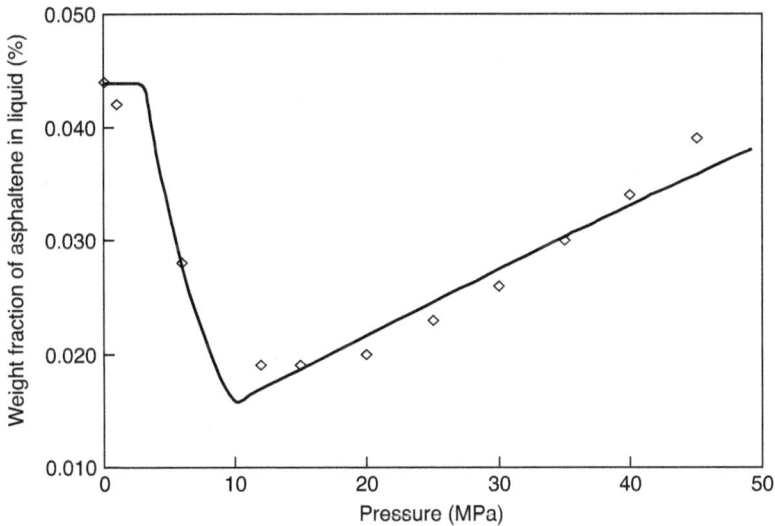

Figure 4.29

Asphaltene solubility as a function of pressure (from Szewczyk 1997 [111]). The line is calculated using the Peng Robinson EoS with specific mixing rules [120]).

Model Recommendations

Many different attempts have been made to describe asphaltene deposits. It is not the purpose of this book to describe them in detail. Four, very different, underlying assumptions may be at the basis of the models [107]:

Solubility models

In this case, it is assumed that the deposit is a separate liquid phase. Many papers propose using the Flory-Huggins models (Hirschberg *et al.*, 1984 [115] and Hirschberg and Hermans, 1984 [116], Yaranton *et al.*, 2007 [117, 118]). Nevertheless, equations of state are also very valuable tools. Szewczyk and Behar [119] use a simple cubic equation of state with the Abdoul *et al.* [120] group contribution mixing rules. Even the SAFT equation of state, which has a term to explicitly describe association phenomena, has been used with some success [121-123]. The model of Buckley's team ([124, 125]) assumes a liquid-liquid equilibrium with a solubility parameter correlated with the refractive index (RI).

Solid phase models

In this case, it is assumed that the asphaltene is a solid and the equation described in section 3.4.4.1 (p. 222) can be used directly [126, 127].

Colloidal models [128]

Asphaltenes are considered to be particles stabilised in a fluid by resin molecules. The transfer of resins from the colloid to the oil phase destabilises the colloid, resulting in asphaltene flocculation. The model is that of a liquid-liquid equilibrium where only the resin fugacity is considered. In the oil phase, the fugacity is calculated using an equation of state, and in the flocculated phase, the Flory-Huggins method is used. This choice is explained by the fact that both asphaltene and resins can be considered as polymers.

Micellisation models [129, 130]

In this case we are no longer dealing with phase equilibrium, but rather with an unstable mixture of two phases. The continuous phase is oil and the dispersed phases are asphaltene particles kept in suspension thanks to the interfacial tension. Flocculation occurs when the size of the micelles grows larger than a critical micelle concentration.

None of these models can truly be called predictive. In all cases, the characterisation (parameters) of the heavy petroleum fraction must be fitted on experimental data. The best of these models can be used to extrapolate the data to other conditions (pressure, temperature or composition).

4.2.3 Phase equilibrium in presence of H_2 or other supercritical gases

Hydrogen is a very important gas in refining. It is present in almost all upgrading reactions related to heavy oil treatments. In the future, great hopes are built on its use as a fuel. Dihydrogen is the smallest molecule and has a very low critical point ($T_c = 33.19$ K and $P_c = 1.313$ MPa). It is always found in supercritical conditions in industrial use. In fact, there are two different molecules of hydrogen, differing by the spin of its two atoms (the ortho- and para-hydrogen). The relative amount of the two spins depends on the temperature. At 80 K,

50% of each kind is present while at 200 K only 25% of ortho-hydrogen is found. The latter mixture is called normal hydrogen. Pure hydrogen is well described by a modified BWR type equation of state with 32 constants as proposed by Younglove (1982 [131]).

Hydrogen + hydrocarbon mixtures exhibit type III phase behaviour, which means that at low temperature and high pressure a phase split is observed. The low temperature behaviour has no practical consequences, since the operational temperature is generally sufficiently high.

Other gases such as oxygen and nitrogen behave in a similar way in the presence of a liquid hydrocarbon, although more soluble, and can therefore be treated similarly.

Due to the large difference in volatility between hydrogen and hydrocarbons, the phase envelope always reaches very high pressures. The critical point of hydrogen-hydrocarbon mixtures is rarely required in process simulation.

The equilibrium coefficients (representative of their volatility) of different gases in mixture with a representative hydrocarbon (*n*-decane) as a function of temperature, at 1 MPa pressure, are drawn in figure 4.30. The figure clearly shows that hydrogen is the least soluble of all the gases represented (highest volatility or equilibrium coefficient). Its solubility increases with temperature (volatility decreases). Nitrogen is next, with very similar behaviour. The methane solubility shows a minimum (maximum volatility) at a temperature close to 420 K, as observed previously (figure 4.19). The more soluble a gas, the more its solubility decreases with higher temperature.

The effect of the solvent on the solubility is shown in figure 4.31 [132]. It indicates that the molar mass has little effect on hydrogen solubility, but that the chemical family is important: the solubility in olefins and paraffins is much larger than in aromatic components.

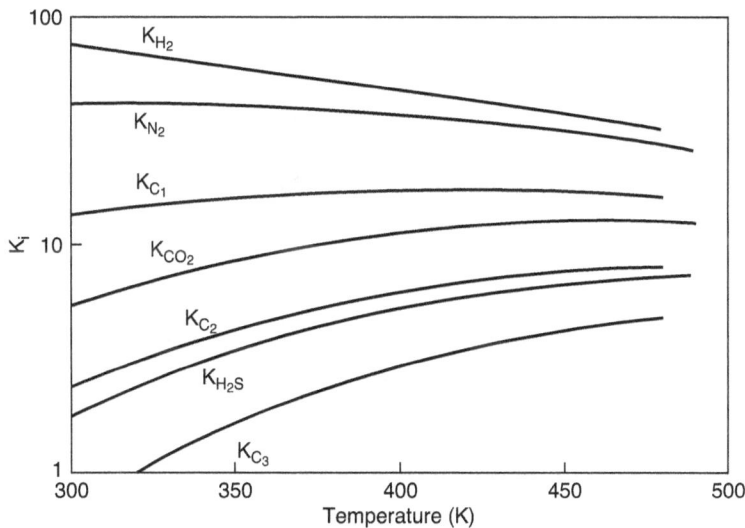

Figure 4.30

Distribution coefficient of a light gas + *n*-decane (as a representative of the hydrocarbon family) mixture. The plot is a result of a Grayson-Streed calculation, at 1 MPa pressure.

These plots are shown in mole fraction. It is important to note that the weight fraction solubility is much smaller because of the relative molar mass. On a mass fraction scale, the gas solubility strongly decreases with increasing solvent molar mass.

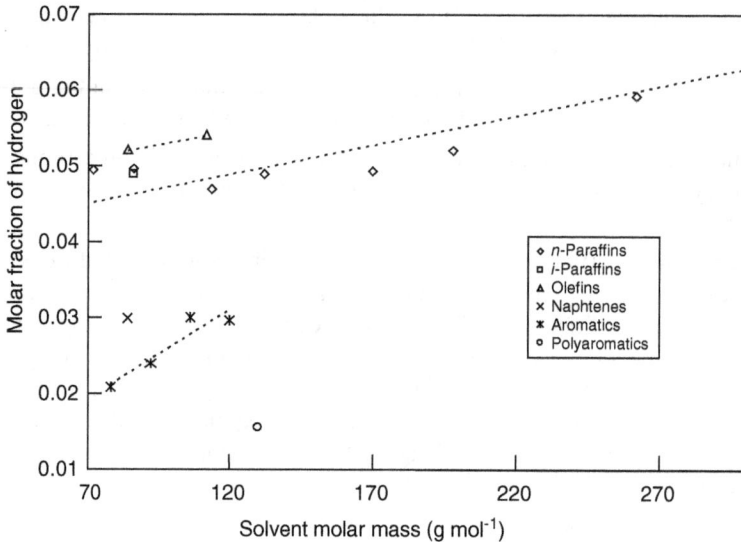

Figure 4.31

Mole fraction solubility of hydrogen in various solvents (Ferrando [132]).

Model Recommendations

Many different models have been proposed to calculate hydrogen – hydrocarbon mixtures:

Grayson and Streed method (GS)

For thermodynamic modelling of hydrogen equilibrium, the traditional method of Chao and Seader (1961 [133]) has been improved by Grayson and Streed (1963 [134]) especially for this purpose. Some more recent modifications have been proposed by Lee, Erbar and Edmister [135]. The method is based on a heterogeneous, asymmetric approach (see section 2.2.3.1.B, p. 68). The distribution coefficient is calculated as follows:

$$K_i = \frac{y_i}{x_i} = \frac{\varphi_i^{L,*}(T,P)\gamma_i(T,\mathbf{x})}{\varphi_i^{V}(T,P,\mathbf{y})} \qquad (4.25)$$

where the three factors are calculated using three different models:

• The pure liquid fugacity coefficient $\varphi_i^{L,*}(T,P)$ is calculated using a specific, corresponding states method, described previously in section 3.4.4.4, page 224.
• The liquid activity coefficient $\gamma_i(T,\mathbf{x})$ is calculated using the regular solution model.
• The vapour phase fugacity coefficient $\varphi_i^{V}(T,P,\mathbf{y})$ is computed from the Redlich Kwong cubic equation of state.

The advantage of this method is that it is predictive, as it requires no interaction parameter data. It also is easy to use for hydrocarbon mixtures, as it requires the knowledge of pure component parameters that can easily be determined also for petroleum pseudo-components (T_c, P_c, ω , liquid molar volume and solubility parameter).

For heavier hydrocarbons however (greater than C_{15}), the predicted hydrogen solubility may be quite different from the experimental values (see figure 4.33). Hence, it is *not recommended* for these components: the solubility increase with the solvent molar mass, observed in figure 4.31, is not correctly taken into account.

Cubic equations of state

It is important to stress the extreme sensitivity of the hydrogen fugacity to its characteristic parameters, and in particular the acentric factor. In all models, make sure to use the values recommended by the authors.

• **Grabowski and Daubert** (1978-1979 [136-138]) adapted the SRK model for various light gases, including hydrogen. This form has been recommended by API for hydrocarbon + hydrogen mixtures.

• **Moysan *et al.*** (1983-1986 [139-141]) calculated interaction parameters k_{ij} for 10 mixtures and observed very strong temperature dependence using SRK and PR with a classic mixing rule. In addition, k_{ij} coefficients are very high (reaching 0.7 at high temperature) indicating that this rule can no longer be considered as a "correction". They also observed however that the interaction parameter of hydrogen is almost independent of the kind of solvent. They therefore propose a simple correlation for both SRK and PR equations of state (hydrogen is component *i* and solvent is component *j*):

$$k_{ij} = 1 + \frac{\mu_{EOS}\left(T_{r,i} - 1\right) - 1}{\sqrt{\alpha_i}} \tag{4.26}$$

with $\mu = 0.06$ for SRK and $\mu = 0.0417$ for PR. Almost 150 points have been used with a resulting root mean square error (RMSE) on k_{ij} between 2 and 10%.

• **Gray *et al.*** (1983 [142]) proposed a different kind of formulation for the interaction coefficients of SRK and PR:

$$k_{ij} = A + \frac{BX^3}{1 + X^3}$$

$$0 \le X = \frac{T_{c,j} - 50}{1000 - T_{c,j}} \le 1 \tag{4.27}$$

with $A_{SRK} = 0.0067$; $B_{SRK} = 0.63375$ and $A_{PR} = 0.0736$; $B_{PR} = 0.58984$. They obtain very good agreement (3% on k_{ij}) with their 12 experimental points below 550 K but a wider average absolute deviation for 18 points above this limit (more than 10%). More recently, (Gao *et al.* 2003 [143]) obtained new results for mixtures of H_2 + *n*-alkanes.

• **Twu *et al.*** 1996 [144] take a different point of view. Instead of using a very large interaction parameter, or modifying the mixing rule, they prefer to modify the "alpha" function of hydrogen.

$$\alpha_{H_2} = T_r^{N(M-1)} \exp\left(L\left(1 - T_r^{NM}\right)\right) \tag{4.28}$$

With this approach, interaction coefficients between H_2 and hydrocarbons are equal to 0.

SAFT

Several authors have investigated the use of the molecular-based SAFT equation of state for calculating the phase equilibrium in hydrogen containing mixtures, with good results. Among others, we may mention Le Thi *et al.*, [145], Tran *et al.* [146] who use a group contribution method so that the equation can be applied to any hydrocarbon. The disadvantage of using SAFT, however, is that to date no method exists for computing parameters for petroleum pseudo-components.

Example 4.7 Comparison of experimental values and different models with H$_2$ + *n*-hexane mixture

Experimental values of VLE of hydrogen + *n*-hexane mixture have been published by Nichols *et al.* (1957 [147]). In figure 4.32, these experimental data at 377.6 K are compared with three models: Grayson-Streed, Peng-Robinson with k_{ij} equal to zero and Peng-Robinson with k_{ij} estimated by Moysan. The cubic EoS without k_{ij} strongly underpredicts the bubble pressure, and is clearly not recommended for hydrogen-containing mixtures. This effect is adequately corrected up to a liquid containing 50% hydrogen with the k_{ij} evaluated with the Moysan procedure. The Grayson-Streed model is adequate for bubble pressure prediction up to 30% of hydrogen for this mixture at this temperature. The Dew line is adequately predicted by the cubic EoS with Moysan k_{ij}. Grayson-Streed is less accurate. The same authors report VLE at other temperatures.

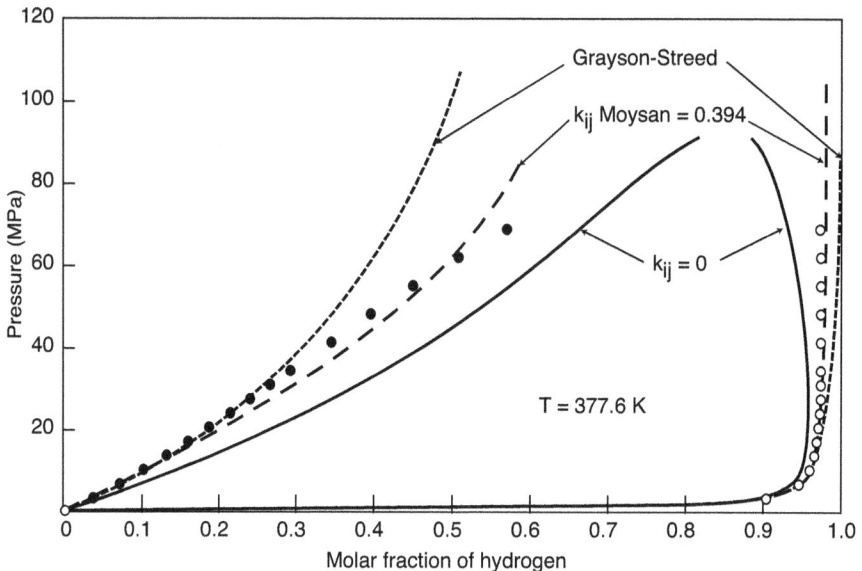

Figure 4.32

Comparison of experimental VLE and different models for a mixture of hydrogen + *n*-hexane at 377.59 K (data from [147]).

A comparison with additional models has been reported by Ferrando and Ungerer (2007 [132]). They have shown that the error in the hydrogen solubility prediction in the liquid phase remains close to 5% for all paraffins, except for **the Grayson-Streed model that overpredicts the hydrogen solubility by more than 20% when the solvent is heavier than C$_{16}$** (figure 4.33). All these models predict the hydrogen concentration in the vapour phase with less than 2% error.

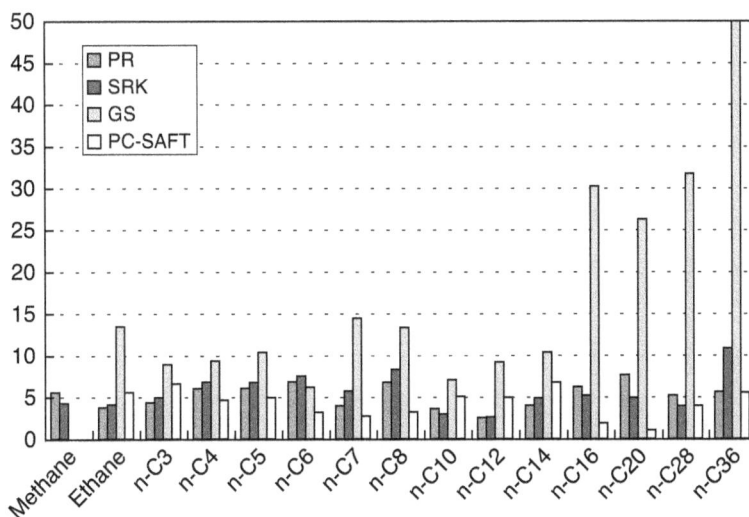

Figure 4.33

Relative error in the prediction of molar composition of hydrogen in the vapour phase in a binary mixture of H$_2$ + n-paraffin [132, 148].

This example is discussed on the website:
http://books.ifpenergiesnouvelles.fr/ebooks/thermodynamics

4.2.4 Phase equilibrium in presence of an aqueous phase

4.2.4.1 Fluid phase equilibrium

Due to its very strong hydrogen bonds, water does not mix with most organic molecules at low temperature. The phase diagrams of water-containing mixtures therefore exhibit type III phase behaviour (Bidart *et al.* [149]) as shown in figure 4.34.

The phase diagram of a water + hydrocarbon mixture forms a heteroazeotrope, at a pressure that is essentially equal to the sum of the vapour pressures of the water and the hydrocarbon.

P

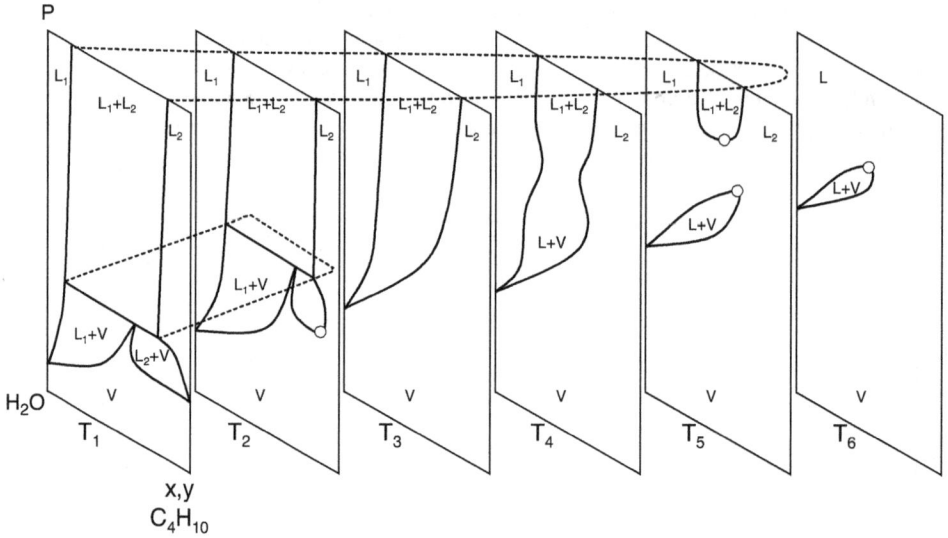

Figure 4.34

Schematic *PTxy* projection for binary mixtures of water(1) + *n*-alkane(2) (Type III).

Model Recommendations

It is generally recognised that describing both aqueous and organic phases simultaneously and accurately is difficult with a single model, which is why heterogeneous methods (different models in each phase) are often proposed. At low concentration of water in both phases, solubility models are preferred (see section 4.2.4.2, p. 297).

Cubic equations of state

Used in a homogeneous approach (same equation in all phases), **cubic equations of state with the classic alpha function and the classic mixing rule are unable to calculate correctly water + hydrocarbon mixtures**.

Nevertheless, by modifying the alpha function (see section 3.4.3.4, p. 207) all cubic equations can represent the water vapour pressure curve:

• specific coefficients of the **Twu** correlation are available [150];
• when the aqueous phase contains NaCl (salt), which is often the case in upstream applications, the model proposed by **Soreide and Whitson** [151] is recommended. This is a modification of the Peng-Robinson EoS where the "salt water" is considered as a single component, described using a specific "alpha" function that takes into account the presence of salt (c_{sw} is the salt molarity). The equation is:

$$\alpha = \left(1 + 0.4530\left(1 - T_r\left(1 - 0.0103c_{sw}^{1.1}\right)\right) + 0.0034\left(T_r^{-3} - 1\right)\right)^2 \qquad (4.29)$$

Specific mixing rules can be used as well. Several such rules have been proposed [152-156]:

• Using the classical mixing rule, only the hydrocarbon phases (vapour or liquid) can be described (water uptake). The value of $k_{ij} = 0.5$ between water and hydrocarbon is often a good choice. **Eubank** [157] proposes an improvement with the empirical correlation:

$$k_{ij} = A_j - B_j T_{rj} - \frac{C_i}{T_{rj}} \qquad (4.30)$$

where T_{rj} is the reduced temperature of the hydrocarbon.

• If an aqueous phase is present, equilibrium can be computed using distinct k_{ij} values in each phase. **Soreide and Whitson** [151] propose, together with the specific equation (4.29) for water:

$$a_{ij}^{NA} = \sum_i \sum_j y_i y_i \left(a_i a_j \right)^{1/2} \left(1 - k_{ij}^{NA} \right)$$

$$a_{ij}^{AQ} = \sum_i \sum_j x_i x_i \left(a_i a_j \right)^{1/2} \left(1 - k_{ij}^{AQ} \right) \qquad (4.31)$$

where the superscripts NA and AQ refer to non-aqueous and aqueous phases.

• A modification to the SRK model has been proposed by Kabadi and Danner [152] and by Panagiotopoulos and Reid [153] for SRK and PR to best fit the vapour pressure of water. In addition, specific mixing rules have been published by Michel *et al.* [154], Daridon *et al.* [155] and Mollerup and Clark [156].

The **Kabadi and Danner** [152] mixing rule is:

$$a(T) = \sum_i \sum_j x_i x_j \sqrt{a_i a_j} \left(1 - k_{ij} \right) + \sum_{i \neq w} a_{wi}'' x_w^2 x_i$$

$$a_{wi}'' = G_i \left(1 - T_{rw}^{0.8} \right) \qquad (4.32)$$

where the index w stands for water and G_i can be calculated from group contributions.

• As shown in section 3.4.3.4.D, p. 211, activity coefficient models can be combined with a cubic equation of state (**G^E-mixing rules**), thus yielding a very powerful mixing rule which can be used for water + hydrocarbon mixtures. Specific interaction parameters for each binary should then be regressed.

Cubic Plus Association (CPA)

The CPA equation of state (section 3.4.3.5.B, p. 218) is now widely recognised as a very powerful method for modelling water + hydrocarbon mutual solubilities [158]. Although several authors have further improved this equation by re-fitting the parameters of organic molecules, it can also be used with the corresponding states parameters for the hydrocarbons [159].

Example 4.8 Prediction of a heteroazeotrope with total liquid immiscibility

Calculate the temperature at which the water(1) + benzene(2) binary mixture form three distinct phases at 0.1 MPa pressure. Give the phase compositions.

When considering only the moderate temperature region (*i.e.* well below the water critical point), a number of simple approximations can be made:

• The hydrocarbons and water are entirely insoluble in each other in the liquid phase.
• The vapour phase is an ideal gas.
• The vapour pressures may be obtained using the Antoine equation:

$$\log P^\sigma = A + \frac{B}{C+T} \tag{4.33}$$

where P^σ is expressed in torr and T in °C, and where the numerical parameters are given in table 4.4.

Table 4.4   Parameters for vapour pressure equation of benzene and water

	A	*B*	*C*
Benzene	6.90565	– 1211.033	220.79
Water	7.96681	– 1668.21	228.0

Analysis:

The Gibbs phase rule states:

$$\Im = \mathcal{N} - \phi + 2$$

There are two components and three phases in the heteroazeotropic point, therefore only one degree of freedom. If pressure is fixed, the solution is defined.

Solution:

The diagram is constructed starting from the pure components. The boiling temperature of benzene is 353.15 K, that of water 373.15 K. At low temperature, two liquid phases are present, whose composition is known:

for the hydrocarbon phase: $x_1^{HC} = 0$, $x_2^{HC} = 1$ and

for the aqueous phase: $x_1^{aq} = 1$, $x_2^{aq} = 0$.

When the vapour is in equilibrium with the aqueous phase, we have

$$y_1 P = P_1^\sigma \tag{4.34}$$

This is the water dew curve.

When the vapour is in equilibrium with the hydrocarbon phase, we have

$$y_2 P = P_2^\sigma \tag{4.35}$$

This is the organic dew curve.

When the vapour is in equilibrium with both aqueous and hydrocarbon liquids, both equation (4.34) and (4.35) are true simultaneously. They can be summed to yield

$$P = P_1^\sigma + P_2^\sigma \tag{4.36}$$

The total three phase pressure is equal to the sum of the two vapour pressures. Only one temperature satisfies this condition: $T = 342.25$ K. At this temperature, the vapour phase composition can be calculated: $y_2 = 0.3$.

Below this temperature, the vapour phase is absent; above this temperature, one of the liquid phases is absent.

The two dew curves (equations 4.34 and 4.35) are thus constructed. They meet at the three-phase point. The diagram is shown in figure 4.35.

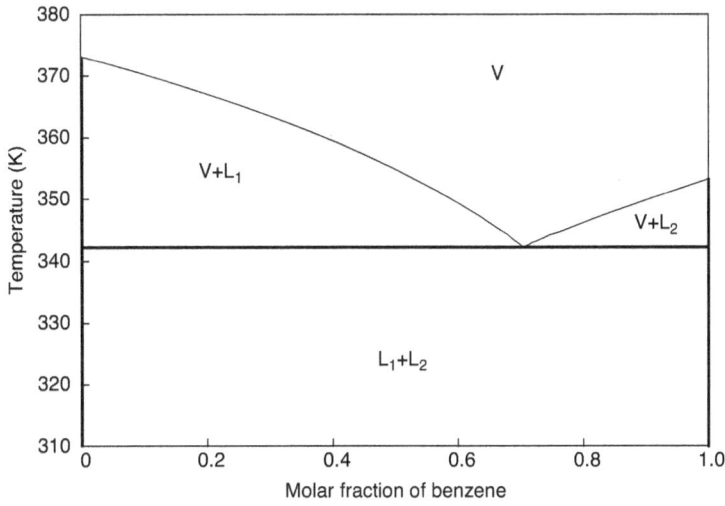

Figure 4.35

Benzene + water heteroazeotrope *Txy* diagram at 0.1 MPa.

This example is discussed on the website:
http://books.ifpenergiesnouvelles.fr/ebooks/thermodynamics

As a consequence of the heteroazeotropic behaviour, the temperature at which the first liquid drop forms upon cooling (the dew point) is lower than the condensation temperature of both pure components. We must always distinguish between the water dew point and the organic dew point since they concern a different liquid phase.

4.2.4.2 Solubilities

In the actual process, it may be important to know in some detail the solubility of water in the organic phase or that of hydrocarbons in the aqueous phase. The two cases will be considered separately below [160].

A. Water solubility in the organic phase

Water solubility in hydrocarbons is an important subject for various reasons, for example corrosion risk upon aqueous phase condensation if the solubility limit is reached due to a change in process conditions.

It is generally observed, under identical pressure and temperature conditions, that water is much more soluble in hydrocarbons than hydrocarbons are in water. This solubility, however, varies greatly depending on whether the organic phase is liquid or gaseous. In a liquid phase, while the solubility is almost independent of pressure, it is highly dependent on temperature. Figure 4.36 shows that the molar solubility in a liquid phase which is almost independent of the hydrocarbon molar mass. The hydrocarbon polarity plays a role: aromatic hydrocarbons absorb the most water.

Water solubility in a gas phase is presented in the GPSA diagram in figure 4.37. Solubility decreases with pressure, and increases with temperature, in accordance with the simple expression of the water distribution coefficient, already introduced in the example 4.8:

$$K_{water} = \frac{P^{\sigma}_{water}}{P} \tag{4.37}$$

It is clear that this expression is a simplification based on low pressure (ideal behaviour of the vapour phase) and water is assumed to be truly pure. This simple rule breaks down either:

- in the presence of salts, where the water volatility decreases with salinity;
- in the presence of acid gases, as they have a strong affinity with water, and especially so at high pressure (the vapour phase is no longer ideal). This phenomenon is further illustrated below (figure 4.45).

In the GPSA diagram, the salt effect is expressed through a correction shown in the inset. The solubility is expressed in mass fraction, resulting in a second correction related to the gas molecular mass or density (second inset).

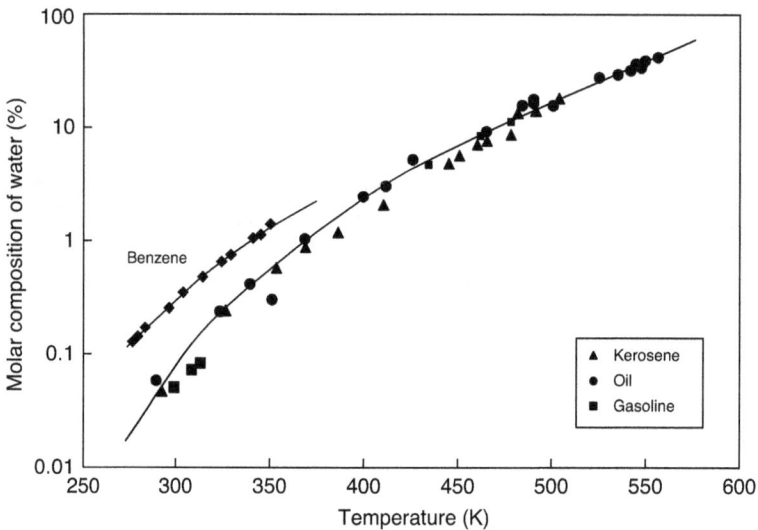

Figure 4.36

Water solubility in liquid hydrocarbons (original from Griswold and Kasch [162] adapted to SI by the authors).

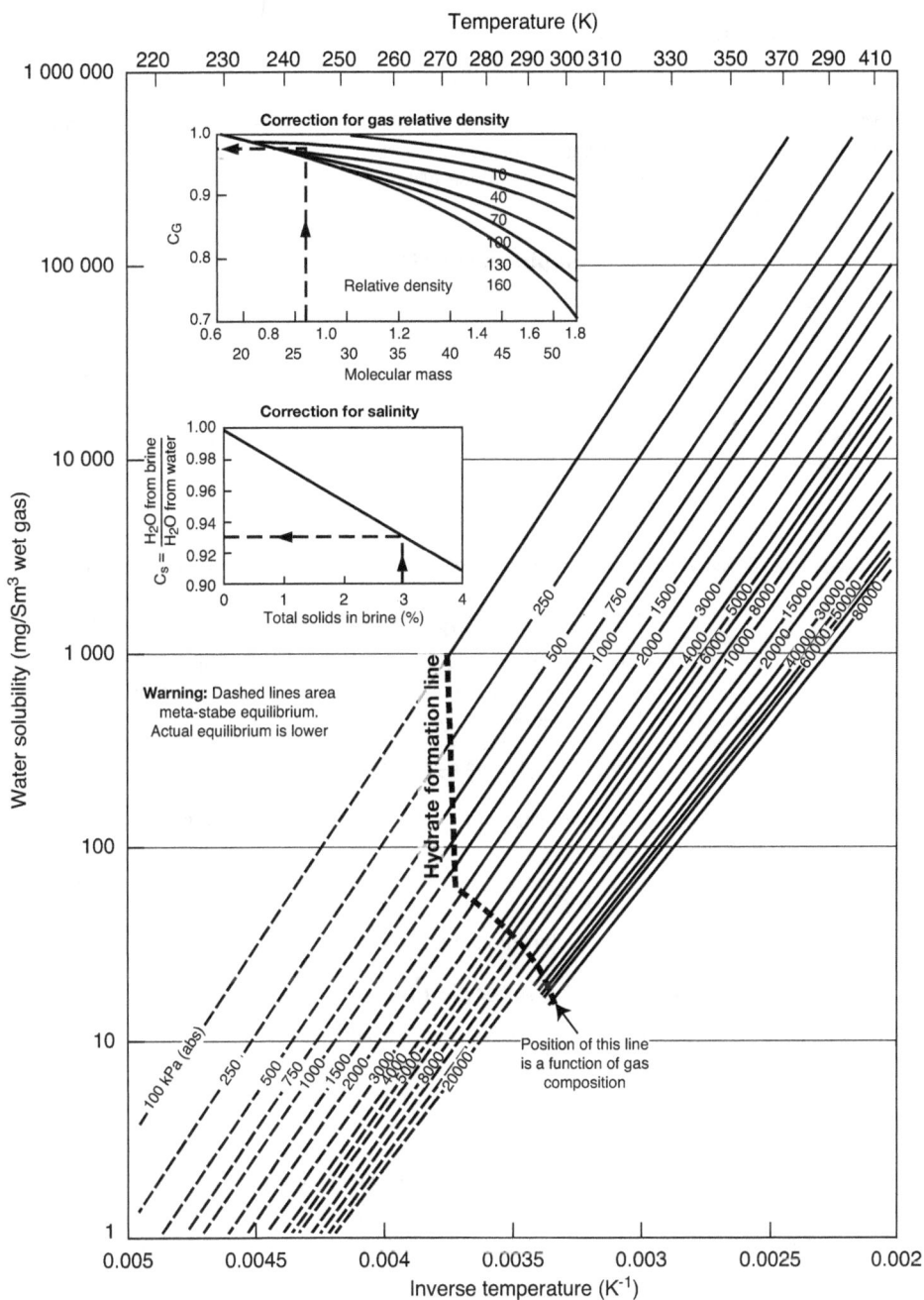

Figure 4.37

Water solubility in natural gas with correction for salinity and relative density (original from McKetta, in Campbell [161], adapted to SI by the authors).

Model Recommendations

Nomographs

Many techniques to evaluate solubility of water in liquid hydrocarbons are based on reading nomographs. If the solubility of water must be estimated in natural gas, the chart of McKetta and Wehe [165] is recommended by GPSA and API (figure 4.37). The work of Griswold and Kash [162] is commonly used as well.

Empirical correlations

• **Tsonopoulos** [163, 164] proposes simple correlations in order to obtain a numerical value of the water solubility in *n*-alkanes as a function of carbon atom number C_N at 298 K, using:

$$\ln(x_W) = \frac{A + BC_N}{C + C_N} \tag{4.38}$$

• **Eubank** [157] has proposed a simple relationship providing the water solubility in *n*-alkanes (written as x_w^0) as a function of temperature:

$$\ln(x_w^0) = -21.2632 + 5.9473 \times 10^{-2}T - 4.0785 \times 10^{-5}T^2 \tag{4.39}$$

• **Brady** [166] notice a relationship between this solubility and the double bond index *I*, written as

$$I = 2\frac{n_D}{n_C} \tag{4.40}$$

where n_D is the number of double bonds and n_C the carbon atoms number. Assuming that the water solubility in hydrocarbons can be calculated using a correction on the solubility in *n*-alkanes x_w^0 , and using Brady's observations, it is possible to write:

$$x_W = x_w^0\left(1 + (13.85 - 0.0267T)I\right) \tag{4.41}$$

• **Tsonopoulos** [163, 164] has investigated extensively all solubility data and proposes a simple correlation:

$$\ln(x_W) = A + \frac{B}{T} \tag{4.42}$$

where *A* and *B* are provided for each hydrocarbon.

Estimate of the water distribution coefficient

In case the water distribution coefficient is required in a vapour-liquid equilibrium, and in *the absence of an aqueous phase*, the equation that can be used is:

$$K_{water} = \frac{P_{water}^\sigma}{Px_w} \tag{4.43}$$

where P_{water}^σ is the water vapour pressure and x_w is the water solubility in the hydrocarbon, as calculated using, for example, (4.41) or (4.42).

In the presence of an aqueous phase, as shown in example 4.8, the water distribution is rather expressed as (4.37).

Equation of state

The models recommended above (cubic EoS or CPA) are well-adapted to calculate the water uptake of hydrocarbons.

B. Hydrocarbon solubility in the aqueous phase

The solubility of hydrocarbon components in water is of great importance for environmental issues. High precision is necessary when predicting these small amounts. Under high pressure and high temperature solubility increases, as illustrated by Dhima (1998 [167]). This effect is observed in natural gas reservoirs where prediction of the amount of gas trapped in water is not negligible.

Figure 4.38 shows how the solubility in water varies with chemical component and temperature. A minimum as a function of temperature is almost always observed in the range between 273 K to 293 K:

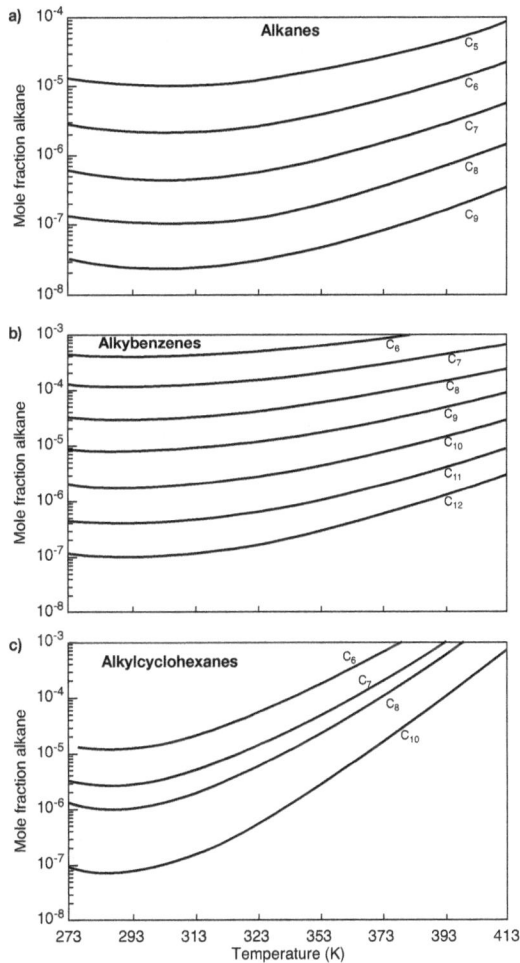

Figure 4.38

Solubility of different families of hydrocarbons in water adapted from Tsonopoulos [163, 164].

Reprinted from [163, 164], with permission from Elsevier.

Model Recommendations

Very few models are able to correctly predict very small hydrocarbon solubilities in water. The equations proposed in the general section (4.2.4.1, p. 293) provide acceptable results, but equations below should be used if accuracy is required.

Empirical solubility correlations

• **Tsonopoulos** [163]

For *n*-alkanes, the solubility at 298,15 K is given as: (regression valid for $4 < C_N < 10$).

$$\ln(x_{hc}) = -3.9069 - 1.51894 C_N \tag{4.44}$$

For alkylbenzenes, alkylcyclohexanes and 1-alkenes, the expressions are respectively [168]:

$$\ln(x_{hc-alkylbenzenes}) = 4.25097 - 1.559463 C_N - \frac{14.89081}{C_N}$$

$$\ln(x_{hc-alkylcyclohexanes}) = -3.74419 - 1.27431 C_N \tag{4.45}$$

$$\ln(x_{hc-1-alkenes}) = -2.24515 - 1.54381 C_N$$

• **Maczynsky et al.** [169-177]

In a recent series of papers, Maczynsky, Wisniewska, Goral and some other contributors obtained a good representation of hydrocarbon solubilities in water. The equation they propose is for alkanes, unsaturated hydrocarbons and ethers (except aromatics):

$$\ln(x) = \ln(x_{min}) + C\left[\frac{T_{min}}{T} - \ln\left(\frac{T_{min}}{T}\right) - 1\right] \tag{4.46}$$

This equation indicates that ln(x) is a linear function of $\left[\dfrac{T_{min}}{T} - \ln\left(\dfrac{T_{min}}{T}\right) - 1\right]$. Both coefficients of the straight line have been observed to be a function of the SRK critical excluded volume $b = 0.08664 R T_c / P_c$.

$$\ln(x_{min}) = c_1 + c_2 b + c_\pi L$$
$$C = c_3 + c_4 b \tag{4.47}$$

where *L* is the number of double bonds in the molecule (*L* = 0 for alkane, *L* = 1 for alkynes, *L* = 2 for alkadienes and alkynes, *L* = 4 for alkadiynes). For aromatics the equation to use is:

$$\ln(x) = \ln(x_{min}) + C\left[\frac{T_{min}}{T} \ln\left(\frac{T_{min}}{T}\right) + 1 - \frac{T_{min}}{T}\right] \tag{4.48}$$

The critical constants and the coordinate of the minimum are given for many components.

Henry's constants

Several authors have proposed the use of Henry constants to calculate solubilities in water. It allows a more theoretical approach (in particular extrapolation with pressure). The use of Henry's constant is explained in section 2.2.3.1.A (p. 66).

• **Dhima et al.** [167, 178, 179] regressed Henry constants at the water vapour pressure, using following correlations:

$$\ln H_i^\sigma(T) = A_i' + \frac{B_i'}{T} + \frac{C_i'}{T^2} \tag{4.49}$$

They also propose to correlate the infinite dilution partial molar volume to obtain pressure dependence of the Henry constant:

$$H_{i,w}(P,T) = H_{i,w}^\infty \cdot \exp \frac{\overline{v_i^\infty}(T)\left(P - P_w^\sigma\right)}{RT}$$ (4.50)

where the partial molar volume at infinite dilution is calculated from Lyckmann *et al.* [180]

• **Plyasunov *et al.*** [181-184] proposed a model to calculate infinite dilution properties in water, using a group contribution approach. Their approach is attractive in that they use the fundamental thermodynamic relations in order to relate all infinite dilution properties (volume, enthalpy, Henry constant).

• **Harvey** [185] proposes a relatively simple correlation which provides Henry constant values up to the critical point:

$$\ln H_i = \ln P_{water}^\sigma + \frac{A}{T^*} + \frac{B\left(1-T^*\right)^{0.355}}{T^*} + C \frac{\exp\left(1-T^*\right)}{T^{*0.41}}$$ (4.51)

where $T^* = T/T_{c,water}$.

C. Hydrocarbon solubility in salt solutions

When the water contains salts, it is well-known [186-188] that the solubility of hydrocarbons decreases. Figure 4.39 illustrates this "salting-out" effect on the solubility of benzene with sodium chloride. However, Bradley *et al.* [189] and Fedushkin *et al.* [190] indicate that some ions (Ag^+ and K^+ in particular) form complexes with benzene and toluene. Thus, their presence in the aqueous phase can increase significantly the solubility of these aromatics, resulting in a "salting-in" effect.

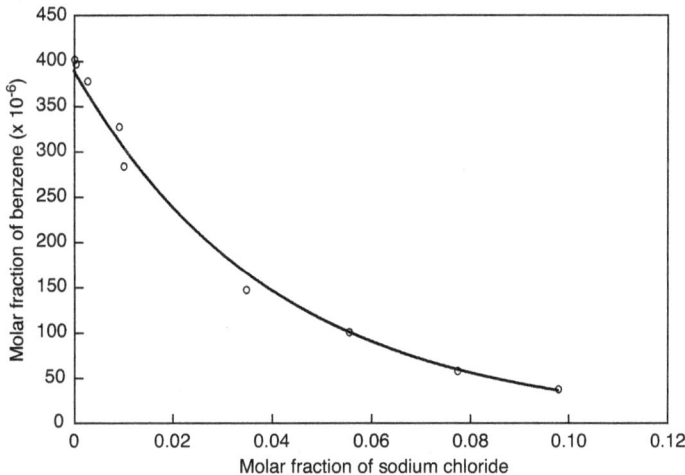

Figure 4.39

Salting out effect on solubility of benzene in presence of water + sodium chloride [189].

Model Recommendations

Equations of state

The salting out behaviour is correctly described by the **Soreide and Whitson** model, although it only considers NaCl as a salt. An equivalent salinity must therefore be defined if another salt is used.

Electrolyte activity coefficient models

The most accurate approach is the use of an electrolyte model, such as the **Pitzer** model, which is often used when complex salt mixtures are present (in particular in geochemical applications). These models have been briefly described in section 3.4.2.5 (p. 186).

Empirical solubility correlations

The simplest approach is the use of the **Setchenow** constants to calculate the decrease in solubility [191]:

$$x_i = x_i^0 \exp\left(-K c_s\right) \tag{4.52}$$

where c_s is the salt concentration and K depends on the type of salt. They propose to calculate the K coefficients as a function of ionic and gas contributions.

$$K = \sum_i \left(h_i + h_g\right) n_i \tag{4.53}$$

where n_i is the stoichiometry of the ion i in the salt, and the contributions h_i and h_g (for gas) are tabulated.

4.2.4.3   Fluid-Solid phase equilibrium (ice and hydrates)

As shown in the phase diagram figure 4.40, when the temperature is low enough, water can exist in two distinct solid phases.

The most well-known is ice, which can be treated essentially in the same way as any pure component solid. Nevertheless, when water and some light components coexist (light hydrocarbons, CO_2, H_2S, N_2), so-called water hydrates can be formed [64]. This is a specific crystalline water structure physically resembling ice, in which small non-polar molecules (typically gases) are trapped inside "cages" of hydrogen bonded water molecules. Without the support of the trapped molecules, the lattice structure of hydrate (or clathrate) would collapse into conventional ice crystal structure or liquid water. Most low molar mass gases (including O_2, H_2, N_2, CO_2, CH_4, H_2S, Ar, Kr, and Xe) as well as some higher hydrocarbons and freons will form hydrates at suitable temperatures and pressures. These crystalline structures are non stoichiometric, as the number of gas molecules per water molecule may vary within limits, depending on pressure and temperature. Three structures have so far been identified: structure I, structure II and structure H, in which the size and distribution of large and small cages are different. Hydrate formation will occur at moderate temperature (up to around 300 K) and high pressure (several MPa). Consequently, the risk of solid deposit is of industrial interest in gas production and transportation, due to possible plugging of pipes.

Figure 4.40 shows a typical *PT* phase diagram. Its shape depends on the fluid composition. This example is given for the case with excess water, and a hydrocarbon fluid whose

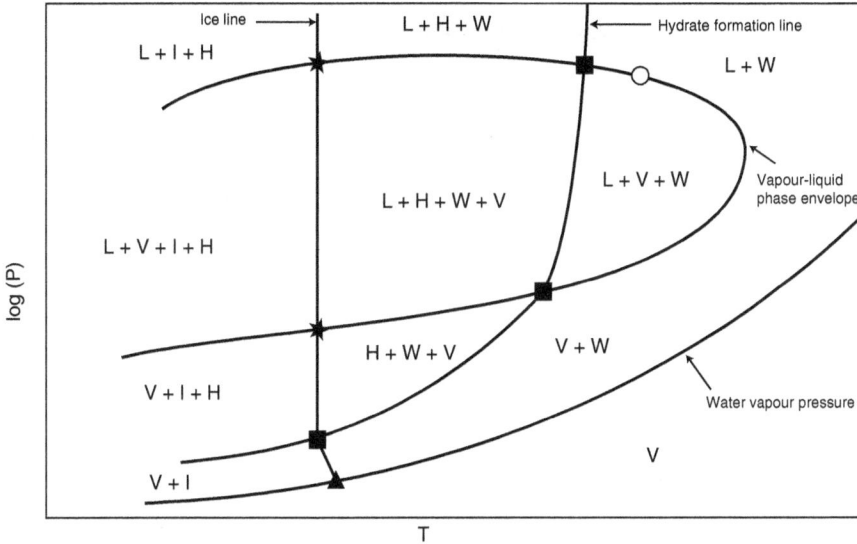

Figure 4.40

PT phase diagram of a hydrocarbon mixture in the presence of water, showing the region of hydrate formation. The phase labels are as follows: V: Vapour; L: Liquid hydrocarbon; W: Aqueous liquid; I: Ice; H: Hydrate. ▲ = Triple point; ■ = Quadruple point; ★ = Quintuple point.

phase envelope is shown as well as its critical point. All lines represent boundaries where a new phase appears. Where two lines cross, all phases that exist in the neighbouring zones coexist. On figure 4.40, one can see a triple point (three coexisting phases, V-I-W), three quadruple points and two quintuple points. Due to the possibility of having several different hydrate structures, the phase diagram for solids may be more complex than that presented here. The position of the hydrate formation line may be affected by the presence of inhibitors or enhancers. Compounds as methanol or glycols (mono ethylene glycol, di ethylene glycol, etc.) are well-known inhibitors [192], but dissolved salts have the same effect.

The books of Caroll (2004 [194]) and Sloan (2008 [64]) give a complete overview of hydrate problems from a process engineer's point of view and are recommended for further reading. Munck's paper (1989 [195]) is also a classical reference.

Model Recommendations

Empirical approaches

These approaches are designed only to calculate the hydrate phase boundary for methane hydrates.

One of the first relationships was proposed by Hammerschmidt in 1934 [196]. In this practical equation the hydrate dissociation pressure is expressed in psia and temperature in °F.

$$T = 8.9P^{0.285} \tag{4.54}$$

Van der Waals-Platteeuw model

As suggested by Sloan [197, 198], all process simulators dealing with calculations concerning the formation of water hydrates currently use a compositional model of the Van der Waals and Platteeuw (1959 [199]) that was further refined by Parrish and Prausnitz in 1972 [200]. It is described in some detail in section 3.4.4.2 (p. 223). Equilibrium is based on the equality of the water chemical potential (or fugacity) between all phases, taking into account the adsorbed components. The main variations between the various models based on the Van der Waals and Platteuw theory, concern the way the reference term and the mixing term are calculated. Similarly, the differences also come from the choice of how to model the other (non hydrate) phases (vapour, liquid hydrocarbon and aqueous). Sloan's team have proposed a complete formulation including several kinds of hydrates and ice formation for high ranges of pressure and temperature (Ballard, 2002 [201]).

Example 4.9 Formation of hydrates

To illustrate the fact that various models can lead to the same description of thermodynamic behaviour, figure 4.41 shows the pressure dissociation curve of methane hydrates as a function of temperature. Two models are compared with the data of Verma (1974 [202]). Both models use the Kihara parameters in the Van der Waals and Platteeuw theory but, for the vapour phase, Parrish & Prausnitz [200] use a modified Rechlich-Kwong EoS, while Youssef *et al.* [203] employ the CPA EoS that explicitly takes into account the auto-association of water. In the pressure range examined, the two models are in good agreement with experimental data.

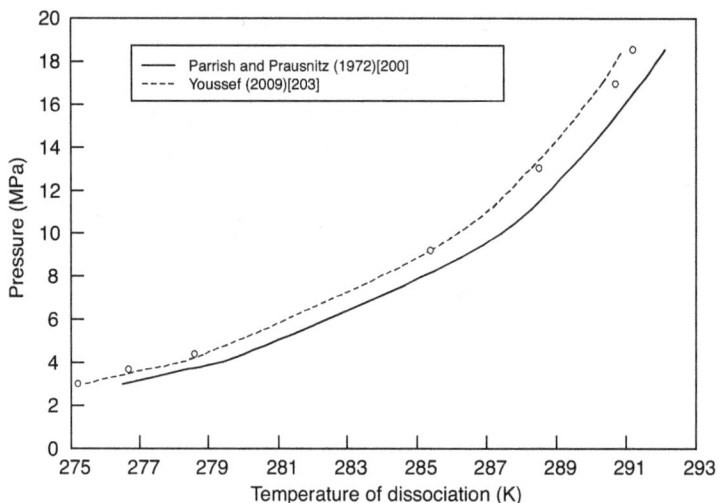

Figure 4.41

Dissociation conditions in *PT* diagram for methane hydrate (Youssef 2009 [203]), data from [202].

This example is discussed on the website:
http://books.ifpenergiesnouvelles.fr/ebooks/thermodynamics

4.2.5 Phase equilibrium in presence of CO_2/H_2S

Almost all natural gas systems contain greater or lesser amounts of CO_2 or H_2S. Although these gases are often considered to have a behaviour close to hydrocarbons, carbon dioxide has a strong quadrupole and hydrogen disulphide forms hydrogen bonds similar to those of water [204]. Their phase behaviour is described in some detail by de Hemptinne and Behar [205]. Carrol and Mather have made a significant contribution to this topic ([206-208]).

4.2.5.1 CO_2 and H_2S with hydrocarbons at high pressures

Due to extensive studies on CO_2 reinjection in reservoirs, many high pressure data have been published. An important compilation can be found by Wichterle *et al.* ([209, 210]). Less data exist for H_2S mixtures. The general appearance of vapour-liquid phase behaviour with light alkanes can be appreciated in figure 4.42.

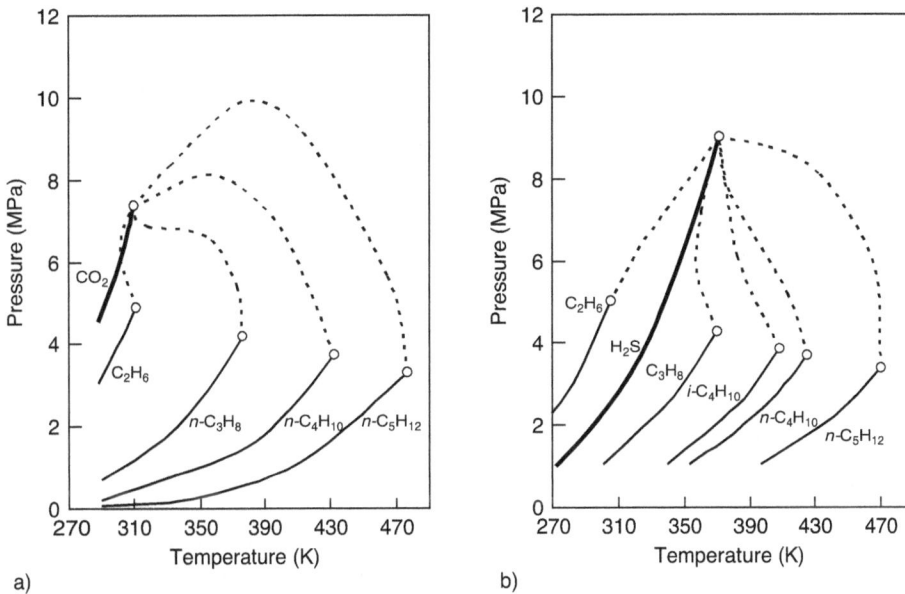

Figure 4.42

Critical loci of CO_2 [212] and H_2S [211] with various hydrocarbon solvents.
Reprinted with permission from [211, 212]. © 1989, American Chemical Society.

The vapour-liquid behaviour of CO_2 with a number of heavier alkanes is summarised in figure 4.43 (a). The vapour pressures of all pure components, including that of CO_2, is shown as solid lines, and the critical point loci as dotted lines.

For light alkanes (lower than n-C_7), only vapour-liquid equilibria are observed. The location of the CO_2 critical point (304.2 K and 7.383 MPa) is such that in a mixture with methane, CO_2 is the heavier component, while in a mixture with propane, it is the lighter component. As the critical temperatures of ethane and CO_2 are almost identical, the CO_2 + ethane phase behaviour is rather unusual. The corresponding *Pxy* phase envelopes are shown in figure 4.43 (b): up to 290 K, regular azeotropic behaviour is observed. Above this temperature, and below the critical temperature of either component, the azeotropic diagram is interrupted and two critical points can be found.

When the number of carbon atoms increases beyond n-C_7, a liquid phase split is observed as illustrated in figure 4.43 (a). A type II diagram is observed, meaning that in addition to the regular vapour-liquid behaviour, a liquid-liquid phase split is observed at low temperature.

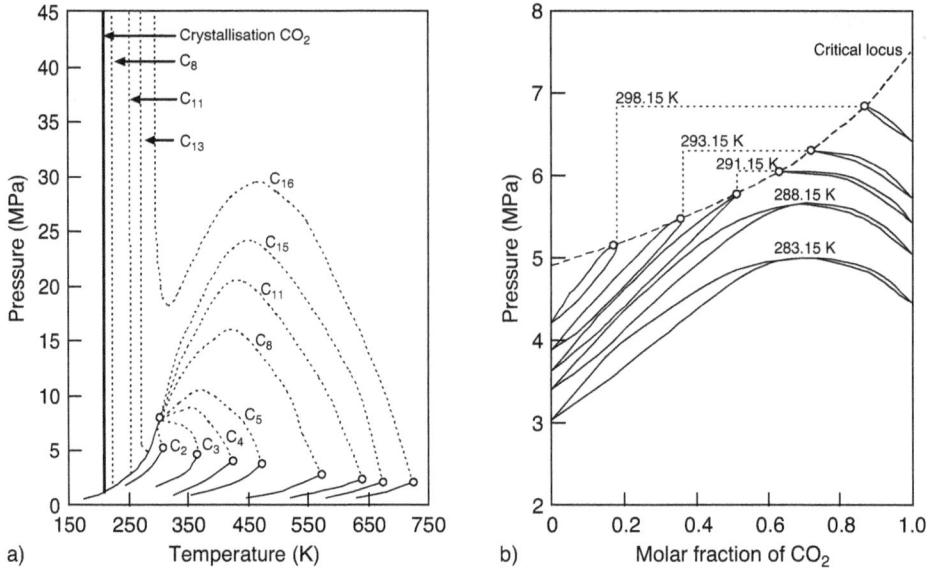

Figure 4.43

(a) *PT* projections for a number of CO_2 + n-alkane and (b) CO_2 + ethane binary mixtures ([212] and [213] respectively mentioned in [214]).

For the CO_2 + n-C_{13} binary mixture, the vapour-liquid and the liquid-liquid critical lines intersect forming a type IV phase diagram. From n-C_{14} onward, the two lines merge into a single continuous line. This is the onset of type III behaviour. The three-phase line still exists. It is invisible on the scale of the figure, being located slightly below the CO_2 vapour pressure line.

Model Recommendations

Cubic EoS with classic mixing rules

Cubic equations of state can correctly reproduce mixtures of CO_2 or H_2S with hydrocarbons. A binary interaction parameter is essential.

For mixtures with CO_2, the value of $k_{ij} = 0.13$ is generally recognised as a good choice [215] for paraffins; for aromatics, the value varies between 0.1 and 0.13 [216]. Kato [215] even proposes a correlation with the acentric factor:

$$k_{ij} = a'(T - b')^2 + c' \tag{4.55}$$

where

$$a' = -0.70421 \times 10^{-5} \log(\omega_i) - 0.132 \times 10^{-7} \tag{4.56}$$

$$b' = 301.58\omega_i + 226.57 \tag{4.57}$$

$$c' = -0.0470356\left[\log(\omega_i) + 1.08884\right]^2 + 0.1304 \tag{4.58}$$

For equilibrium with H_2S, the k_{ij} is highly dependent on the hydrocarbon molar mass. Carroll and Mather [207] propose several correlations, using the hydrocarbon molar mass, acentric factor or boiling temperature. Considering how easy it is to use the acentric factor, we present it here:

$$k_{ij} = 0.088 - 7.82 \times 10^{-2}\omega_i - 7.94 \times 10^{-4}\omega_i^2 \tag{4.59}$$

PPR78

The group contribution calculation of the binary interaction parameter, proposed by Jaubert's group, also works well for acid gases with hydrocarbons (section 3.4.3.4.E, p. 214).

4.2.5.2 CO_2 and H_2S solubilities in liquid hydrocarbons

The solubility of these two light components in various types of hydrocarbon is found at 298.15 K and atmospheric pressure in the data compilation of Fogg and Gerrard [69]. For hydrogen sulphide, authors have observed a molar solubility between 3% and 6%.

Solubility of carbon dioxide is smaller and varies between 0.6% and 1.5% molar at the same temperature 298.15 K and atmospheric pressure. For alkanes, a linear trend line is proposed by Fogg and Gerrard as a function of the number of carbon atoms.

We may state that alkanes are generally better solvents than the other hydrocarbon families shown, except for small aromatics that dissolve H_2S better. It is difficult to extract more information from these plots (figure 4.44).

The molar fraction solubility of both CO_2 and H_2S in hydrocarbon solvents decreases with temperature [69], like the other highly soluble gases shown in figure 4.30.

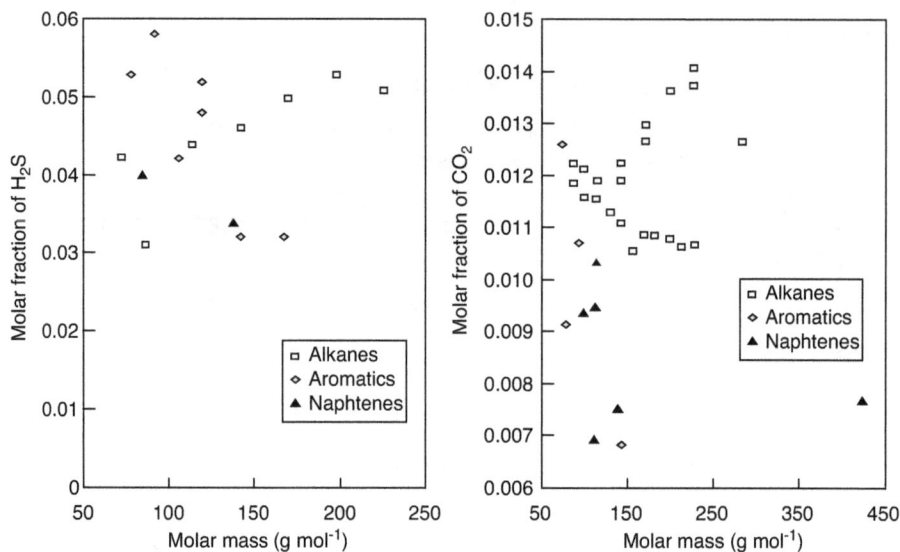

Figure 4.44

Solubilities of H_2S and CO_2 in various hydrocarbon solvents [214], at 298.15 K and atmospheric pressure.

4.2.5.3 Phase equilibria of industrial solvents with CO_2 and H_2S

Acid gas treatment consists in absorbing the acid gas (*i.e.* CO_2 and H_2S) in a solvent. Many alternative solvents are under investigation [217], but at present, only two families of solvent are mainly used on an industrial scale: physical solvents and chemical solvents. A combination of both is however possible.

A. Physical solvents (water and alcohols)

Physical solvents are liquids that exhibit high acid gas solubility. Most often, a mixture of alcohols and water is used. The $H_2O + CO_2$ phase behaviour is illustrated in figure 4.45. It features vapour-liquid, liquid-liquid and a three-phase vapour-liquid-liquid equilibrium line.

The figure 4.46 compares the water uptake of a number of gases as a function of pressure, at similar temperatures. For acid gases such as CO_2 and H_2S, the polarity is such that at higher density, their hydrophobicity is lowered. Hence, a minimum is observed with increasing pressure, and high pressure acid gases are known to contain large amount of water. The water solubility in methane, much lower, is also shown in figure 4.37.

The relatively hydrophilic character of hydrogen sulphide and carbon dioxide is even more visible in the liquid phase (Gillespie and Wilson [221]): as a result of the strong polarity of the dense phase, water is more soluble in liquid hydrogen sulphide and carbon dioxide than in the vapour phase of these components. This is visible in figure 4.45, where the *Txy* cuts clearly show on the three phase lines that the liquid composition is richer in water than the vapour composition.

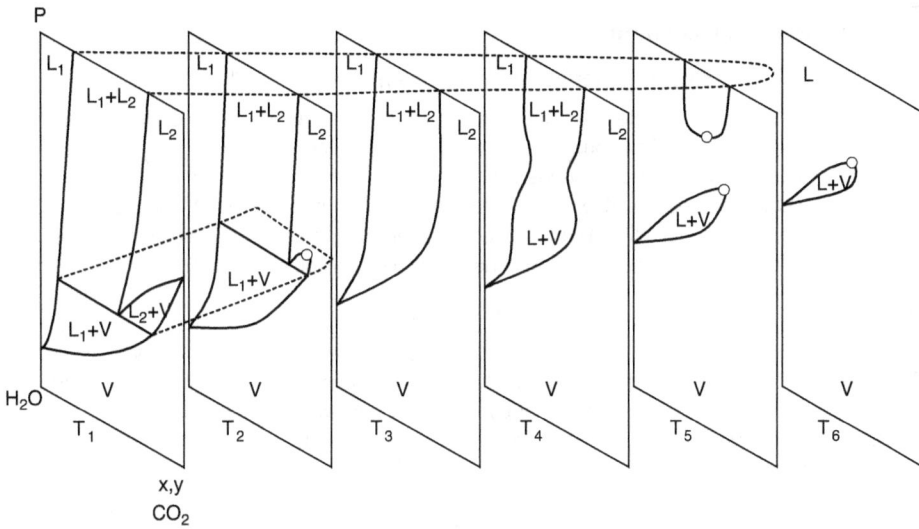

Figure 4.45

TPxy phase diagram for the $H_2O + CO_2$ mixture (from Tödheide and Franck, 1963 [222]).

Figure 4.45 is qualitatively similar to figure 4.34, with the difference that the three-phase pressure is now in between both vapour pressures of the pure components. When more complex mixtures are considered, the three-phase line becomes a three-phase region.

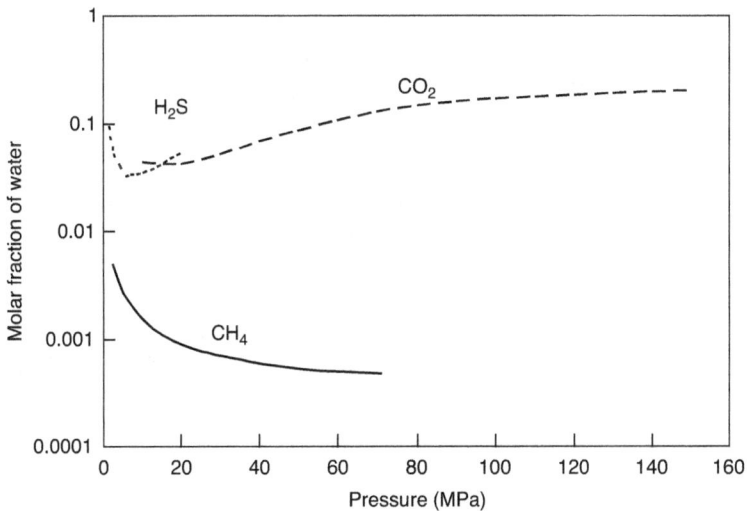

Figure 4.46

Solubility of water in acid gases and methane (data from CO_2 at 383 K [218], H_2S at 377 K [219], CH_4 at 373 K [220]).

Model Recommendations

Heterogeneous (activity coefficient) models

Due to the presence of potentially supercritical components, an asymmetric approach is required to calculate the liquid phase properties with a heterogeneous model (the reference state for gases is infinite dilution in the solvent, while for solvents, it is pure components). Henry constants are then used to calculate the gas solubility in each solvent, and a suitable activity coefficient model (NRTL) can be used provided that the correct reference states are taken into consideration. This has been done but results in a complex approach.

Cubic EoS

To describe the full phase diagram in the presence of water + alcohol mixtures, it is no longer possible to use a classical mixing rule. If a cubic equation of state is chosen, it must be combined with a G^E mixing rule that contains a suitable activity coefficient model. The NRTL model, as used for drawing figure 4.47 is appropriate provided that all binary parameters are adjusted.

CPA

The CPA equation has also been used to describe this system [223, 224]. The major advantage of this equation is that for all hydrocarbons contained in the mixture, the simple SRK cubic equation of state is recovered. No new binary interaction parameter needs therefore to be determined and the pseudo-component correlations can be used. For alcohol and water molecules, specific parameters have been proposed by Kontogeorgis, and interaction parameters have also been proposed by de Hemptinne *et al.* [159]. For mixtures with H_2S, Ruffine *et al.* [204] proposed another set of parameters.

Example 4.10 Example of a vapour-liquid-liquid equilibrium of an acid gas system in the presence of water

Analysis:

In this example, we illustrate how some selected models are capable to predict the phase envelope of a quaternary system composed of H_2O, H_2S, CH_4, CO_2 (50/40/5/5 molar percent). This system has been examined experimentally by Robinson *et al.* (1982) [226] and figure 4.47 shows its phase diagram. This system is quite interesting as it shows both non-ideal mixing behaviour as high pressure effects, in a small range of temperature and pressure.

At high temperature and low pressure, the system is a vapour phase. When the temperature and pressure are large (upper right corner of the diagram), the water forms an aqueous phase and a vapour-liquid domain is observed. For lower temperatures, the hydrogen sulphide forms a second liquid phase and we obtain a three phase equilibrium. Finally, at lower temperatures and high pressure, the vapour phase disappears and we have only liquid-liquid equilibrium.

The chosen models are presented in table 4.5, along with the number of empirical binary parameters that have been adjusted. The fitting used binary VLE data, *i.e.* other than those presented in figure 4.47: calculation of the phase diagram is performed in a predictive mode. The origin of the parameters is also provided in the table. The fewer binary parameters, the better the theoretical foundation of the model must be.

Table 4.5 Parameters depending on selected model

Name	Detail of the model	Number of binary parameters	Origin of binary parameters
SW (Soreide-Witson)	Peng-Robinson EoS with $k_{ij} \neq 0$ for aqueous phase Peng-Robinson EoS with $k_{ij} \neq 0$ for organic phase	12 (6 for each EoS)	Aqueous: Soreide-Witson Organic: state of the art published values
PRH (Peng-Robinson Huron-Vidal)	Peng-Robinson EoS with Huron-Vidal mixing rules with NRTL-V G^E approach	12 (2 for each binary)	Parameters fitted on each binary systems
PSRK	Soave-Redlich-Kwong EoS with MHV1 mixing rules with UNIFAC G^E approach	–	UNIFAC matrix
CPA	Cubic Plus Association with k_{ij} parameters	6 (1 for each binary)	Parameters fitted on each binary systems

Results:

The phase diagrams predicted by these models are shown on figure 4.47.

The water dew curve (lowest curve) is well described by all models: the deviation between experimental and calculated pressure is less than 2.5%.

The prediction of the three phase equilibrium is more difficult. Since we deal with a multicomponent mixture, this three-phase equilibrium appears in a zone (it would have been a line for a binary mixture) that is limited by a hydrocarbon dew curve on the bottom and a hydrocarbon bubble curve on top. A vapour-liquid critical point (not drawn) is found at the limit between dew and bubble lines. The dew curve corresponding to the appearance of the hydrogen sulphide-rich liquid phase upon pressure increase is close to the pure H_2S vapour pressure curve and is well predicted by all models (except PSRK).

On the other hand, the bubble curve of the three phase equilibrium is difficult to estimate and there are large differences between the models. For PRH and SW, agreement with experimental data is close to perfection. This result illustrates that it is possible to describe this type of complex mixture phase diagrams knowing its binary sub-systems.

PSRK and CPA produce the worst results: bubble temperature is underestimated by up to 40 K for PSRK and up to 10 K for CPA; bubble pressure is over-estimated by 4 MPa for PSRK and 3 MPa for CPA. We must bear in mind that PSRK is used in a totally predictive manner and that the equation of state gives a good qualitative idea of the phase behaviour. For CPA, the result is very interesting since it uses fewer parameters. It may even produce better results if the binary $H_2O + H_2S$ is described with more detail as in [227].

To conclude this example, we can note that in opposition with the other models, SW uses a heterogeneous approach (hence no continuity between the aqueous and the hydrocarbon phases). As a result, SW is not able to calculate the critical point between the aqueous and a hydrocarbon phase. This has no real impact on the figure since the only critical point seen is that between the liquid and the vapour hydrocarbon phases.

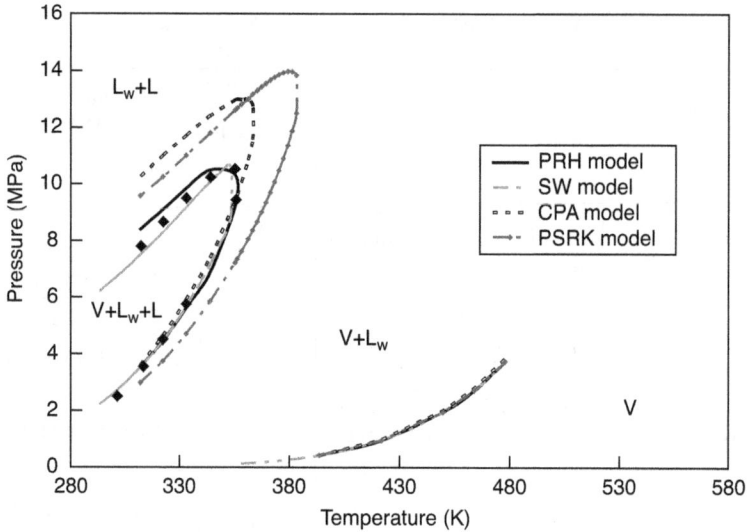

Figure 4.47

Phase envelope of the water + hydrogen sulphide + methane + carbon dioxide (data from [226]). L_w for aqueous liquid and L is the another liquid phase.

This example is discussed on the website:
http://books.ifpenergiesnouvelles.fr/ebooks/thermodynamics

B. Chemical solvents

CO_2 and H_2S are two acid components (their respective acidity constants pKa are about 6.4 and 7.0 at 298.15 K) and this property can be used in order to find a good solvent. Basic solvents such as solutions of aqueous alkanolamines (monoethanolamine, di-ethanol amine, etc.) have a high absorption capacity because the amine component reacts chemically with the acid gases. For example, with CO_2, we have:

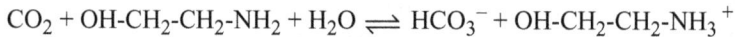

$$CO_2 + OH\text{-}CH_2\text{-}CH_2\text{-}NH_2 + H_2O \rightleftharpoons HCO_3^- + OH\text{-}CH_2\text{-}CH_2\text{-}NH_3^+$$

This is why this kind of solvent is called a chemical solvent.

Studies of these systems involve simultaneous resolution of the phase equilibrium and the reaction equilibrium (section 2.2.4, p. 86). Specific algorithms should be developed to solve this type of problem. The main purpose of the modelling work is to reproduce the absorption isotherm as shown in figure 4.48. This graph gives the partial pressure of acid gas as a function of its loading in the liquid phase (defined as the ratio of the number of moles of CO_2 and the number of moles of amine). We see that the partial pressure increases with temperature. Note also that the pressure range covers 5 decades. This behaviour is typical of chemical systems.

Considering the industrial importance of these mixtures, a large number of data have been measured, but not all have been published. Barreau *et al.* [228] and Blanchon Le Bouhelec *et al.* [229] report experimental measurements and compare them with theoretical model predictions (electrolyte NRTL) as shown in figure 4.48. We should also mention the works of Garst and Lawson (1972-1976 [230-232]), Lee and Mather (1977 [233]), Jou *et al.* (1982-2000 [234-236]), Carroll *et al.* (1993; 2002 [206, 237]), and the work of Anderko *et al.* (2002 [238, 239]) based on the previous works of Li and Mather (1994 [240, 241]).

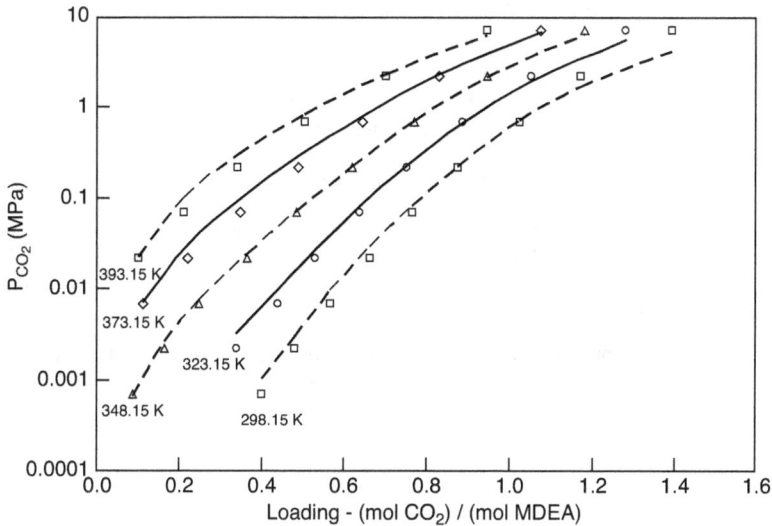

Figure 4.48

Carbon dioxide + MDEA + water (comparison of data of Li *et al.* [241] and Blanchon Le Bouhelec's NRTL electrolyte model [242]).

Model Recommendations

Considering that the increased solubility is due to the chemical equilibrium between ionic species, the model must take into account this chemical equilibrium. For this reason, both chemical and physical equilibria must be considered simultaneously.

Semi – empirical approach

The first compositional approach was developed by Kent and Eisenberg (1976) [243]. They solve the mass balance with the mass action laws of the chemical reactions. In their model, non-ideality of the liquid phase is included into the chemical equilibrium constant, which is apparent. This kind of approach allows good representation once fitted on a data set, but extrapolation is not possible.

Electrolyte activity coefficient model

A more rigorous approach consists in solving the same chemical equilibrium relations, but the non-ideality of the liquid phase is calculated using an activity coefficient model (*i.e.* concentrations are replaced by activities). The activity coefficient model must explicitly take into account the charge effect of the ionic species (as discussed in section 3.4.2.5, p. 186). The classical activity models (NRTL, UNIQUAC, etc.) allow this type of extension and there is also a specific approach for ionic species such as Pitzer's activity model. The NRTL-electrolyte approach has been used by the team of Rochelle (Austgen 1989 [244], Posey 1996 [245]) and is available in some process simulators. A description of this approach is given by Blanchon Le Bouhelec (2007 [228, 229]) and this model is used in figure 4.48. An approach such as this not only results in good modelling of phase equilibria, but also in good prediction of the heat of absorption, which is a second important aspect of these systems. For more details on the model, refer to Barreau *et al.* [228].

Electrolyte equation of state model

It is also possible to model these systems with an equation of state including a specific ionic term. Vallée [246, 247] used the electrolyte equation of state of Fürst and Renon [248] to describe water + alcanolamine + acid gases systems. However, this approach is still under development and should be considered as an exploratory avenue.

In all cases, it is important to stress the need for a large number of good quality experimental data to regress the numerous parameters.

4.2.6 Phase equilibrium in the presence of molecules containing heteroatoms (e.g. oxygenated)

Biomolecules contain large amounts of oxygen, which results in polar molecules or the formation of hydrogen bonds. All possible non-idealities can then be observed. The expected phase behaviour can be explained using the simple rule that the stronger the inter-actions among molecules of the same kind, the larger the positive deviation from ideality and the larger the risk for liquid-liquid phase split. Two types of interaction can be identified: polar (weaker) or hydrogen bonding (stronger type of interaction).

Two major families of anti-solvents can be identified for oxygenated molecules:

- one has a very strong polar and hydrogen-bonding capacity (water, for example),
- the other exhibits very non-polar behaviour (e.g. long chain *n*-alkanes).

As an example, it has been documented that phenol forms liquid-liquid equilibrium with water, but also with alkanes heavier than $n\text{-}C_6$.

The phase diagrams found are often of type II (with an upper critical end point, as discussed in section 4.2.1, p. 265).

4.2.6.1 Polar interactions

The presence of a heteroatom results in the formation of permanent dipoles (examples are esters, ketones, ethers, etc.). Polar molecules interact more strongly with other polar molecules, which explains why an azeotropic phase diagram is observed when acetates (for example butanal) and an alkane are mixed (see figure 4.49). Azeotropy is always positive in this case.

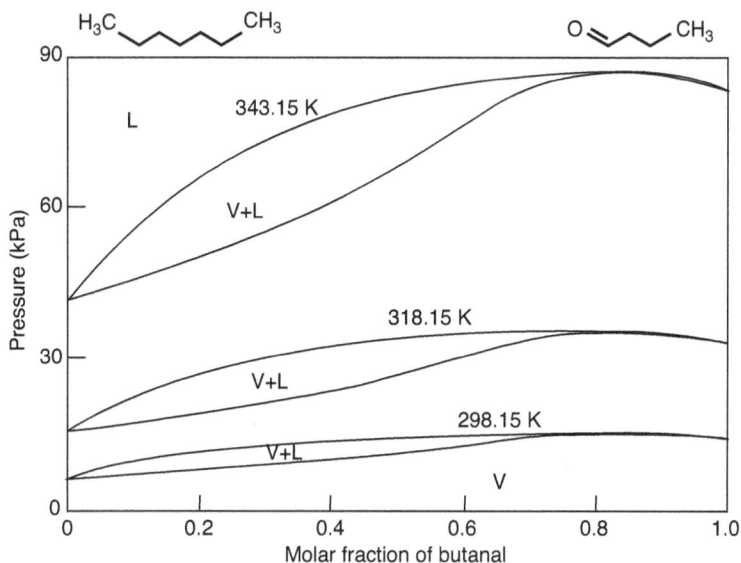

Figure 4.49

Azeotropic behaviour of butanal + *n*-heptane (Nguyen-Huynh [249]).

When the polarities are highly different, liquid-liquid phase splits may be observed at low temperature. This is the case between polar molecules and long-chain alkanes. As an example, the isobaric phase diagram between acetophenone and *n*-decane is shown in figure 4.50.

4.2.6.2 Hydrogen bonds: auto-association

Hydrogen bonds can be formed when a lone electron pair (two are present on all oxygen atoms) neighbours an electropositive site (proton donor or electron acceptor). This configuration is typically encountered on the hydrogen atom of, for example, an OH group (alcohol, water, acid, etc.). These bonds have a high bonding energy (typically 15 kJ mol^{-1} [251]). In case of auto-association, positive deviations are observed (e.g. alcohols with alkanes, as shown in figure 4.51).

The case of vapour-liquid equilibria of alcohols with hydrocarbons has been extensively investigated by the group of Goral and Skrzecz [252-254]. These systems also lead to liquid-liquid phase splits at low temperatures, as exemplified in figure 4.52. Note that for light alkanes, an azeotrope is observed, leading to a heteroazeotrope at higher pressures, while for

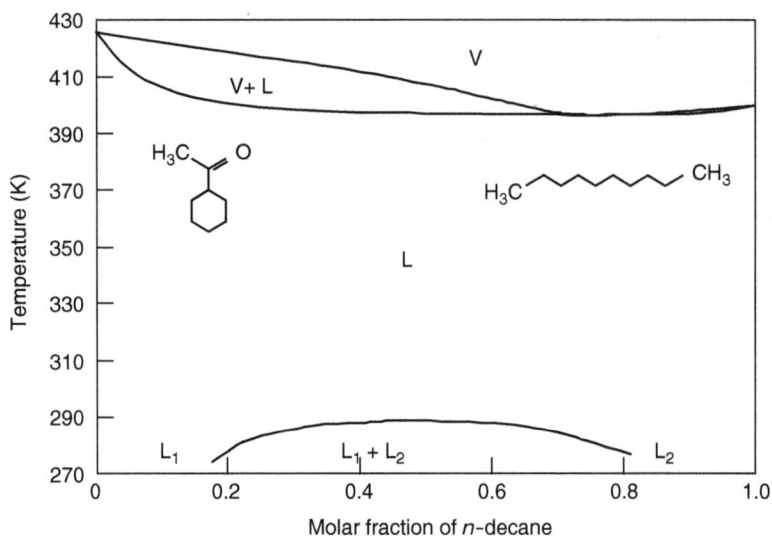

Figure 4.50

Liquid-liquid equilibrium between *n*-decane and acetophenone at atmospheric pressure (liquid-liquid data from [250] and model NRTL for VLE equilibrium).

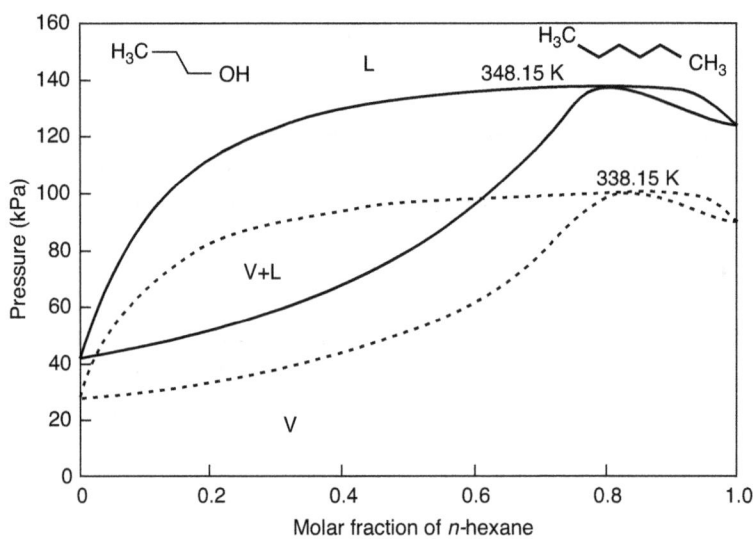

Figure 4.51

1-Propanol + *n*-hexane binary vapour-liquid equilibrium, from [256].

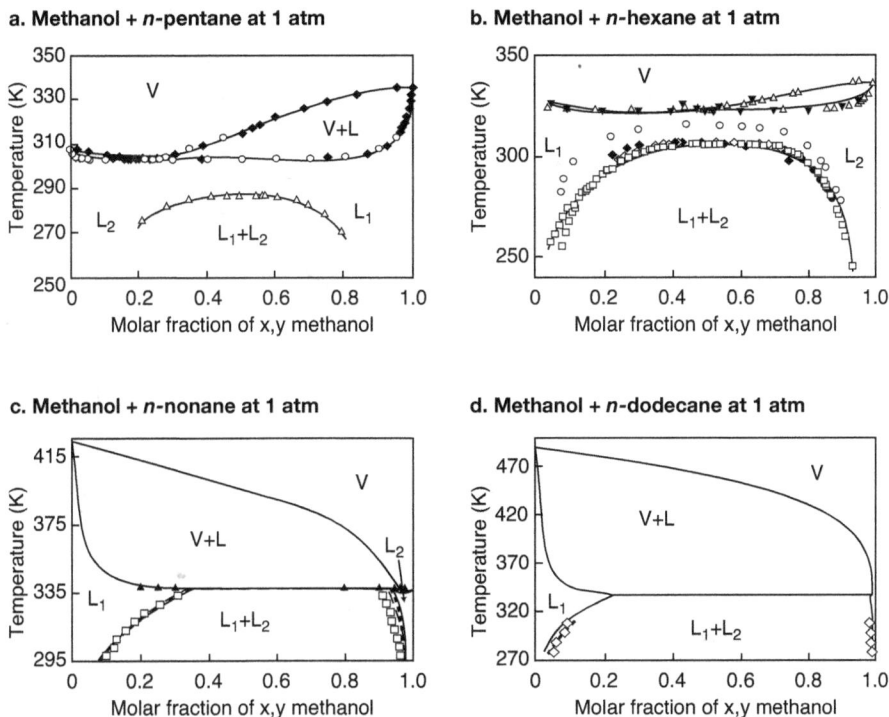

Figure 4.52

Methanol + *n*-alkane binary vapour-liquid and liquid-liquid equilibrium at atmospheric pressure, from [257] and [258]. Symbols are experimental data. Lines are calculated using PPC-SAFT.

heavy alkanes, the difference in boiling temperature becomes large enough so that the VLE phase diagram no longer shows azeotropic behaviour. In that case, a vapour-liquid-liquid three phase line is observed.

Ethanol also forms liquid-liquid phase splits with *n*-alkanes, but only with much longer hydrocarbon chains (starting from *n*-C$_{12}$ [255]).

Water (strongly auto-associating) + hydrocarbon phase splits are extreme examples of this same phenomenon. They have already been discussed in section 4.2.4, p. 293.

4.2.6.3 Effect of cross-association

Cross-association results in strong molecular interactions between unlike molecules. Hence, depending on the relative strength of auto- and cross-association, it may produce negative deviation from the ideal behaviour, with negative azeotropes. Examples can be found when polar but non auto-associating molecules (which often contain a lone pair of electrons on an oxygen atom, that act as proton acceptor) are mixed with molecules containing a proton donor. An illustrative example is provided in figure 4.53 with acetone and chloroform.

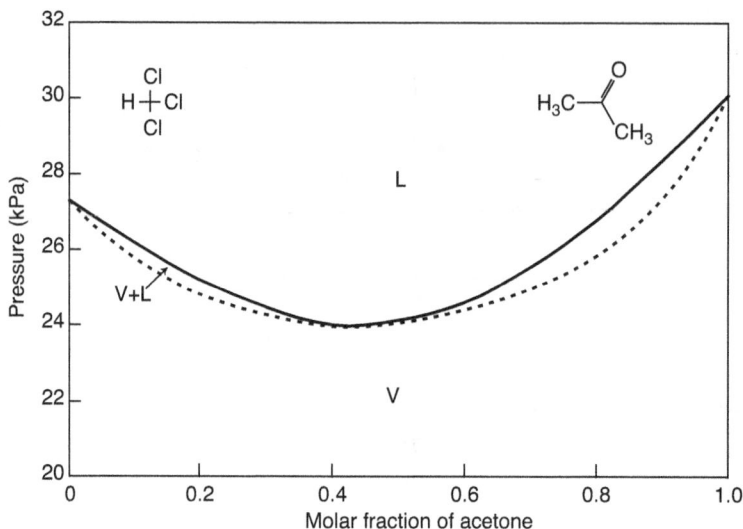

Figure 4.53

Negative azeotrope with the acetone + chloroform mixture at 298.15 K (data from Litinov, 1952 [259]).

Cross-association may also exist with liquid-liquid phase split: when a long chain ester or ether is in the presence of water, the very strong phase split between water and hydrocarbon is reduced. This is illustrated in figure 4.54. In the lower part of the figure (below 10^{-4} molar fraction) organic molecule solubility in water is shown, and in the upper part of the plot, water solubility in the organic component is shown. It is clear that alkanes are the most hydrophobic of all components. The presence of an oxygenated group clearly enhances the mutual solubility, but in all cases solubility of water in the organic component is several orders of magnitude higher than the solubility of the organic component in water. The order is almost identical in both phases:

$$\text{alkanes} < \text{ether} < \text{ester} < \text{alcohol}$$

The solubility of the organic component in water does not show the typical minimum observed for *n*-alkanes in water. Instead, solubility decreases with increasing temperature. This is most likely due to the cross-association phenomenon which decreases while temperature increases, resulting in a lowering of solubility.

Cross-association may also lead to quite unusual liquid-liquid equilibrium phenomena when the ratio of cross-association with respect to auto-association increases with decreasing temperature. The example in figure 4.55 is taken from the water + tetrahydrofuran mixture. At high temperature (above 380 K), the strong auto-association results in phase separation, with an upper critical end point. When the temperature is decreased, cross-association becomes more important with respect to auto-association thus resulting in improved miscibility, eventually closing the liquid-liquid phase diagram with a lower critical end point.

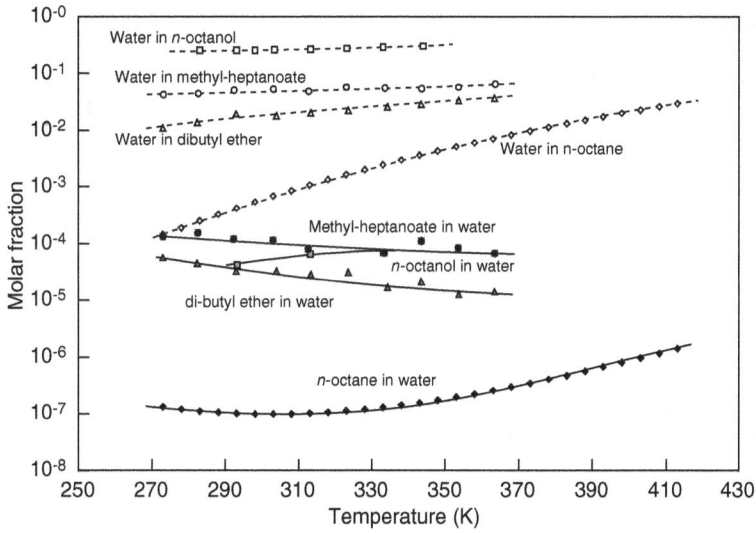

Figure 4.54

Mutual solubilities of water and chains containing 8 carbon atoms and an oxygenated group (data from [71]).

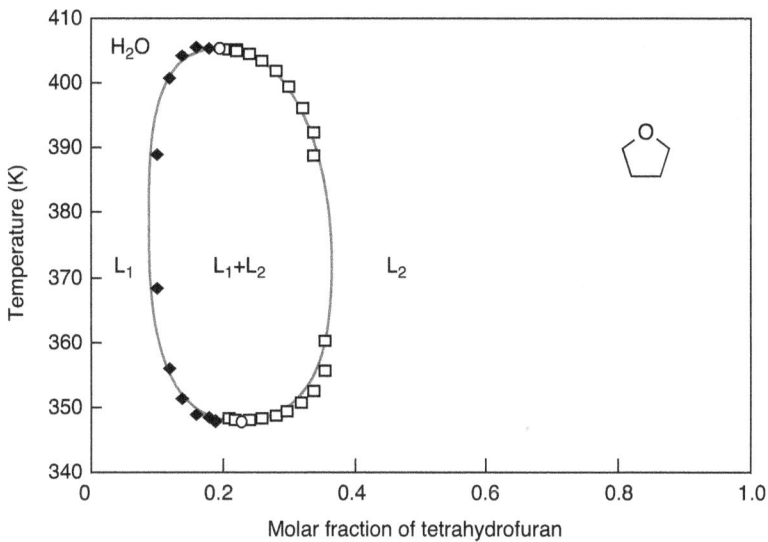

Figure 4.55

Liquid-liquid equilibrium data for the water + tetrahydrofuran mixture at 6 MPa [260].

Model Recommendations

G^E models

Except for some acids, such as acetic acids which form dimers in the vapour phase, the vapour can generally be considered as an ideal gas (process conditions are often at low pressure). Strong non-ideality is handled by using an activity coefficient model, as NRTL or UNIQUAC, provided that the parameters have been validated on experimental data. It may be important to consider temperature-dependent parameters, thus increasing the number of data needed.

In the absence of experimental data, UNIFAC can be used at low pressure; PSRK or VTPR at high pressure. These predictive models are not adequate for aqueous phases.

Association models (SAFT/CPA/GCA)

Following the recent developments in molecular equations of state, use of the **CPA** [261] can now be recommended for strongly associating mixtures. No systematic study of all types of mixtures is available, however.

Although we have not discussed this model in detail, several papers [262, 263] indicate that the **GCA** equation of state, whose parameters also are accessible through group contributions, provides very encouraging results (see 3.4.3.5.B.b, p. 219).

Due to its stronger theoretical basis, however, **SAFT** is probably better suited to describe these types of mixtures. Several groups [264, 265] have shown how a balanced use of a polar term together with an association term can represent correctly the complex phase behaviour observed here very well. Group contribution methods make it possible to use this equation in a predictive manner.

Example 4.11 VLE and LLE calculation of the methanol + *n*-hexane mixture

Methanol is a polar and associating molecule. It forms a strongly non-ideal mixture with hydrocarbons: both liquid-liquid phase split and azeotropic behaviour are encountered. A heteroazeotrope is observed at low pressure. In figure 4.56, the ability of several models to represent this complex system is shown. We draw the reader's attention to several points:

Properties used in the fit: the NRTL parameters have been fitted separately on VLE and LLE data, resulting in two different models. When NRTL is fitted on VLE, it represents these data correctly, but results in a total mismatch for LLE (except possibly at low temperature, but this may be due to pure luck). When NRTL is fitted on LLE, its representation of the VLE curve is seriously impaired, although the azeotropic behaviour is still visible.

The parameters fitted on vapour-liquid equilibrium may not be suitable for liquid-liquid equilibrium calculations, and vice-versa.

The parameters of the GP-PPC-SAFT EoS shown in the plot have been obtained from group contributions. This model is entirely predictive.

Extrapolation of properties: for some applications, it may be important to have good predictions of Henry's constants, in addition to the azeotropic behaviour. In figure 4.56, the quality of this property prediction can be evaluated by looking at the slopes of the bubble temperatures with respect to composition. It is clear that none of the models fitted on the azeotropic VLE is capable of calculating this property correctly. The model most suited for this type of extrapolation is that using a Wertheim association term (SAFT).

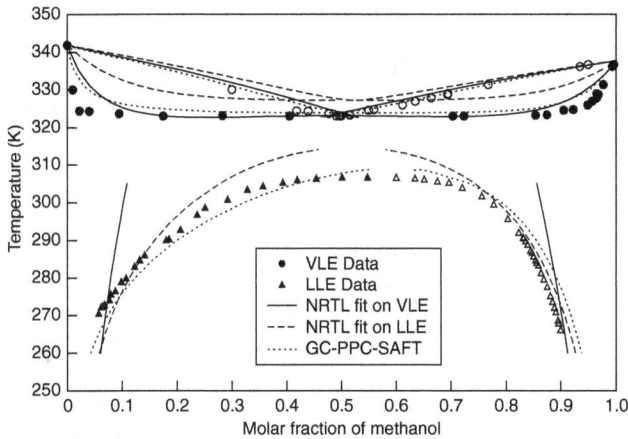

Figure 4.56

Vapour-liquid and liquid-liquid equilibria of the methanol + *n*-hexane mixture at atmospheric pressure (data from Raal *et al.* 1972 [266]).

Very often, no experimental data are available for fitting parameters. In that case, a fully predictive model should be used. UNIFAC is very often considered as the most appropriate method. Recently, the group contribution polar PC-SAFT (GC-PPC-SAFT) EoS has been developed whose aim is to provide a predictive approach for phase equilibrium calculations of polar and associating molecules [267, 268]. Figure 4.57 compares both UNIFAC and GC-PPC-SAFT on the same mixture as above. It shows that UNIFAC better predicts the bubble temperature of the mixture, but is unable to calculate the liquid-liquid phase split correctly.

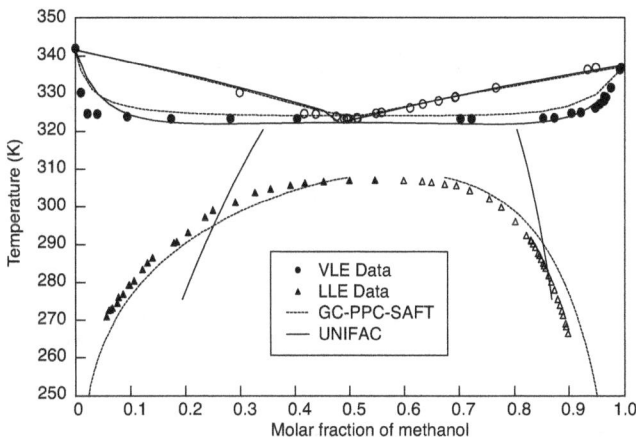

Figure 4.57

Vapour-liquid and liquid-liquid equilibria of the methanol + *n*-hexane mixture at atmospheric pressure (data from Raal *et al.* 1972 [266]). The GC-PPC-SAFT parameters are from Mourah [258].

This example is discussed on the website:
http://books.ifpenergiesnouvelles.fr/ebooks/thermodynamics

4.2.7 Other systems of industrial interest

It is impossible to cover all systems of industrial interest in this work. Specific recommendations nevertheless exist in a number of cases, and some suitable references for the interested reader are proposed.

4.2.7.1 Polymers

Polymer mixtures may exhibit both vapour-liquid and liquid-liquid equilibria, in a fashion similar to that of light and heavy alkanes: the particularity of these types of mixtures is that the liquid-liquid phase split may show an upper critical solution temperature below the lower critical solution temperature: these are type IV phase diagrams. A very good example is provided by the polystyrene + acetone system shown in figure 4.58. It illustrates how increasing polymer molar mass results in lower mutual miscibility: when the molar mass is not too large, two immiscibility regions are found, with a full miscibility region at intermediate temperatures; when the molar mass is high, a typical hourglass-shape immiscibility region is observed. Solubility decreases with increasing pressure [60].

The phenomena that explain this type of behaviour is briefly mentioned in section 3.4.2.3.D (p. 181): an entropic deviation from ideality is observed whose driving force is found in the large difference in free volume in the two pure components. Either lattice-fluid or SAFT-type equations of state are adapted for this type of problems [270] (section 3.4.3, p. 189).

Far more should be detailed for these kinds of mixture. A very good theoretical explanation of the phenomena occuring in polymer blends is provided by Prausnitz *et al.* [60] or Sanchez and Panayiotou [94].

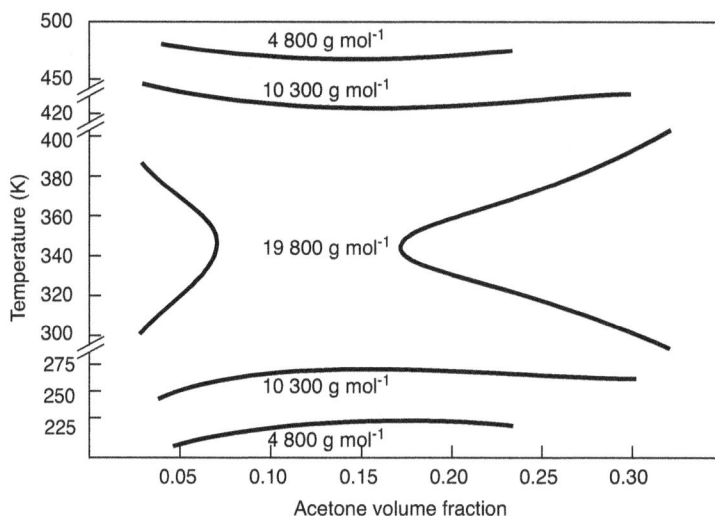

Figure 4.58

Isobaric phase equilibrium of the polystyrene + acetone mixture, depending on the polymer molar mass [269].

Reprinted with permission from [269]. © 1972, American Chemical Society.

4.2.7.2 Ionic liquids

Ionic liquids are a new kind of solvent receiving considerable attention in the research community. These solvents are in fact organic salts which remain liquid at room temperature. Like all salts, they are composed of a cation and an anion. The difference from the usual salts is that these ions consist of organic chains, which may have different molar masses. Due to their ionic character, they do not evaporate, and therefore remain in the liquid phase. A liquid-liquid phase split may be observed, however, especially in the presence of water or organic material. A large number of possible cation-anion combinations exist, so they can be customised to the intended application.

It is out of the scope of this book to discuss this topic. Several review papers exist in the literature, and the interested reader is invited to consult the relevant papers directly [271, 272]. It may be of use to mention that an UNIFAC method has been dedicated to ionic liquids [273].

4.3 CONCLUSION: HOW TO CHOOSE A MODEL

4.3.1 The right questions

The practicing engineer requires a simple, unique model that will predict all possible types of fluid behaviour. Hopefully, the discussions in this book will have convinced him that the state of knowledge today does not allow such a simple approach.

In the absence of a simple model, the engineer will request a clear answer on which model to choose for his/her application: some kind of expert system that provides a unique answer, obviously including the appropriate parameters. Again, he/she may be faced with disappointment: often, several methods are possible, and no best choice can be identified.

> The only universal recommendation that can and should be made is that any model must be compared with experimental data: these are to be taken as a guideline for the choice of any theoretical approach.

In the same way as has been proposed in the past [274], this book suggests several entry points to answer the question:

4.3.1.1 What is the property of interest?

This is extensively discussed in chapter 2. We may identify two types of properties:

- Single phase properties, discussed in the first section of this fourth chapter. Figure 1.9 page 18, in the very first chapter of this book also illustrates which model can be used depending on the process pressure-temperature location with respect to the phase envelope.
- Phase equilibrium properties, discussed in the second part of this fourth chapter. This second type of property is generally much more difficult to calculate accurately.

Note that the simulation tools generally offer "property packages" that contain a pre-defined selection of property-model combinations. An example of such combinations is provided in table 4.7 hereinafter. This combination is often thermodynamically inconsistent as it is rarely possible to have the same accuracy for different properties using the same model. Yet, this situation is most often acceptable.

4.3.1.2 What is the fluid composition?

This issue is extensively discussed in the third chapter of this book, since the choice of model parameters is directly related to the system composition. More specifically, one must be able to distinguish entropic deviations from ideality (related to size asymmetry) and enthalpic deviations (which are due to differing polarities and/or hydrogen bonding). In this fourth chapter, a number of typical systems that may be of industrial interest have been investigated.

4.3.1.3 What are pressure and temperature conditions of the process?

There are two main families of fluid phase models: activity coefficient and equation of state). Although equations of state can describe all fluid phase conditions (vapour, liquid, supercritical), the use of activity coefficient models may have some practical advantages. As discussed in section 2.2.2 (p. 52), the use of activity coefficients implies a heterogeneous approach for calculating distribution coefficients, while a homogeneous approach is used with equations of state. Table 4.6, taken from de Hemptinne and Behar [275], summarises some advantages and disadvantages of both approaches.

Table 4.6   Heterogeneous and homogeneous approaches for vapour – liquid distribution coefficients calculation (taken from [275])

$K_i = \dfrac{y_i}{x_i}$	Homogeneous approach	Heterogeneous approach
	$K_i = \dfrac{\varphi_i^L}{\varphi_i^V}$	$K_i = \dfrac{\wp_i(P,T)P_i^\sigma(T)\,\varphi_i^\sigma(T)\gamma_i(x,T)}{P}\;\dfrac{}{\varphi_i^V(y,P,T)}$
Advantages	• High pressures. • Phase envelope calculation, including the near-critical region	• Improved accuracy because of the large choice of models, in particular for non-ideal mixtures
Disadvantages	• Limited by the choice of an appropriate mixing rule	• Limited to pressures below 1.0 MPa* • No phase envelope calculation • Requires the asymmetric convention in case of supercritical components • Unable to take into account free-volume effects in polymer-solvent systems

* there is no truly theoretical reason for this limit to 1.0 MPa: use of a Poynting correction can enlarge the validity domain up to 1.5 MPa, but above this pressure, one of the components is often supercritical, resulting in the need to use the asymmetric convention.

The use of the **heterogeneous approach**, although very powerful for non-ideal mixtures, is in principle limited to low pressure applications. It requires the use of:

• Pure component property correlations. For database components, accurate correlations exist. Otherwise, other methods should be employed, as discussed in section 3.1.1.2 (p. 109). A clear difficulty exists when one of the components is supercritical Henry's law should be used in this case, leading to the much more complex asymmetric convention as explained in section 2.2.3.1.A (p. 63).

- An activity coefficient model. In this case, a distinction is made between predictive models (less accurate, but can be used without data, like the regular solution or the Flory models) and correlative models (require many experimental data, or, equivalently, a well-furnished parameter database, e.g. NRTL or UNIQUAC). This is discussed in section 3.4.2.6 (p. 188). Note that UNIFAC is a very powerful alternative when no data exist, and that COSMO-RS or COSMO-SAC can now be used to create quite accurate pseudo-experimental values for liquid phase activity coefficients.

Today, there is a clear trend in favour of the **homogeneous approach**, both because it allows a coherent description of all fluid phases (and thus describes critical points), and because new, powerful equations of state have become available:

- The traditional limitations of the cubic equations of state have been overcome by the G^E-based mixing rules (section 3.4.3.4.E.b, p. 214). This approach requires a large number of data for fitting and validation. The PSRK method, which uses UNIFAC can be considered predictive. The PPR78 approach, which uses a Van Laar type G^E model is also predictive but can only be applied to mixtures of non-polar compounds and supercritical gases.
- Molecular based equations offer improved predictivity (SAFT is discussed as an example here, in section 3.4.3.5, p. 216). Its use with group contributions is now extensively studied by several research groups, and it is expected that this method will become increasingly used in the future.

4.3.2 Decision tree

Nevertheless, it may be useful to provide a decision tree for vapour-liquid phase equilibrium calculations, as has already been attempted by other authors [275, 276]. It might help the beginning engineer in his search. In the figure 4.59, we aim at finding an approach that is suited for pre-studies. For more advanced studies, a full analysis of key components as discussed in chapter 3, yielding a more accurate method, is needed.

In figure 4.59, the reader will be guided through a number of key questions, labelled with a greek letter, and further explained below (comments to figure 4.59). At the end of each selection process, the letter helps positioning the reader in table 4.7 for additional elements regarding the choice of the thermodynamic method. This table provides some complementary information regarding the choice of single-phase properties. For cubic EoS (CEOS), the importance to verify either the "alpha" function or the mixing rule parameters is stressed using either the "α" or "k_{ij}" symbol followed by an exclamation mark. Some cases, that are particularly difficult to model, are indicated with a red arrow inviting the engineer to search for additional data.

The meaning of the colours is as follows: yellow means heterogeneous approach, green means homogeneous approach (with an equation of state). Generally, the high pressure (equation of state, green) approaches are more complex, but also applicable at low pressure.

The significance of the keywords used, both in table 4.7 and in figure 4.59, is provided in table 4.8.

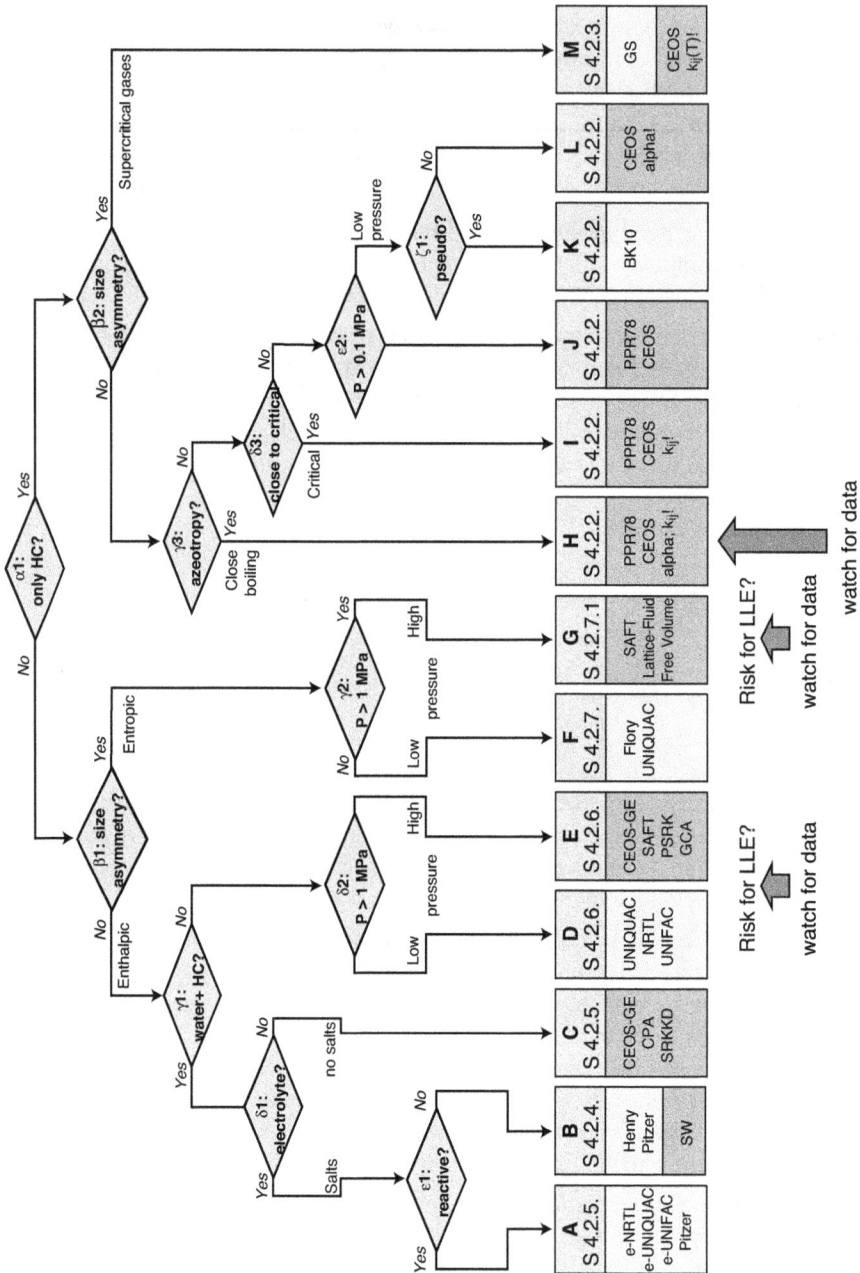

Figure 4.59

Decision tree for vapour-liquid equilibrium calculations.

Question	Label	Explanation
α1: only HC?		The first, basic, question is related to the **presence of other than hydrocarbons** (except gases) in the mixture. The branch on the right corresponds to non-polar systems where the three parameter corresponding states principle can be applied. This branch is also well adapted to petroleum cuts. Non-hydrocarbons gases (CO_2, H_2S...) may also be described on the "hydrocarbon" branch. For mixtures that contain non-hydrocarbons (that include heteroatoms, *i.e.* O, S, N...), the other branch should be chosen.
β1: size asymmetry?		When it is attested that the mixture is non-ideal, a distinction should be made between **enthalpic and entropic deviations** from ideality (discussion in section 3.4.2.1, page 173). Entropic deviations results from mixtures with molecules of strongly different sizes or shapes (size asymmetry); enthalpic deviation is related to different interaction energies between the species.
β2: size asymmetry?	M	This question refers to the **presence of supercritical gases** in the system (see section 4.2.3, page 288). The answer is yes when a significant amount of H_2, N_2... is present in the system. If so, the traditional wisdom is to recommend the Grayson-Streed (GS) method, but it is generally safer (especially for equilibria with heavy ends) to use a cubic EoS on the condition that temperature-dependent interaction parameters (k_{ij}) are available. Methane can in some cases be considered as strongly supercritical (especially at high temperature), but may also lead to close-to-critical conditions, as discussed in "G" below, when temperature is low.
γ1: water+ HC?		**Water – hydrocarbon** mixtures are complex mixtures because of their very different chemical affinities. The properties of the aqueous phase require specific models (choose yes), as explained in detail in section 4.2.4 (p. 293). If other strongly polar components are present, mutual solubilities between the hydrocarbon and aqueous liquids strongly increase, and the right branch should be chosen (choose no), as discussed in section 4.2.6 (p. 316).
γ2: P > 1 MPa	F	For mixtures that exhibit **entropic non-idealities and moderate pressures**, below 1 MPa, the Flory theory provides a fully predictive scheme, but it is most often combined with a local composition model, as in the UNIQUAC theory (which requires interaction parameters). The equations of state in item **G** is also adequate in this case.
γ2: P > 1 MPa	G	For mixtures that exhibit **entropic non-idealities at high pressure** conditions, an equation of state that takes into account free volume is required, as liquid-liquid phase split may be observed. The SAFT or Lattice-Fluid EoS (Sanchez-Lacombe) are designed for these systems.
γ3: azeotropy?	H	Close boiling systems often lead to the **presence of azeotropes** (useful or not). This is discussed in section 4.2.2.1 (p. 271) and requires validation on experimental data. The cubic EoS, which can be recommended in this case, must be well-tuned to both pure component vapour pressures (the so-called alpha function) and mixture bubble pressures (key binaries), as discussed in some details in section 3.4.3.4 (p. 204).
δ1: electrolyte?	B	The presence of **electrolytes (salts)** will strongly affect the result: in the presence of these ionic species, electrolyte models, briefly introduced in section 3.4.2.5 (p. 186), can be used with some success (Pitzer is most often used for complex electrolyte systems). At high pressures, an equation of state should be used. The Soreide and Whitson equation (SW) is here recommended although it can only handle NaCl salt, and no other polar component than water.
δ1: electrolyte?	C	**An aqueous phase is present**, that contain no other polar component, and no salt. In some applications, it may be considered pure water (use a "decant" option). This will still let water dissolve in the other phases, but the aqueous phase is approximated as a pure component. In some cases, it may be important to know the true composition of the aqueous phase, in which case the empirical or Henry-constant models discussed in section 4.2.4 (p. 293) can be used. The CPA EoS is well-suited for this problem. Yet, any cubic EoS can also be used here, provided that the alpha function is validated and a complex mixing rule (at least two binary interaction parameters: kij and kji as in the SRKKD Eos or a GE-based mixing rule as discussed in section 3.4.3.4.E.b (p. 214). These models are also used for physical treatment of acid gases (section 4.2.5.3, p. 310).

Comments to figure 4.59.

Question	Label	Explanation
δ2: P > 1 MPa	D	In the presence of **a mixture of very polar components** (as for example discussed in section 4.2.6, p. 316), it is generally recommended to use an activity coefficient model, with well-documented binary interaction parameters, on the condition that pressure is not too high. UNIQUAC and NRTL are often used indifferently, and UNIFAC (section 3.4.2.4.C, p. 184: the Dortmund version is considered better) is preferred when no parameters are available. Note that it is often possible to fill automatically the UNIQUAC or NRTL parameter matrix using UNIFAC. In case of risk of liquid-liquid phase split, it is essential to validate the model with experimental data, because a slight change in conditions may have a great effect on LLE. The vapour phase is often considered ideal, but the Hayden-O'Connell virial coefficient can be needed for hydrogen-bonding systems. Sometimes, special approaches are required (hexamer forming of HF, for example). The equations of state discussed under label **E** are increasingly used for low pressure calculations.
δ2: P > 1 MPa	E	In the presence of **a mixture of very polar components** (as for example discussed in section 4.2.6, p. 316), the activity coefficient models are limited in pressure. At pressures higher than 1 MPa, an equation of state should be used. A cubic EoS can do the job if an appropriate activity coefficient model is included in the GE-type mixing rule (as in the case of the PSRK EoS). The SAFT model is increasingly used in this context.
δ3: close to critical	I	The process may be focusing on components that are **close to their critical points** (either gases, as CO_2 or H_2S, or light hydrocarbons). Only a homogeneous method with an equation of state can be used. Cubic EoS are particularly well-suited because their very construction makes that the pure component critical point is exact. For mixtures, the prediction should be used with caution. The PPR78 EoS has been designed for this type of problems. The other properties (enthalpic or density), require a high quality virial EoS (as recommended by NIST, see also section 3.4.3.3, p. 199).
ε1: reactive?	A	In some cases, **chemical reactions in the aqueous phase** strongly affect the vapour-liquid equilibrium (e.g. amine treatment). Specific packages (that include simultaneous phase and chemical equilibrium calculations) must be used for these reacting systems. They very often use the electrolyte version of NRTL or UNIQUAC as discussed in section 3.4.2.5 (p. 186). Some simulators propose a special "sour gas" package for acid gas + water mixtures.
ε2: P > 0.1 MPa	J	If pressure is larger than 0.1 MPa, and no specific azeotropic condition is feared, the traditional cubic EoS can be used without great danger. This is the case for stabilization columns, for example.
ζ1: pseudo?	K	In some cases, very low pressure calculations are needed, which means that the liquid is a heavy component. Very often, petroleum pseudo-components are used (typically for vacuum distillation). In this case, the automated API nomograph BK10 (which is nothing but an improved Raoult's law) is traditionally recommended.
ζ1: pseudo?	L	In the case the heavy end is paraffinic, for example (wax treatment), it is recommended to use a cubic EoS, for which the validity of the alpha function has been extended to heavy components (as in PR78).

Comments to figure 4.59 (cont'd).

Table 4.7 Model Choice depending on the Result of the Selection Process in Figure 4.59

	Fugacity (equilibrium)	Enthalpy/entropy	Density	Comments
A	e-NRTL, e-UNIQUAC	Correlation + include excess	Specific	Watch for parameters
	e-UNIFAC			Predictive
	Special Sour Packages	Specific		
B	Pitzer		Specific	No other polar component than water
	SW			Only NaCl; no other polar component than water
C	CEOS -GE	CEOS -GE	Aqueous phase: use water; hydrocarbon phase: use API	Requires parameter fit
	CPA	CPA		Good for mixed solvents
	CEOS with specific mixing rules	CEOS with specific mixing rules		Most simulators offer specific solutions (e.g. SRKKD) for water + hydrocarbons
D	UNIQUAC, NRTL	Correlation + include excess properties	Correlation	Watch for parameters, especially when using excess enthalpy
	UNIFAC	Correlation		
E	CEOS -GE	CEOS -GE		Requires parameter fit
	SAFT	SAFT	SAFT	Group contribution methods may make this model predictive
	GCA		GCA	Often used for biofuel applications
	VTPR	VTPR	VTPR	Predictive, hence less accurate; more recent version of PSRK
	PSRK			Predictive, hence less accurate
F	UNIQUAC, Flory	Correlation	Correlation	No LLE
G	SAFT	SAFT	SAFT	Group contribution methods may make this model predictive
	Lattice Fluid (Sanchez-Lacombe)		Correlation	Group contribution methods may make this model predictive
	UNIFAC Free Volume			Predictive

Table 4.7   Model Choice depending on the Result of the Selection Process in Figure 4.59 (cont'd)

	Fugacity (equilibrium)	Enthalpy/entropy	Density	Comments
H	CEOS	Lee Kesler	Lee Kesler (API)	Check alpha parameter on vapour pressures; Evaluate mixing rules (k_{ij}) on mixture data
	PPR78	Lee Kesler	Lee Kesler (API)	
I	CEOS alpha, k_{ij}	Lee Kesler, MBWR	Lee Kesler MBWR	Close to the critical point, it is extremely difficult to have accurate predictions of the single phase properties Check alpha parameter on vapour pressures; Evaluate mixing rules (k_{ij}) on mixture data
J	CEOS, PPR78	Lee Kesler	Lee Kesler (API)	
K	BK10	Lee Kesler	Lee Kesler (API)	For pseudo-components only
L	PR (SRK), alpha PPR78	Lee Kesler	Lee Kesler (API)	Check alpha parameter on vapour pressures
M	PR (SRK), k_{ij}	Lee Kesler	Lee Kesler	Evaluate mixing rules (k_{ij}) on mixture data (a temperature dependent k_{ij} is often essential)
	GS	Lee Kesler	Lee Kesler	Not for hydrogen solubility in heavy components

Table 4.8   Keywords for Models

Keyword	Description	Section of this book	Page
API	API correlations: these are very similar to those presented in section 3.1.1.2	3.1.1.2	109
Aqueous phase: use water	Specific water density correlation, as proposed by NIST (steam tables)	3.4.3.3.C	200
BK10	Improved Antoine correlations		
CPA	Cubic Plus Association	3.4.3.5.B	218
e-NRTL,	Electrolyte NRTL	3.4.2.5	186
e-UNIFAC	Electrolyte UNIFAC	3.4.2.5	186
e-UNIQUAC	Electrolyte UNIQUAC	3.4.2.5	186
Flory	Flory activity coefficient model	3.4.2.3.B	180
GS	Grayson Streed	4.2.3	290
GCA	Group Contribution with Association equation of state	3.4.3.5.B	218
Lattice Fluid (Sanchez-Lacombe)	Lattice fluid Equation of State	3.4.3.6	219
Lee Kesler	Lee & Kesler method	3.4.3.3.C	202
Correlation	When considering the liquid phase, it may be interesting to use library correlations for the pure component, with an ideal mixing rule (see section 2.2.2.2, p. 60). Most simulators offer well-validated correlations	3.1.1	109
Correlation + include excess (UNIQUAC, NRTL)	In addition to using library correlations, for strongly non-ideal mixtures, the excess property must be added, as calculated using NRTL or UNIQUAC and eq (2.101), p. 62		
MBWR	Modified Benedict Webb & Rubin EoS	3.4.3.3.C	203
NRTL	Activity coefficient model	3.4.2.2.C	177
Pitzer	Activity coefficient model	3.4.2.5	186
PPR78	Predictive Peng-Robinson with temperature dependent k_{ij}. (Van Laar introduced in the PR EoS, using a g^E-mixing rule).	3.4.3.4.E	214
CEOS	Cubic Equation of State: Peng Robinson or Soave-Redlich Kwong equations of state are used indifferently	3.4.3.4	204
CEOS alpha	It is important to validate the alpha function of cubic equations of state. The "Twu" correlation generally improves the results	3.4.3.4.B	206
CEOS k_{ij}	A simple mixing rule can do it, but well-adjusted binary interaction parameters are needed	3.4.3.4.E	212

Table 4.8 Keywords for Models (cont'd)

Keyword	Description	Section of this book	Page
CEOS -GE	A complex mixing rule is needed, possibly using several k_{ij}, but a G^E-type mixing rule is best	3.4.3.4.E	212
PSRK	Predictive SRK (UNIFAC introduced in the SRK EoS, using a G^E mixing rule)	3.4.3.4.E	212
SAFT	Statistical Associating Fluid Theory	3.4.3.5.A	216
Soreide&Whitson	Equation of State	4.2.4	294
special Sour packages	Special, simulator-dependent, packages		
specific	Special, simulator-dependent, packages		
SRK	Soave Redlich Kwong (see above)	3.4.3.4	204
SRKKD	SRK with the Kabadi-Danner mixing rule	4.2.4	295
UNIFAC	Activity coefficient model	3.4.2.4.C	184
UNIFAC Free Volume	Activity coefficient model		
UNIQUAC	Activity coefficient model	3.4.2.4.B	183
VTPR	Volume Translated Peng Robinson	3.4.3.4.E	212

REFERENCE LIST

[1] de Swaan Arons, J. and de Loos, T. "Phase Behavior: Phenomena, Significance and Models"; Ed. Sandler, S. I.; Marcel Dekker, Inc., **1994**.

[2] Kiran, E. and Levelt Sengers, J. M. H. "Supercritical Fluids; Fundamentals for Applications"; Kluwer Academic Publishers, **1994**.

[3] Sadus, R. J. "High Pressure Phase Behaviour of Multicomponent Fluid Mixtures"; Elsevier Science Publisher: Amsterdam, **1992**.

[4] Poling, B. E., Prausnitz, J. M. and O'Connell, J. P. "The Properties of Gases and Liquids" 5th Ed.; McGraw-Hill: New York, **2000**.

[5] Reid, R. C., Prausnitz, J. M. and Sherwood, T. K. "The Properties of Gases and Liquids"; McGraw-Hill Book Company: New York, **1977**.

[6] Reid, R. C., Prausnitz, J. M. and Poling, B. E. "The Properties of Gases and Liquids" 4th Ed. McGraw-Hill Book Company: New York, **1987**.

[7] Domb, C. "Phase Transitions and Critical Phenomena", Academic Press, **2000**.

[8] Anisimov, M. A. and Sengers, J. V. "Critical Region" in *Equations of State for Fluids and Fluid Mixtures*; Ed. Sengers, J. V., Kayser, R. F., Peters, C. J. and White, H. J. Jr; Experimental Thermodynamics; Elsevier: Amsterdam, **2000**.

[9] Maxwell, J. B. "Data Book on Hydrocarbons"; Van Nostrand Company, Inc: London – New York – Toronto, **1950**.

[10] *API Technical Data Book* – "Petroleum Refining", 3rd Ed. 6; American Petroleum Institute: Washington, D.C., **1976**.

[11] Wuithier, P. "Le Pétrole: Raffinage et Génie Chimique (Tome 1)" Ed. Technip: Paris, **1972**.

[12] Katz, D. L., Cornell, D., Kobayashi, R., Poettmann, F. H., Vary, J. A., Elenbaas, J. R. and Weinaug, C. F. "Handbook of Natural Gas Engineering"; McGraw-Hill Book Company: New York, Toronto, London, **1959**.

[13] Danner, R. P. and Daubert, T. E. "Manual for Predicting Chemical Process Design"; American Institute of Chemical Engineers (AIChE): New York, **1983**.

[14] Mollier, R. "Neue Tabellen und Diagramme für Wasserdampf"; Springer: Berlin, **1906**.

[15] Yaws, C. L. "Chemical Properties Handbook"; McGraw-Hill: New York, **2005**.

[16] Colina, C. M. and Muller, E. A. "Joule-Thomson Inversion Curves by Molecular Simulation" *Molecular Simulation* **1997**, **19**, 4, 237-246.

[17] Colina, C. M. and Olivera-Fuentes, C. "Prediction of the Joule-Thomson Inversion Curve of Air from Cubic Equations of State" *Cryogenics* **1998**, **38**, 7, 721-728.

[18] Castillo, M. G., Colina, C. M., Dubuc, J. E. and Olivera-Fuentes, C. G. "Three-Parameter Corresponding-States Correlations for Joule-Thomson Inversion Curves" *International Journal of Thermophysics* **1999**, **20**, 6, 1737-1751.

[19] Colina, C. M. and Olivera-Fuentes, C. "Predicted Inversion Curve and Third Virial Coefficients of Carbon Dioxide at High Temperatures" *Industrial & Engineering Chemistry Research* **2002**, **41**, 5, 1064-1068.

[20] Colina, C. M., Turrens, L. F., Gubbins, K. E., Olivera-Fuentes, C. and Vega, L. F. "Predictions of the Joule-Thomson Inversion Curve for the *n*-Alkane Series and Carbon Dioxide from the Soft-SAFT Equation of State" *Industrial & Engineering Chemistry Research* **2002**, **41**, 5, 1069-1075.

[21] Castro-Marcano, F., Olivera-Fuentes, C. G. and Colina, C. M. "Joule-Thomson Inversion Curves and Third Virial Coefficients for Pure Fluids from Molecular-Based Models" *Industrial & Engineering Chemistry Research* **2008**, **47**, 22, 8894-8905.

[22] Kortekaas, W. G., Peters, C. J. and Arons, J. D. "Joule-Thomson Expansion of High-Pressure-High-Temperature Gas Condensates" *Fluid Phase Equilibria* **1997**, **139**, 1-2, 205-218.

[23] Kortekaas, W. G., Peters, C. J. and Arons, J. D. "High Pressure Behavior of Hydrocarbons – Joule-Thomson Expansion of Gas Condensates" *Revue de l'Institut Français du Pétrole* **1998**, **53**, 3, 259-263.

[24] Segura, H., Kraska, T., Mejia, A. S., Wisniak, J. and Polishuk, I. "Unnoticed Pitfalls of Soave-Type Alpha Functions in Cubic Equations of State" *Industrial & Engineering Chemistry Research* **2003**, **42**, 22, 5662-5673.

[25] Nichita, D. V. and Leibovici, C. F. "Calculation of Joule-Thomson Inversion Curves for Two-Phase Mixtures" *Fluid Phase Equilibria* **2006**, **246**, 1-2, 167-176.

[26] Nichita, D. V., Bessieres, D. and Daridon, J. L. "Calculation of Joule-Thomson Inversion Curves for Multiphase Systems with Waxy Solid-phase Precipitation" *Energy & Fuels* **2008**, **22** 6, 4012-4018.

[27] Kiselev, S. B. "Cubic Crossover Equation of State" *Fluid Phase Equilibria* **1998**, 147, 7-23.

[28] de Hemptinne, J. C. and Ungerer, P. "Accuracy of the Volumetric Predictions of Some Important Equations of State for Hydrocarbons, Including a Modified Version of the Lee-Kesler Method" *Fluid Phase Equilibria* **1995**, **106**, 1-2, 81-109.

[29] Span, R. and Wagner, W. "A New Equation of State for Carbon Dioxide Covering the Fluid Region from the Triple-Point Temperature to 1100 K at Pressures up to 800 MPa" *Journal of Physical and Chemical Reference Data* **1996**, **25**, 6, 1509-1596.

[30] Jacobsen, R. T., Penoncello, S. G., Lemmon, E. and Span, R. "Multiparameter Equations of State" in *Equations of State for Fluids and Fluid Mixtures*; Ed. Sengers, J. V., Kayser, R. F., Peters, C. J, and White, H. J. Jr; Experimental Thermodynamics; Elsevier: Amsterdam, **2000**.

[31] Span, R. and Wagner, W. "Equations of State for Technical Applications. I. Simultaneously Optimized Functional Forms for Nonpolar and Polar Fluids" *International Journal of Thermophysics* **2003**, **24**, 1, 1-39.

[32] Span, R. and Wagner, W. "Equations of State for Technical Applications. II. Results for Nonpolar Fluids" *International Journal of Thermophysics* **2003**, **24**, 1, 41-109.

[33] Span, R. and Wagner, W. "Equations of State for Technical Applications. III. Results for Polar Fluids" *International Journal of Thermophysics* **2003**, **24**, 1, 111-162.

[34] Van Konynenburg, P. H. and Scott, R. L. "Critical Lines and Phase Equilibria in Binary van der Waals Mixtures" *Philosophical Transacions of the Royal Society* **1980**, 298, 495-540.

[35] Stanley, E. "Introduction to Phase Transitions and Critical Phenomena"; Oxford University Press: New York, Oxford, **1971**.

[36] Papon, P., Leblond, J., Meijer and de Gennes P. G. "Physique des Transitions de Phases – Concepts et Applications"; Dunod: Paris, **2002**.

[37] O'Connell, J. P. and Haile, J. M. "Thermodynamics: Fundamentals for Applications" 1st Ed.; Cambridge University Press, **2005**.

[38] Scott, R. L. "Van der Waals-Like Global Phase Diagrams" *Physical Chemistry Chemical Physics* **1999**, **1**, 18, 4225-4231.

[39] Scott, R. L. and Konynenburg, P. H. "Static Properties of Solutions. Van der Waals and Related Models for Hydrocarbon Mixtures" *Discussion of the Faraday Society* **1970**, 49, 87-97.

[40] Van Konijnenburg, P. H. and Scott, R. L. "Critical Lines and Phase Equilibria in Binary van der Waals Mixtures" *Philosophical Transactions of the Royal Society A* **1980**, **298**, 1442, 495-540.

[41] Schneider, G. M. "The Continuity and Family Concepts: Useful Tools in Fluid Phase Science" *Physical Chemistry Chemical Physics* **2004**, 6, 2285-2290.

[42] Clark, G. N. I., Galindo, A., Jackson, G., Rogers, S. and Burgess, A. N. "Modeling and Understanding Closed-Loop Liquid – Liquid Immiscibility in Aqueous Solutions of Poly(Ethylene Glycol) Using the SAFT-VR Approach with Transferable Parameters" *Macromolecules* **2008**, **41**, 17, 6582-6595.

[43] Smits, P. J., Peters, C. J. and De Swaan Arons, J. D. "High Pressure Phase Behaviour of $\{x_1CF_4+x_2NaCl+(1-x_1-x_2)H_2O\}$" *Journal of Chemical Thermodynamics* **1997**, **29**, 4, 385-393.

[44] Smits, P. J., Smits, R. J. A., Peters, C. J. and De Swaan Arons, J. "Measurement of the High Pressure – High Temperature Fluid Phase Behavior of the Systems CF_4+H_2O, $CF_4+H_2O+NaCl$, CHF_3+H_2O and $CHF_3+H_2O+NaCl$" *Fluid Phase Equilibria,* **sept 1998**, 150-151, 745-751.

[45] Schneider, G. M. "High-Pressure Phase Equilibria and Spectroscopic Investigations up to 200 MPa on Fluid Mixtures Containing Fluorinated Compounds: a Review" *Fluid Phase Equilibria* **2002**, 199, 1-2, 307-317.

[46] Peters, C. J "Multiphase Equilibria in Near-Critical Solvents" in *Supercritical Fluids; Fundamentals for Application*; Ed. Kiran, E. and Levelt Sengers, J. M. H.; Series E: Applied Sciences Vol 273; Kluwers: Amsterdam, **1993**.

[47] Schneider, G. M. "Physicochemical Aspects of Fluid Extraction" *Fluid Phase Equilibria* **1983**, **10**, 2-3, 141-157.

[48] Wang, J. L. and Sadus, R. J. "Phase Behaviour of Binary Fluid Mixtures: a Global Phase Diagram Solely in Terms of Pure Component Properties" *Fluid Phase Equilibria* **2003**, **214**, 1, 67-78.

[49] Wang, J. L. and Sadus, R. J. "Global Phase Diagram for Anisotropic Binary Fluid Mixtures: Reverse Type IV Behaviour" *Molecular Physics* **2003**, **101**, 14, 2211-2217.

[50] Aparicio-Martinez, S. and Hall, K. R. "Use of PC-SAFT for Global Phase Diagrams in Binary Mixtures Relevant to Natural Gases. 3. Alkane + Non-Hydrocarbons" *Industrial & Engineering Chemistry Research* **2007**, **46**, 1, 291-296.

[51] Brandt, F. S., Broers, P. M. A. and de Loss, T. W. "High Pressure Phase Equilibria in the Systems Ammonia + Potassium Iodide, + Sodium Iodide, + Sodium Bromide, and + Sodium Thiocyanate: Solid-Liquid-Vapor and Liquid-Liquid-Vapor Equilibria" *International Journal of Thermophysics* **2001**, **22**, 4, 1045-1055.

[52] Schneider, G. M. "High-Pressure Investigations on Fluid Systems: a Challenge to Experiment, Theory, and Application" *Journal of Chemical Thermodynamics* **1991**, **23**, 4, 301-326.

[53] Shaw, J. M. and Behar, E. "SLLV Phase Behavior and Phase Diagram Transitions in Asymmetric Hydrocarbon Fluids" *Fluid Phase Equilibria* **2003**, *209*, 2, 185-206.

[54] Yelash, L. V. and Kraska, T. "The Global Phase Behaviour of Binary Mixtures of Chain Molecules: Theory and Application" *Physical Chemistry Chemical Physics* **1999**, *1*, 18, 4315-4322.

[55] Brunner, E., Thies, M. C. and Schneider, G. M. "Fluid Mixtures at High Pressures: Phase Behavior and Critical Phenomena for Binary Mixtures of Water with Aromatic Hydrocarbons" *Journal of Supercritical Fluids* **2006**, *39*, 2, 160-173.

[56] de Loos, T. W. "Understanding Phase Diagrams" in *Supercritical Fluids; Fundamentals for Application*; Ed. Kiran, E. and Levelt Sengers, J. M. H.; Series E: Applied Sciences **273**; Kluwers: Amsterdam, **1993**.

[57] Sorensen, J. M., Magnussen, T., Rasmussen, P. and Fredenslund, A. "Liquid-Liquid Equilibrium Data – Their Retrieval, Correlation and Prediction .1. Retrieval" *Fluid Phase Equilibria* **1979**, *2*, 4, 297-309.

[58] Sorensen, J. M., Magnussen, T., Rasmussen, P. and Fredenslund, A. "Liquid-Liquid Equilibrium Data – Their Retrieval, Correlation and Prediction .2. Correlation" *Fluid Phase Equilibria* **1979**, *3*, 1, 47-82.

[59] Papon, P., Leblond, J., Meijer and P.H.E. *Physique des Transitions de Phases – Concepts et Applications*; Dunod: Paris, **2002**.

[60] Prausnitz, J. M., Lichtenthaler, R. N. and Gomes de Azevedo, E. *Molecular Thermodynamics of Fluid Phase Equilibria* 3rd Ed.; Prentice Hall Int., **1999**.

[61] Shaw, J. and Behar, E. "SLLV Phase Behavior and Phase Diagram Transitions in Asymmetric Hydrocarbon Fluids" *Fluid Phase Equilibria* **2003**, 209, 185-206.

[62] Zemaitis, J. F., Clark, D. M., Rafal, M. and Scrivner, N. C. "Handbook of Aqueous Electrolyte Thermodynamics"; Design Institute for Physical Properties (DIPPR), **1986**.

[63] Rafal, M., Bethold, J. W. and Scrivner, N. C. "Models for Electrolyte Solutions" in *Models for Thermodynamics and Phase Equilibria Calculations*; Ed. Sandler, S. I.; Marcel Dekker, Inc, NY: New York, **1994**.

[64] Sloan, E. D. and Koh, C. A. "Clathrate Hydrates of Natural Gases" 3rd Ed.; CRC/Taylor & Francis, **2008**.

[65] Peters, C. J "Multiphase Equilibria in Near-Critical Solvents" in *Supercritical Fluids; Fundamentals for Application*; Ed. Kiran, E. and Levelt Sengers, J. M. H.; Series E: Applied Sciences **273**; Kluwers: Amsterdam, **1994**.

[66] Polishuk, I., Wisniak, J. and Segura, H. "Prediction of the Critical Locus in Binary Mixtures Using Equation of State I. Cubic Equations of State, Classical Mixing Rules, Mixtures of Methane-Alkanes" *Fluid Phase Equilibria* **1999**, *164*, 1, 13-47.

[67] Polishuk, I., Wisniak, J., Segura, H., Yelash, L. V. and Kraska, T. "Prediction of the Critical Locus in Binary Mixtures Using Equation of State – II. Investigation of van der Waals-Type and Carnahan-Starling-type Equations of State" *Fluid Phase Equilibria* **2000**, *172*, 1, 1-26.

[68] Arnaud, J. F., Ungerer, P., Behar, E., Moracchini, G. and Sanchez, J. "Excess Volumes and Saturation Pressures for the System Methane + *n*-Tetracosane at 374K. Representation by Improved EOS Mixing Rules" *Fluid Phase Equilibria* **1996**, *124*, 1-2, 177-207.

[69] Fogg, P. G. T. and Gerrard, W. "Solubility of Gases in Liquids"; J. Wiley & Sons, Chichester, **1991**.

[70] Zuliani, M., Barreau, A., Vidal, J., Alessi, P. and Fermeglia, M. "Measurements, Correlation and Prediction of Henry Constants of Light Alkanes in Model Heavy Hydrocarbons and Petroleum Fractions" *Fluid Phase Equilibria* **1993**, 82, 141-148.

[71] Dechema "Detherm" v 1.4.1.0.1, **2005**.

[72] Boldyrev, A. V., Komarov, V. M. and Krichevtov, V. K. "Liquid-Vapor Equilibrium in the System Water-Furfinal Alcohol" *J. Appl. Chem. USSR,* **1973**, *46*, 10, 2487-2488.

[73] Soave, G. "Equilibrium Constants for a Modified Redlich-Kwong Equation of State" *Chemical Engineering Science* **1972**, 27, 1197-1203.

[74] Peng, D. Y. and Robinson, D. B. "A New Two-Constant Equation of State" *Industrial & Engineering Chemistry Fundamentals* **1976**, 15, 59-64.

[75] Robinson, D. B. "The Characterization of The Heptanes and Heavier Fractions for the GPA Peng-Robinson Programs" GPA, RR-28, **1978**.

[76] Nasrifar, K., Bolland, O. and Moshfeghian, M. "Predicting Natural Gas Dew Points from 15 Equations of State" *Energy & Fuels* **2005**, 19, 2, 561-572.

[77] Nasrifar, K. and Bolland, O. "Prediction of Thermodynamic Properties of Natural Gas Mixtures Using 10 Equations of State Including a New Cubic Two-Constant Equation of State" *Journal of Petroleum Science and Engineering* **2006**, 51, 3-4, 253-266.

[78] Pfohl, O., Giese, T., Dohrn, R. and Brunner, G. "1. Comparison of 12 Equations of State with Respect to Gas-Extraction Processes: Reproduction of Pure-Component Properties when Enforcing the Correct Critical Temperature and Pressure" *Industrial & Engineering Chemistry Research* **1998**, 37, 8, 2957-2965.

[79] Kordas, A., Magoulas, K., Stamataki, S. and Tassios, D. "Methane Hydrocarbon Interaction Parameters Correlation for the Peng-Robinson and the T-Mpr Equation of State" *Fluid Phase Equilibria*, **1995**, 112, 1, 33-44.

[80] Jaubert, J. N. and Mutelet, F. "VLE Predictions with the Peng-Robinson Equation of State and Temperature Dependent kij Calculated Through a Group Contribution Method" *Fluid Phase Equilibria* **2004**, 224, 285-304.

[81] Tamouza, S., Passarello, J. P., Tobaly, P. and de Hemptinne, J. C. "Group Contribution Method with SAFT EOS Applied to Vapor Liquid Equilibria of Various Hydrocarbon Series" *Fluid Phase Equilibria* **2004**, 222-223, 67-76.

[82] Tamouza, S., Passarello, J. P., Tobaly, P. and de Hemptinne, J. C. "Application to Binary Mixtures of a Group Contribution SAFT EOS" *Fluid Phase Equilibria* **2005**, 228-229, 409-419.

[83] Tihic, A., Kontogeorgis, G. M., von Solms, N. and Michelsen, M.L. "Applications of the Simplified Perturbed-Chain SAFT Equation of State using an Extended Parameter Table" *Fluid Phase Equilibria* **2008**, 248, 1, 29-43.

[84] Tihic, A., Kontogeorgis, G. M., von Solms, N., Michelsen, M.L. and Constantinou, L. "A Predictive Group-Contribution Simplified PC-SAFT Equation of State: Application to Polymer Systems." *Industrial & Engineering Chemistry Research* **2008**, 47, 15, 5092-5101.

[85] Tihic, A., von Solms, N., Michelsen, M.L., Kontogeorgis, G. M. and Constantinou, L. "Application of sPC-SAFT and Group Contribution sPC-SAFT to Polymer Systems-Capabilities and Limitations" *Fluid Phase Equilibria,* **2009**, 281, 1, 70-77.

[86] Lymperiadis, A., Adjiman, C. S., Galindo, A. and Jackson, G. "A Group Contribution Method for Associating Chain Molecules Based on the Statistical Associating Fluid Theory (SAFT-gamma)" *Journal of Chemical Physics,* **2007**, 127, 23.

[87] Lymperiadis, A., Adjiman, C. S., Jackson, G. and Galindo, A. "A Generalisation of the SAFT-gamma Group Contribution Method for Groups Comprising Multiple Spherical Segments" *Fluid Phase Equilibria* **2008**, 274, 1-2, 85-105.

[88] Reamer, H. H., Sage, B. H. and Lacey, W. N. "Phase Equilibria in Hydrocarbon Systems. Volumetric and Phase Behavior of the Ethane-*n*-Pentane System" *Journal of Chemical & Engineering Data* **1960**, 5, 1, 44-50.

[89] Nguyen-Huynh, D., Passarello, J. P. and Tobaly, P. "*In Situ* Determination of Phase Equilibria of Methyl Benzoate plus Alkane Mixtures Using an Infrared Absorption Method. Comparison with Polar GC-SAFT Predictions" *Journal of Chemical and Engineering Data* **2009**, 54, 6, 1685-1691.

[90] Lin, H. M., Sebastian, H. M., Simnick, J. J. and Chao, K. C. "Gas-Liquid Equilibrium in Binary Mixtures of Methane with *N*-Decane, Benzene and Toluene" *Journal of Chemical & Engineering Data* **1979**, 24, 2, 146-149.

[91] Susarev, M. P. and Chen, S. T. "Calculation of the Liquid-Vapor Equilibrium in Ternary Systems from Data for Binary Systems. The Benzene – *n*-Hexane – Cyclo-Hexane System" *Zhurnal Fizicheskoi Khimii* **1963**, **37**, 8, 1739-1744.

[92] Polishuk, I, Wisniak, J. and Segura, H. "Estimation of Liquid – Liquid – Vapour Equilibria in Binary Mixtures of *n*-Alkanes" *Industrial & Engineering Chemistry Research* **2004**, **43**, 18, 5957-5964.

[93] Luks, K. D. and Kohn, J. P "The Topography of Multiphase Equilibria Behavior: What can it tell the Design Engineer" GPA, **1984**.

[94] Sanchez, I. C. and Panayiotou, C. G. "Equation of State Thermodynamics of Polymer and Related Solutions" in *Models for Thermodynamic and Phase Equilibria Calculations*; Ed. Sandler, S. I.; Chemical Industries; Marcel Dekker, Inc.: New York, **1994**.

[95] Radfarnia, H. R., Kontogeorgis, G. M., Ghotbi, C. and Taghikhani, V. "Classical and Recent Free-Volume Models for Polymer Solutions: A Comparative Evaluation" *Fluid Phase Equilibria* **2007**, **257**, 1, 63-69.

[96] de Hemptinne, J. C. "Benzene Crystallization Risks in the LIQUEFIN Natural Gas Process" *Process Safety Progress* **2005**, **24**, 3, 203-212.

[97] Won, K. W. "Thermodynamics for Solid Solution-Liquid-Vapor Equilibria – Wax Phase Formation from Heavy Hydrocarbon Mixtures" *Fluid Phase Equilibria* **1986**, 30, 265-279.

[98] Vidal, J. "Thermodynamics: Applications in Chemical Engineering and the Petroleum Industry"; Editions Technip: Paris, **2003**.

[99] Ungerer, P., Faissat, B., Leibovici, C., Zhou, H., Behar, E., Moracchini, G. and Courcy, J. P. "High Pressure High Temperature Reservoir Fluids: Investigation of Synthetic Condensate Gases containing a Solid Hydrocarbon" *Fluid Phase Equilibria* **1995**, **111**, 2, 287-311.

[100] Coutinho, J. A. P. "Predictive UNIQUAC: A New Model for the Description of Multiphase Solid-Liquid Equilibria in Complex Hydrocarbon Mixtures" *Industrial & Engineering Chemistry Research* **1998**, **37**, 12, 4870-4875.

[101] Coutinho, J. A. P., Mirante, F. and Pauly, J. "A New Predictive UNIQUAC for Modeling of Wax Formation in Hydrocarbon Fluids" *Fluid Phase Equilibria* **2006**, **247**, 1-2, 8-17.

[102] Esmaeilzadeh, F., Fathi Kaljahi, J. and Ghanaei, E. "Investigation of Different Activity Coefficient Models in Thermodynamic Modelling of Wax Precipitation" *Fluid Phase Equilibria* **2006**, 248, 7-18.

[103] Ji, H-Y, Tohidi, B., Danesh, A. and Todd, A. C. "Wax Phase Equilibria: Developing a Thermodynamic Model using a Systematic Approach" *Fluid Phase Equilibria* **2004**, **216**, 2, 201-217.

[104] Pauly, J., Daridon, J. L., Coutinho, J. A. P., Lindeloff, N. and Andersen, S. I. "Prediction of Solid-Fluid Phase Diagrams of Light Gases-Heavy Paraffin Systems up to 200 MPa using an Equation of State-GE Model" *Fluid Phase Equilibria* **2000**, **167**, 2, 145-159.

[105] Jensen, M. R., Ungerer, P., de Weert, B. and Behar, E. "Crystallisation of Heavy Hydrocarbons from Three Synthetic Condensate Gases at High Pressure" *Fluid Phase Equilibria* **2003**, **208**, 1-2, 247-260.

[106] Pedersen, K. S., Fredenslund, Aa. and Thomassen, P. "Properties of Oils and Natural Gases" Gulf Publishing Company: Houston, TX, **1989**.

[107] Pina, A., Mougin, P. and Behar, E. "Characterization of Asphaltenes and Modelling of Flocculation – State of the Art" *Oil & Gas Science and Technology-Revue de l'Institut Français du Pétrole* **2006**, **61**, 3, 319-343.

[108] Werner, A., Behar, F., de Hemptinne, J. C. and Behar, E. "Thermodynamic Properties of Petroleum Fluids during Expulsion and Migration from Source Rocks" *Organic Geochemistry* **1996**, **24**, 10-11, 1079-1095.

[109] Shaw, J. M. "Toward Common Generalized Phase Diagrams-for Ashpaltene Containing Hydrocarbon Fluids" *Abstracts of Papers of the American Chemical Society* **2002**, 224, U276-U276.

[110] Mutelet, F., Ekulu, G., Solimando, R. and Rogalski, M. "Solubility Parameters of Crude Oils and Asphaltenes" *Energy & Fuels* **2004**, **18**, 3, 667-673.

[111] Szewczyk, V. "Modélisation thermodynamique compositionnelle de la flocculation des bruts asphalténiques" Institut National Polytechnique de Lorraine, **1997**.

[112] Buckley, J. S. "Predicting the Onset of Asphaltene Precipitation from Refractive Index Measurements" *Energy & Fuels* **1999**, **13**, 2, 328-332.

[113] Buckley, J. S. and Wang, J. X. "Crude Oil and Asphaltene Characterization for Prediction of Wetting Alteration" *Journal of Petroleum Science and Engineering* **2002**, **33**, 1-3, 195-202.

[114] Maham, Y., Chodakowski, M. G., Zhang, X. and Shaw, J. "Asphaltene Phase Behavior: Prediction at a Crossroads" *Fluid Phase Equilibria* **2005**, **227**, 2, 177-182.

[115] Hirschberg, A., de Jong, L. N. G., Schipper, B. A. and Meijer, J. G. "Influence of Pressure and Temperature on Asphaltene Flocculation" *Society of Petroleum Engineers Journal* **1984**, 24 (June), 283-293.

[116] Hirschberg, A. and Hermans, L. "Asphaltene Phase Behaviour: A Molecular Thermodynamic Model. International Symposium" in *Characterization of Heavy Crude Oils and Petroleum Residues*; Centre National de la Recherche Scientifique. Elf Aquitaine. Institut Français du Pétrole. Total, Compagnie Française de Raffinage: Lyon, **1984**.

[117] Yarranton, H. W., Sztukowski, D. M. and Urrutia, P. "Effect of Interfacial Rheology on Model Emulsion Coalescence – I. Interfacial Rheology" *Journal of Colloid and Interface Science* **2007**, **310**, 1, 246-252.

[118] Yarranton, H. W., Urrutia, P. and Sztukowski, D. M. "Effect of Interfacial Rheology on Model Emulsion Coalescence – II. Emulsion Coalescence" *Journal of Colloid and Interface Science* **2007**, **310**, 1, 253-259.

[119] Szewczyk, V. and Behar, E. "Compositional Model for predicting Asphaltene Flocculation" *Fluid Phase Equilibria* **1999**, 158-160 (June), 459-469.

[120] Abdoul, W., Rauzy, E. and Peneloux, A. "Group-Contribution Equation of State for Correlating and Predicting Thermodynamic Properties of Weakly Polar and Non-Associating Mixtures; Binary and Multicomponent Systems" *Fluid Phase Equilibria* **1991**, 68, 47-102.

[121] Ting, P. D., Hirasaki, G. J. and Chapman, W. G. "Modeling of Asphaltene Phase Behavior with the SAFT Equation of State" *Petroleum Science and Technology* **2003**, **21**, 3-4, 647-661.

[122] Buckley, J. S., Hirasaki, G. J., Liu, Y., Von Drasek, S., Wang, J. X. and Gil, B. S. "Asphaltene Precipitation and Solvent Properties of Crude Oils" *Petroleum Science and Technology* **1998**, **16**, 3-4, 251-285.

[123] Gonzales, D. L., Ting, P. D., Hirasaki, G. J. and Chapman, W. G. "Prediction of Asphaltene Instability under Gas Injection with the PC-SAFT Equation of State" *Energy & Fuels* **2005**, **19**, 4, 1230-1234.

[124] Wang, J. X. and Buckley, J. S. "A Two-Component Solubility Model of the Onset of Asphaltene Flocculation in Crude Oils" *Energy & Fuels* **2001**, **15**, 5, 1004-1012.

[125] Wang, J. X., Buckley, J. S., Burke, N. E. and Creek, J. L. "A Practical Method for Anticipating Asphaltene Problems" *SPE Production & Facilities* **2004**, **19**, 3, 152-160.

[126] Nghiem, L., Hassam, M. and Nutakki, R. "Efficient Modelling of Asphaltene Precipitation" SPE Annual Technical Conference and Exhibition, **1993**, Houston, Texas.

[127] Thomas, F., Bennion, D., Bennion, D. and Hunter, B. "Experimental and Theoretical Studies of Solids Precipitation from Reservoir Fluids" *Journal of Canadian Petroleum Technology* **1992**, **31**, 1, 22-31.

[128] Leontaritis K. and Mansoori, G. A. "Asphaltene Flocculation during Oil Production and Processing. A Thermodynamic-Colloidal Model" SPE International Symposium on Oilfield Chemistry, **1987**, San Antonio, Texas.

[129] Victorov, A. I. and Firoozabadi, A. "Thermodynamics of Asphaltene Precipitation in Petroleum Fluids by a Micellisation Model" *AIChE Journal* **1996**, 42, 1753-1764.

[130] Pan, H. and Firoozabadi, A. "A Thermodynamic Micellization Model for Asphaltene Precipitation: Part I: Micellar Size and Growth" SPE 36741; Annual Technical Conference and Exhibition Denver, Co, **1996**.

[131] Younglove, B. A. "Thermophysical Properties of Fluids. 1. Argon, Ethylene, Parahydrogen, Nitrogen, Nitrogen Trifluoride and Oxygen" *Journal of Physical and Chemical Reference Data* **1982**, 11, Supplement 1.

[132] Ferrando, N. and Ungerer, P. "Hydrogen/Hydrocarbon Phase Equilibrium Modelling with a Cubic Equation of State and a Monte Carlo Method" *Fluid Phase Equilibria* **2007**, 254, 1-2, 211-223.

[133] Chao, K. C. and Seader, J. D. "A General Correlation of Vapor-Liquid Equilibria in Hydrocarbon Mixtures" *AIChE Journal* **1961**, 7, 4, 598-605.

[134] Grayson, H. G. and Streed, C. W. "Vapor-Liquid Equilibria for High Temperature, High Pressure Hydrogen-Hydrocarbon Systems" 6th World Congress for Petroleum, Frankfurt, **1963**.

[135] Lee, B. I., Erbar, J. H. and Edmister, W. C. "Prediction of Thermodynamic Properties for Low Temperature Hydrocarbon Process Calculations" *AIChE Journal* **1973**, 19, 2, 349-356.

[136] Graboski, M. S. and Daubert, T. E. "Modified Soave Equation of State for Phase-Equilibrium Calculations .1. Hydrocarbon Systems" *Industrial & Engineering Chemistry Process Design and Development* **1978**, 17, 4, 443-448.

[137] Graboski, M. S. and Daubert, T. E. "Modified Soave Equation of State for Phase-Equilibrium Calculations .2. Systems Containing CO_2, H_2S, N_2, and Co" *Industrial & Engineering Chemistry Process Design and Development* **1978**, 17, 4, 448-454.

[138] Graboski, M. S. and Daubert, T. E. "Modified Soave Equation of State for Phase-Equilibrium Calculations .3. Systems Containing Hydrogen" *Industrial & Engineering Chemistry Process Design and Development* **1979**, 18, 2, 300-306.

[139] Moysan, J. M. P., Huron, M. J., Paradowski, H. and Vidal, J. "Prediction of the Solubility of Hydrogen in Hydrocarbon Solvents Through Cubic Equations of State" *Chemical Engineering Science* **1983**, 38, 7, 1085-1092.

[140] Moysan, J. M., Paradowski, H. and Vidal, J. "Correlation Defines Phase-Equilibria for H_2, CH_4 and N_2 Mixes" *Hydrocarbon Processing* **1985**, 64, 7, 73-76.

[141] Moysan, J. M. P., Huron, M. J., Paradowski, H. and Vidal, J. "Prediction of Phase Behaviour of Gas-Containing Systems with Cubic Equations of State" *Chemical Engineering Progress* **1986**, 41, 8, 2069-2074.

[142] Gray, R. D., Heidman, J. L., Hwang, S. C. and Tsonopoulos, C. "Industrial Applications of Cubic Equations of State for VLE Calculations, with Emphasis on H_2 Systems" *Fluid Phase Equilibria* **1983**, 13, 59-76.

[143] Gao, W., Robinson, R. L. and Gasem, K. A. M. "Alternate Equation of State Combining Rules and Interaction Parameter Generalizations for Asymmetric Mixtures" *Fluid Phase Equilibria* **2003**, 213 (1-2), 19-37.

[144] Twu, C. H., Coon, J. E., Harvey, A. H. and Cunningham, J. R. "An Approach for the Application of a Cubic Equation of State to Hydrogen Hydrocarbon Systems" *Industrial & Engineering Chemistry Research* **1996**, 35, 3, 905-910.

[145] Le Thi, C., Tamouza, S., Passarello, J. P., Tobaly, P. and de Hemptinne, J. C. "Modeling Phase Equilibrium of H_2 + *n*-Alkane and CO_2 + *n*-Alkane Binary Mixtures Using a Group Contribution Statistical Association Fluid Theory Equation of State (GC-SAFT-EOS) with a k_{ij} Group Contribution Method" *Industrial & Engineering Chemistry Research* **2006**, 45, 20, 6803-6810.

[146] Tran, T. K. S., Nguyen-Huynh, D., Ferrando, N., Passarello, J. P., de Hemptinne, J. C. and Tobaly, P. "Modeling VLE of H_2 + Hydrocarbon Mixtures Using a Group Contribution SAFT with a kij Correlation Method Based on London's Theory" *Energy & Fuels* **2009**, 23, 3, 2658-2665.

[147] Nichols, W. B., Reamer, H. H. and Sage, B. H. "Volumetric and Phase Behavior in the Hydrogen-*n*-Hexane System" *AIChE Journal* **1957**, 3, 2, 262-267.

[148] Ferrando, N. "Comparison of Different Models with Experimental Points for H_2 + Alkane Mixtures", Personal Communication, **2009**.

[149] Bidart, C., Segura, H. and Wisniak, J. "Phase Equilibrium Behavior in Water (1) + *n*-Alkane (2) Mixtures" *Industrial & Engineering Chemistry Research* **2007**, *46*, 3, 947-954.

[150] Twu, C. H., Watanasiri, S. and Tassone, V. "Methodology for Predicting Water Content in Supercritical Gas Vapor and Gas Solubility in Aqueous Phase for Natural Gas Process" *Industrial & Engineering Chemistry Research* **2007**, *46*, 22, 7253-7259.

[151] Soreide, I. and Whitson, C. "Peng-Robinson Predictions for Hydrocarbons, CO_2, N_2, and H_2S with Pure Water and NaCl brine" *Fluid Phase Equilibria* **1992**, *77*, 217-240.

[152] Kabadi, V. N. and Danner, R. P. "A Modified Soave-Redlich-Kwong Equation of State for Water Hydrocarbon Phase-Equilibria" *Industrial & Engineering Chemistry Process Design and Development* **1985**, *24*, 3, 537-541.

[153] Panagiotopoulos, A. Z. and Reid, R. C. "New Mixing Rule for Cubic Equations of State for Highly Polar, Asymmetric Systems" *ACS Symposium Series* **1986**, 300, 571-582.

[154] Michel, S., Hooper, H. H. and Prausnitz, J. M. "Mutual Solubilities of Water and Hydrocarbons from an Equation of State. Need for an Unconventional Mixing Rule" *Fluid Phase Equilibria* **1989**, *45*, 2-3, 173-189.

[155] Daridon, J. L., Lagourette, B., Saint-Guirons, H. and Xans, P. "A Cubic Equation of State Model for Phase Equilibrium Calculation of Alkane + Carbon Dioxide + Water Using a Group Contribution kij" *Fluid Phase Equilibria* **1993**, 91, 31-54.

[156] Mollerup, J. M. and Clark, W. M. "Correlation of Solubilities of Gases and Hydrocarbons in Water" *Fluid Phase Equilibria* **1989**, 51, 257-268.

[157] Eubank, P. T. "Measurements and Predictions of Three Phase Water/Hydrocarbon Equilibria" *Fluid Phase Equilibria* **1994**, 102, 181-203.

[158] Kontogeorgis, G. M., Folas, G. K., Muro-Sune, N., von Solms, N., Michelsen, M. L. and Stenby, E. H. "Modelling of Associating Mixtures for Applications in the Oil & Gas and Chemical Industries" *Fluid Phase Equilibria* **2007**, *261*, 1-2, 205-211.

[159] de Hemptinne, J. C., Mougin, P., Barreau, A., Ruffine, L., Tamouza, S. and Inchekel, R. "Application to Petroleum Engineering of Statistical Thermodynamics – Based Equations of State" *Oil & Gas Science and Technology-Revue de l'Institut Français du Pétrole* **2006**, *61*, 3, 363-386.

[160] de Hemptinne, J. C., Dhima, A. and Zhou, H. "The Importance of Water-Hydrocarbon Phase Equilibria during Reservoir Production and Drilling Operations" *Revue de l'Institut Français du Pétrole* **1998**, *53*, 3, 283-301.

[161] Campbell, J. M. "Gas Conditioning and Processing" 2nd Ed.; Published by John M. Campbell, PO Box 869: Norman, OK, **1972**.

[162] Griswold, J. and Kasch, J. E. "Hydrocarbon-Water Solubilities at Elevated Temperatures and Pressures" *Industrial & Engineering Chemistry* **1942**, *34*, 7, 804-806.

[163] Tsonopoulos, C. "Thermodynamic Analysis of the Mutual Solubilities of Normal Alkanes and Water" *Fluid Phase Equilibria* **1999**, 156, 21-33.

[164] Tsonopoulos, C. "Thermodynamic Analysis of the Mutual Solubilities of Hydrocarbons and Water" *Fluid Phase Equilibria* **2001**, 186, 185-206.

[165] McKetta, J. J. and Wehe, A. H. "Use This Chart for Water Content of Natural Gases" *Petroleum Refiner* **1958**, 37 (August), 153.

[166] Brady, C. J., Cunningham, J. R. and Wilson, G. M. "Water-Hydrocarbon Liquid-Liquid-Vapor Equilibrium Measurements to 530 °F" GPA Research Report 62, **1982**.

[167] Dhima, A., de Hemptinne, J. C. and Moracchini, G. "Solubility of Light Hydrocarbons and their Mixtures in Pure Water under High Pressure" *Fluid Phase Equilibria* **1998**, *145*, 1, 129-150.

[168] Tsonopoulos, C. and Ambrose, D. "Vapor-Liquid Critical Properties of Elements and Compounds. 8. Organic Sulfur, Silicon, and Tin Compounds (C + H plus S, Si, and Sn)" *Journal of Chemical and Engineering Data* **2001**, *46*, 3, 480-485.

[169] Maczynski, A., Goral, M., Wisniewska-Goclowska, B., Skrzecz, A. and Shaw, D. "Mutual Solubilities of Water and Alkanes" *Monatshefte fur Chemie* **2003**, **134**, 5, 633-653.

[170] Maczynski, A., Wisniewska-Goclowska, B. and Goral, M. "Recommended Liquid-Liquid Equilibrium Data. Part 1. Binary Alkane-Water Systems" *Journal of Physical and Chemical Reference Data* **2004**, **33**, 2, 549-577.

[171] Goral, M., Maczynski, A. and Wisniewska-Goclowska, B. "Recommended Liquid-Liquid Equilibrium Data. Part 2. Unsaturated Hydrocarbon-Water Systems" *Journal of Physical and Chemical Reference Data* **2004**, **33**, 2, 579-591.

[172] Goral, M., Wisniewska-Goclowska, B. and Maczynski, A. "Recommended Liquid-Liquid Equilibrium Data. Part 3. Alkylbenzene-Water Systems" *Journal of Physical and Chemical Reference Data* **2004**, **33**, 4, 1159-1188.

[173] Maczynski, A., Shaw, D. G., Goral, M., Wisniewska-Goclowska, B., Skrzecz, A., Maczynska, Z., Owczarek, I., Blazej, K., Haulait-Pirson, M. C., Kapuku, F., Hefter, G. T. and Szafranski, A. "IUPAC-NIST Solubility Data Series. 81. Hydrocarbons with Water and Seawater – Revised and Updated. Part 1. C-5 Hydrocarbons with Water" *Journal of Physical and Chemical Reference Data* **2005**, **34**, 2, 441-476.

[174] Maczynski, A., Shaw, D. G., Goral, M., Wisniewska-Goclowska, B., Skrzecz, A., Owczarek, I., Blazej, K., Haulait-Pirson, M. C., Hefter, G. T., Maczynska, Z., Szafranski, A., Tsonopoulos, C. and Young, C. L. "IUPAC-NIST Solubility Data Series. 81. Hydrocarbons with Water and Seawater – Revised and Updated. Part 2. Benzene with Water and Heavy Water" *Journal of Physical and Chemical Reference Data* **2005**, **34**, 2, 477-552.

[175] Maczynski, A., Shaw, D. G., Goral, M., Wisniewska-Goclowska, B., Skrzecz, A., Owczarek, I., Blazej, K., Haulait-Pirson, M. C., Hefter, G. T., Maczynska, Z., Szafranski, A. and Young, C. L. "IUPAC-NIST Solubility Data Series. 81. Hydrocarbons with Water and Seawater – Revised and Updated. Part 3. C_6H_8-C_6H_{12} Hydrocarbons with Water and Heavy Water" *Journal of Physical and Chemical Reference Data* **2005**, **34**, 2, 657-708.

[176] Maczynski, A., Shaw, D. G., Goral, M., Wisniewska-Goclowska, B., Skrzecz, A., Owczarek, I., Blazej, K., Haulait-Pirson, M. C., Hefter, G. T., Kapuku, F., Maczynska, Z. and Young, C. L. "IUPAC-NIST Solubility Data Series. 81. Hydrocarbons with Water and Seawater – Revised and Updated. Part 4. C_6H_{14} Hydrocarbons with Water" *Journal of Physical and Chemical Reference Data* **2005**, **34**, 2, 709-753.

[177] Maczynski, A., Shaw, D. G., Goral, M., Wisniewska-Goclowska, B., Skrzecz, A., Owczarek, I., Blazej, K., Haulait-Pirson, M. C., Hefter, G. T., Kapuku, F., Maczynska, Z., Szafranski, A. and Young, C. L. "IUPAC-NIST Solubility Data Series. 81. Hydrocarbons with Water and Seawater – Revised and Updated. Part 5. C_7 Hydrocarbons with Water and Heavy Water" *Journal of Physical and Chemical Reference Data* **2005**, **34**, 3, 1399-1487.

[178] Dhima, A., de Hemptinne, J. C. and Jose, J. "Solubility of Hydrocarbons and CO_2 Mixtures in Water under High Pressure" *Industrial & Engineering Chemistry Research* **1999**, **38**, 8, 3144-3161.

[179] de Hemptinne, J. C., Dhima, A. and Shakir, S. "The Henry Constant for 20 Hydrocarbons, CO_2 and H_2S in Water as a Function of Pressure and Temperature" 14th Symposium on Thermophysical Properties, Boulder, CO, **2000**.

[180] Lyckman, E. W., Eckert, C. A. and Prausnitz, J. M. "Generalized Reference Fugacities for Phase Equilibrium Thermodynamics" *Chemical Engineering Science* **1965**, 20, 685-691.

[181] Plyasunov, A. V. and Shock, E. L. "Group Contribution Values of the Infinite Dilution Thermodynamic Functions of Hydration for Aliphatic Noncyclic Hydrocarbons, Alcohols, and Ketones at 298.15 K and 0.1 MPa" *Journal of Chemical and Engineering Data* **2001**, **46**, 5, 1016-1019.

[182] Plyasunova, N. V., Plyasunov, A. V. and Shock, E. L. "Group Contribution Values for the Thermodynamic Functions of Hydration at 298.15 K, 0.1 MPa. 2. Aliphatic Thiols, Alkyl Sulfides, and Polysulfides" *Journal of Chemical and Engineering Data* **2005**, **50**, 1, 246-253.

[183] Plyasunov, A. V., Plyasunova, N. V. and Shock, E. L. "Group Contribution Values for the Thermodynamic Functions of Hydration at 298.15 K, 0.1 MPa. 4. Aliphatic Nitriles and Dinitriles" *Journal of Chemical and Engineering Data* **2006**, **51**, 5, 1481-1490.

[184] Plyasunov, A. V., Plyasunova, N. V. and Shock, E. L. "Group Contribution Values for the Thermodynamic Functions of Hydration at 298.15 K, 0.1 MPa. 3. Aliphatic Monoethers, Diethers, and Polyethers" *Journal of Chemical and Engineering Data* **2006**, **51**, 1, 276-290.

[185] Harvey, A. H. "Semiempirical Correlation for Henry's Constants Over Large Temperature Ranges" *AIChE Journal* **1996**, 42, 1491-1494.

[186] Price, L. C. "Aqueous Solubility of Petroleum as Applied to its Origin and Primary Migration" *American Association of Petroleum Geologists (AAPG) Bulletin* **1976**, **60**, 2, 213-244.

[187] Groves, F. R. "Solubility of Cycloparaffins in Distilled Water and Salt Water" *Journal of Chemical & Engineering Data* **1988**, 33, 136-138.

[188] Keeley, D. F., Hoffpauir, M. A. and Meriwether, J. R. "Solubility of Aromatic Hydrocarbons in Water and Sodium Chloride Solutions of Different Ionic Strengths: Benzene and Toluene" *Journal of Chemical & Engineering Data* **1988**, 33, 87-89.

[189] Bradley, R. S., Dew, M. J. and Munro, D. C. "The Solubility of Benzene and Toluene in Water and Aqueous Salt Solutions Under Pressure" *High Temperatures- High Pressures* **1973**, 5, 169-176.

[190] Fedushkin, I. L., Lukoyanov, A. N., Hummert, M. and Schumann, H. "Coordination of Benzene to a Sodium Cation" *Russian Chemical Bulletin* **2007**, **56**, 9, 1765-1770.

[191] Weisenberger, S. and Schumpe, A. "Estimation of Gas Solubilities in Salt Solutions at Temperatures from 273K to 363K" *AIChE Journal* **1996**, **42**, 1, 298-300.

[192] Cohen, J. M., Wolf, P. F. and Young, W. D. "Enhanced Hydrate Inhibitors: Powerful Synergism with Glycol Ethers" *Energy & Fuels* **1998**, **12**, 2, 216-218.

[193] Mohammadi, A. H., Chapoy, A., Richon, D. and Tohidi, B. "Experimental Measurement and Thermodynamic Modeling of Water Content in Methane and Ethane Systems" *Industrial & Engineering Chemistry Research* **2004**, **43**, 22, 7148-7162.

[194] Carroll, J. J. "An Examination of the Prediction of Hydrate Formation Conditions in Sour Natural gas" GPA Europe, Spring Meeting, Dublin, Ireland.

[195] Munck, J., Skjold-Joergensen, S. and Rasmussen, P. "Computations of the Formation of Gas Hydrates" *Chemical Engineering Science* **1988**, **43**, 10, 2661-2672.

[196] Hammerschmidt, E. G. "Formation of Gas Hydrates in Natural Gas Transmission Lines" *Industrial & Engineering Chemistry* **1934**, **26**, 8, 851-855.

[197] Sloan, E. D. "Clathrate Hydrates: The Other Common Solid Water Phase" *Industrial & Engineering Chemistry Research* **2000**, **39**, 9, 3123-3129.

[198] Sloan, E. D. "Fundamental Principles and Applications of Natural Gas Hydrates" *Nature* **2003**, **426**, 6964, 353-359.

[199] Van der Waals, J. H. and Platteeuw, J. C. "Clathrate Solutions" *Advances in Chemical Physics* **1959**, **2**, 1, 1-57.

[200] Parrish, W. R. and Prausnitz, J. M. "Dissociation Pressure of Gas Hydrates Formed by Gas Mixtures" *Industrial & Engineering Chemistry Process Design and Development* **1972**, **11**, 1, 26-35.

[201] Ballard, A. L. "A Non-Ideal Hydrate Solid Solution Model for a Multi-phase Equilibria Program", Colorado School of Mines, Golden, **2002**.

[202] Verma, V. K. "Gas Hydrates from Liquid Hydrocarbon-Water Systems", University of Michigan, Ann Arbor, **1974**.

[203] Youssef, Z. "Etude thermodynamique de la formation d'hydrates en absence d'eau liquide: mesures et modélisation", Université Claude Bernard, Lyon, **2009**.

[204] Ruffine, L., Barreau, A. and Mougin, P. "How to Represent H₂S within the CPA EOS" *Industrial Engineering Chemistry Research* **2006**, 45, 7688-7699.

[205] de Hemptinne, J. C. and Behar, E. "Propriétés thermodynamiques de systèmes contenant des gaz acides; Etude bibliographique" *Oil & Gas Science and Technology* **2000**, **55**, 6, 617-637.

[206] Carroll, J. J. "Phase Equilibria Relevant to Acid Gas Injection: Part 2 – Aqueous Phase Behaviour" *Journal of Canadian Petroleum Technology* **2002**, **41**, 7, 39-43.

[207] Carroll, J. J. and Mather, A. E. "A Generalized Correlation for the Peng-Robinson Interaction Coefficients for Paraffin-Hydrogen Sulfide Binary Systems" *Fluid Phase Equilibria* **1995**, 105, 221-228.

[208] Carroll, J. J. and Mather, A. E. "Phase-Equilibrium in the System Water Hydrogen-Sulfide – Hydrate-Forming Conditions" *Canadian Journal of Chemical Engineering* **1991**, **69**, 5, 1206-1212.

[209] Wichterle, I., Linek, J., Wagner, Z. and Kehiaian, H. "Vapor-Liquid Equilibrium Bibliographic Database" ELDATA: International Electronic *Journal of Physico-Chemical Data* **1993**.

[210] Wichterle, I., Linek, J., Wagner, Z. and Kehiaian, H. "Vapor-Liquid Equilibrium Bibliographic Database. Supplement to 1993" ELDATA: International Electronic Journal of Physico-Chemical Data, ELDATA: Int. Electron. *Journal of Physical and Chemistry Data* **1995**, 5.

[211] Leu, A. D. and Robinson, D. B. "Equilibrium Phase Properties of the *n*-butane-Hydrogen Sulfide and Isobutane-Hydrogen Sulfide Binary Systems" *Journal of Chemical and Engineering Data* **1989**, **34**, 3, 315-319.

[212] Poetmann, F. H. and Katz, D. L. "Phase Behaviour of Binary Carbon Dioxide – Paraffin Systems" *Industrial & Engineering Chemistry* **1945**, **37**, 9, 847-853.

[213] Ohgaki, K. and Katayama, T. "Isothermal Vapor-Liquid Equilibrium Data for the Ethane – Carbon Dioxide System at High Pressures" *Fluid Phase Equilibria* **1977**, **1**, 1, 27-32.

[214] de Hemptinne, J. C. and Behar, E. "Propriétés thermodynamiques de systèmes contenant des gaz acides; Etude bibliographique" *Oil & Gas Science and Technology-Revue de l'Institut Français du Pétrole* **2000**, **55**, 6, 617-637.

[215] Kato, K., Nagahama, K. and Hirata, M. "Generalized Interaction Parameter for the Peng-Robinson Equation of State: Carbon Dioxide-*n*-Paraffin Binary Systems" *Fluid Phase Equilibria* **1981**, 7, 219-231.

[216] Yau, J. S. and Tsai, F. N. "Correlation of Solubilities of Carbon Dioxide in Aromatic Compounds" *Fluid Phase Equilibria* **1992**, 73, 1-25.

[217] Lecomte, F., Broutin, P. and Lebas, E. "Le captage du CO_2: des technologies pour réduire les émissions de gaz à effet de serre" 1st Ed.; Editions Technip: Paris, **2010**.

[218] Takenouchi, S. and Kennedy, G. G. "The Binary System H_2O-CO_2 at High Temperatures and Pressures" *American Journal of Science* **1964**, 262, 1055-1074.

[219] Selleck, F. T. "Phase Behavior in the Hydrogen Sulfide-Water System" *Industrial & Engineering Chemistry Research* **1952**, **44**, 9, 2219-2226.

[220] Ugrozov, V. V. "Equilibrium Compositions of Vapor-Gas Mixtures over Solutions" *Zhurnal Fizicheskoi Khimii* **1996**, **70**, 7, 1328-1329.

[221] Gillespie, P. C. and Wilson, G. M. "Vapor-Liquid and Liquid-Liquid Equilibria: Water-Methane; Water-Carbon Dioxide; Water-Hydrogen Sulfide; Water-*n*-Pentane; Water-Methane-*n*-Pentane" GPA, Research Report 48, **1982**.

[222] Todheide, K. and Franck, E. U. "Das Zweiphasengebiet und die Kritische Kurve im System Kohlendioxid-Wasser Bis zu Drucken von 3500 bar" *Zeitschrift für Physikalische Chemie Neue Folge* **1963**, 37, 387-401.

[223] Kontogeorgis, G. M., Michelsen, M. L., Folas, G. K., Derawi, S., von Solms, N. and Stenby, E. H. "Ten years with the CPA (Cubic-Plus-Association) Equation of State. Part 1. Pure Compounds and Self-Associating Systems" *Industrial & Engineering Chemistry Research* **2006**, **45**, 14, 4855-4868.

[224] Kontogeorgis, G. M., Michelsen, M. L., Folas, G. K., Derawi, S., von Solms, N. and Stenby, E. H. "Ten Years with the CPA (Cubic-Plus-Association) Equation of State. Part 2. Cross-Associating and Multicomponent Systems" *Industrial & Engineering Chemistry Research* **2006**, **45**, 14, 4869-4878.

[225] Huc, A.Y. "Heavy Crude Oils: from Geology to Upgrading, an Overview; Ed. Technip: Paris, **2011**.

[226] Robinson, D. B., Huang, S. H., Leu, A. D. and Ng, H. J. "The Phase Behavior of Two Mixtures of Methane, Carbon Dioxide, Hydrogen Sulfide and Water" GPA, Research Report 57, **1982**.

[227] Tsivintzelis, I. *et al.* "Modeling Phase Equilibria for Acid Gas Mixtures using the CPA Equation of State". 1. Mixtures with H_2S, *AIChE Journal* **2010**, **56**, 11, 2965-2982.

[228] Barreau, A., Blanchon Le Bouhelec, E., Tounsi, K. N. H., Mougin, P. and Lecomte, F. "Absorption of H_2S and CO_2 in Alkanolamine Aqueous Solution: Experimental Data and Modelling with the Electrolyte-NRTL Model" *Oil & Gas Science and Technology-Revue de l'Institut Français du Pétrole* **2006**, **61**, 3, 345-361.

[229] Blanchon Le Bouhelec, E., Mougin, P., Barreau, A. and Solimando, R. "Rigorous Modeling of the Acid Gas Heat of Absorption in Alkanolamine Solutions" *Energy & Fuels* **2007**, **21**, 4, 2044-2055.

[230] Garst, A. W. and Lawson, J. D. "Solubility of Methane, Ethane and Propane in Aqueous Monoethanolamine and Aqueous Diethanolamine Solutions" Amoco Production Co., F72-P-22, **1972**.

[231] Lawson, J. D. and Garst, A. W. "Gas Sweetening Data – Equilibrium Solubility of Hydrogen-Sulfide and Carbon-Dioxide in Aqueous Monoethanolamine and Aqueous Diethanolamine Solutions" *Journal of Chemical and Engineering Data* **1976**, **21**, 1, 20-30.

[232] Lawson, J. D. and Garst, A. W. "Hydrocarbon Gas Solubility in Sweetening Solutions – Methane and Ethane in Aqueous Monoethanolamine and Diethanolamine" *Journal of Chemical and Engineering Data* **1976**, **21**, 1, 30-32.

[233] Lee, J. I. and Mather, A. E. "Solubility of Hydrogen Sulfide in Water" *Berichte der Bunsen-Gesellschaft für Physikalische Chemie* **1977**, **81**, 10, 1021-1023.

[234] Jou, F. Y., Mather, A. E. and Otto, F. D. "Solubility of H_2S and CO_2 in Aqueous Methyldiethanolamine Solutions" *Industrial & Engineering Chemistry Process Design and Development* **1982**, **21**, 4, 539-544.

[235] Jou, F. Y., Carroll, J. J., Mather, A. E. and Otto, F. D. "Solubility of Mixtures of Hydrogen-Sulfide and Carbon-Dioxide in Aqueous N-Methyldiethanolamine Solutions" *Journal of Chemical and Engineering Data* **1993**, **38**, 1, 75-77.

[236] Jou, F. Y., Otto, F. D. and mather, A. E. "Equilibria of H_2S and CO_2 in Triethanolamine Solutions" *Canadian Journal of Chemical Engineering* **1985**, **63**, 1, 122-125.

[237] Carroll, J. J., Jou, F. Y., Mather, A. E. and Otto, F. D. "The Distribution of Hydrogen-Sulfide Between An Aqueous Amine Solution and Liquid Propane" *Fluid Phase Equilibria* **1993**, 82, 183-190.

[238] Anderko, A., Wang, P. M. and Rafal, M. "Electrolyte Solutions: from Thermodynamic and Transport Property Models to the Simulation of Industrial Processes" *Fluid Phase Equilibria* **2002**, 194, 123-142.

[239] Wang, P. M., Anderko, A. and Young, R. D. "A Speciation-Based Model for Mixed-Solvent Electrolyte Systems" *Fluid Phase Equilibria* **2002**, **203**, 1-2, 141-176.

[240] Li, Y. G. and Mather, A. E. "Correlation and Prediction of the Solubility of Carbon-Dioxide in a Mixed Alkanolamine Solution" *Industrial & Engineering Chemistry Research* **1994**, **33**, 8, 2006-2015.

[241] Li, Y. G. and Mather, A. E. "Correlation and Prediction of the Solubility of N_2O in Mixed-Solvents" *Fluid Phase Equilibria* **1994**, 96, 119-142.

[242] Blanchon le Bouhelec, "Contribution à la thermodynamique de l'absorption des gaz acides H_2S et CO_2 dans les solvants eau-alcanolamine-méthanol: mesures expérimentales et modélisation", Institut National Polytechnique de Lorraine – Nancy, **2006**.

[243] Kent and Eisenberg "Better Data for Amine Creating" *Hydrocarbon Processing* **1976**, **55**, 2, 87-90

[244] Austgen, D. M., Rochelle, G. T., Peng, X. and Chen, C. C. "Model of Vapor-Liquid Equilibria for Aqueous Acid Gas – Alkanolamine Systems Using the Electrolyte-NRTL Equation" *Industrial & Engineering Chemistry Research* **1989**, 28, 1060-1073.

[245] Posey, M. L., Tapperson, K. G. and Rochelle, G. T. "A Simple Model for Prediction of Acid Gas Solubilities in Alkanolamines" *Gas Separation & Purification* **1996**, 10, 3, 181-186.

[246] Vallée, G. "Caractérisation expérimentale et modélisation par une équation d'état de systèmes électrolytiques complexes", École des Mines de Paris, **1998**.

[247] Vallée, G., Mougin, P., Jullian, S. and Furst, W. "Representation of CO$_2$ and H$_2$S Absorption by Aqueous Solutions of Diethanolamine Using an Electrolyte Equation of State" *Industrial & Engineering Chemistry Research* **1999**, 38, 9, 3473-3480.

[248] Fürst, W. and Renon, H. "Representation of Excess Properties of Electrolyte Solutions Using a New Equation of State" *AIChE Journal* **1993**, 39, 2, 335-343.

[249] Nguyen-Huynh, D. "Modélisation thermodynamique de mélanges symétriques et asymétriques de composés polaires oxygénés et/ou aromatiques par GC-SAFT", Université Paris XIII, **2008**.

[250] Krupatkin, I. L. and Basanov, A. N. "Ternary Heterogeneous Liquid Systems with Three Chemical Compounds (m-Cresol – Acetophenone Complexes)" *Russ. J. Phys. Chem.* **1970**, 44, 3, 440-442.

[251] Solomonov, B. N., Novikov, V. B., Varfolomeev, M. A. and Klimovitskii, A. E. "Calorimetric Determination of Hydrogen-Bonding Enthalpy for Neat Aliphatic Alcohols" *Journal of Physical Organic Chemistry* **2005**, 18, 11, 1132-1137.

[252] Goral, M., Oracz, P., Skrzecz, A., Bok, A. and Maczynski, A. "Recommended Vapor-Liquid Equilibrium Data. Part 1: Binary n-Alkanol-n-Alkane Systems" *Journal of Physical and Chemical Reference Data* **2002**, 31, 3, 701-748.

[252] Goral, M., Oracz, P., Skrzecz, A., Bok, A. and Maczynski, A. "Recommended Vapor-Liquid Equilibrium Data. Part 2: Binary Alkanol-Alkane Systems" *Journal of Physical and Chemical Reference Data* **2003**, 32, 4, 1429-1472.

[254] Goral, M., Skrzecz, A., Bok, A., Maczynski, A. and Oracz, P. "Recommended Vapor-Liquid Equilibrium Data. Part 3. Binary Alkanol-Aromatic Hydrocarbon Systems" *Journal of Physical and Chemical Reference Data* **2004**, 33, 3, 959-997.

[255] Dahlmann, U. and Schneider, G. M. "(Liquid + Liquid) Phase-Equilibria and Critical Curves of (Ethanol + Dodecane Or Tetradecane Or Hexadecane Or 2,2,4,4,6,8,8-Heptamethylnonane) from 0.1-MPa to 120.0-MPa" *Journal of Chemical Thermodynamics* **1989**, 21, 9, 997-1004.

[256] Maciel, M. R. W. and Francesconi, A. Z. "Excess Gibbs Free-Energies of (Normal-Hexane + Propane-1-ol) at 338.15 and 348.15-K and of (Normal-Hexane + Propan-2-ol) at 323.15-K, 338.15-K, and 348.15-K" *Journal of Chemical Thermodynamics* **1988**, 20, 5, 539-544.

[257] Lallemand, T. "Modélisation thermodynamique des mélanges méthanol – eau – hydrocarbures", Université Aix-Marseille II, **1998**.

[258] Mourah, M. "Modélisation des équilibres de phases liquide-liquide et liquide-vapeur des mélanges contenant de l'eau, des alcools et des hydrocarbures", Université Paris 13 (Paris Nord), **2009**.

[259] Litvinov, N. D. "Isothermal Vapor-Liquid Equilibrium in the Systems of Three Unlimited Miscible Liquids" *Zhurnal Fizicheskoi Khimii* **1952**, 26, 8, 1144-1151.

[260] Riesco, N. and Trusler, J. P. M. "Novel Optical Flow Cell for Measurements of Fluid Phase Behaviour" *Fluid Phase Equilibria* **2005**, 228, 233-238.

[261] Oliveira, M. B., Teles, A. R. R., Queimada, A. J. and Coutinho, J. A. P. "Phase Equilibria of Glycerol Containing Systems and their Description with the Cubic-Plus-Association (CPA) Equation of State" *Fluid Phase Equilibria* **2009**, 280, 1-2, 22-29.

[262] Ferreira, O., Macedo, E. A. and Brignole, E. A. "Application of the GCA-EoS Model to the Supercritical Processing of Fatty Oil Derivatives" *Journal of Food Engineering* **2005**, 70, 4, 579-587.

[263] Espinosa, S., Diaz, S. and Fornari, T. "Extension of the Group Contribution Associating Equation of State to Mixtures Containing Phenol, Aromatic Acid and Aromatic Ether Compounds" *Fluid Phase Equilibria* **2005**, **231**, 2, 197-210.

[264] Tumakaka, F., Gross, J. and Sadowski, G. "Thermodynamic Modelling of Complex Systems Using PC-SAFT" *Fluid Phase Equilibria* **2005**, 228-229, 89-98.

[265] Nguyen-Huynh, D., Falaix, A., Passarello, J. P., Tobaly, P. and de Hemptinne, J. C. "Predicting VLE of Heavy Esters and Their Mixtures Using GC-SAFT" *Fluid Phase Equilibria* **2008**, 264, 184-200.

[266] Raal, J. D., Code, Russel K. and Best, Donald A. "Examination of Ethanol-*n*-Heptane, Methanol-*n*-Hexane Systems Using New Vapor-Liquid Equilibrium Still" *Journal of Chemical & Engineering Data* **1972**, **17**, 2, 211-216.

[267] Nguyen-Huynh, D., Passarello, J. P., Tobaly, P. and de Hemptinne, J. C. "Modeling Phase Equilibria of Asymmetric Mixtures Using a Group-Contribution SAFT (GC-SAFT) with a k(ij) Correlation Method Based on London's Theory. 1. Application to CO_2 + *n*-Alkane, Methane plus *n*-Alkane, and Ethane plus *n*-Alkane Systems" *Industrial & Engineering Chemistry Research* **2008**, **47**, 22, 8847-8858.

[268] Nguyen-Huynh, D., Passarello, J. P., Tobaly, P. and de Hemptinne, J. C. "Application of GC-SAFT EOS to Polar Systems Using a Segment Approach" *Fluid Phase Equilibria* **2008**, **264**, 1, 62-75.

[269] Siow K.S., Delmas, G. and Patterson, D. "Cloud-Point Curves in Polymer Solutions with Adjacent Upper and Lower Critical Solution Temperatures" *Macromolecules* **1972**, **5**, 1, 29-34.

[270] Lambert, M. L., Song, Y. and Prausnitz, J. M. "Equations of State for Polymer Systems" in *Equations of State for Fluids and Fluid Mixtures*; Ed. Sengers, J. V., Kayser, R. F., Peters, C. J, and White, H. J. Jr; Experimental Thermodynamics; Elsevier: Amsterdam, **2000**.

[271] Marsh, K. N., Boxall, J. A. and Lichtenthaler, R. N. "Room Temperature Ionic Liquids and Their Mixtures – a Review" *Fluid Phase Equilibria* **2004**, 219, 93-98.

[272] Keskin, S., Kayrak-Talay, D., Akman, U. and Hortacsu, O. "A Review of Ionic Liquids Towards Supercritical Fluid Applications" *Journal of Supercritical Fluids* **2007**, **43**, 1, 150-180.

[273] Lei, Z., Zhang, J. and Chen, B. "UNIFAC Model for Ionic Liquids" *Industrial & Engineering Chemistry Research* **2009**, 48, 2697-2704.

[274] Gani, R. and Oconnell, J. P. "A Knowledge Based System for the Selection of Thermodynamic Models" *Computers & Chemical Engineering* **1989**, **13**, 4-5, 397-404.

[275] de Hemptinne, J. C. and Behar, E. "Thermodynamic Modelling of Petroleum Fluids" *Oil & Gas Science and Technology-Revue de l'Institut Français du Pétrole* **2006**, **61**, 3, 303-317.

[276] Carlson, E. "Don't Gamble with Physical Properties for Simulation" *Chemical Engineering Progress* **1996**, 35-46.

5

Case Studies

The choice of a modelling approach is very often guided by the experience on other systems, or by the availability of parameter databases. This is acceptable in the sense that most industrial computations are based on previous cases that worked correctly. Yet, it may also be dangerous when extrapolating conditions, and it is then recommended to submit the problem to a thermodynamic analysis.

Our purpose in this book is to help the engineer in selecting, fitting and validating the appropriate model for his or her particular application. In the website that is related to this work, it is intended to publish case studies showing that, although the problem in each case is quite different, a common path can be identified for reaching the solution.

The philosophy is discussed in the first chapter. In the same spirit as proposed by other authors [1], it is based on a number of observations that have been analysed in detail in the chapters 2 to 4: thermodynamic computations rest on the combined understanding of three major concepts (properties, mixtures and phases):

Properties

The second chapter of this book has stressed the importance of choosing a model in accordance with the properties that are requested. This is important as it may be useful to consider using different models depending on the task. A given model may not be equally accurate for all properties.

Mixtures

In the third chapter, the relationship between components and models was made clear: a large number of models have been discussed, and the physical relevance of their parameters was mentioned. In the best case, the parameters originate from a confrontation with experimental data, but this cannot always be guaranteed. As a result, a given model that works very well in some conditions may perform poorly with other pressures, temperatures or compositions.

Phases

In the fourth chapter, we have insisted on a number of specific phase diagrams that may be encountered in the process industry, using a classification that is based on the system composition.

In this chapter, which is an introduction to the case studies, a first section summarizes the proposed procedure for a thermodynamic analysis, and a second section lists some of the major processes found in the chemical and petroleum industry, along with some key questions that can help the process engineer in his/her analysis.

5.1 PROBLEM SOLVING PROCEDURE

Once the problem has been correctly identified as discussed in the chapters 2 to 4 of this book, the solution procedure can start. A three-step procedure is proposed:
- evaluate **the best model**, including the adequate simplifying assumptions, based on the problem analysis discussed above,
- determine **the data** that exist or that are needed in order to either fit or validate the model,
- **regress and evaluate** the result in order to decide which is the best model/parameter combination.

5.1.1 Evaluation of the most appropriate model(s)

In this first step, the observations that have been made in the problem analysis are used in order to justify possible simplifying assumptions that will guide the choice.

There is generally no single choice for a thermodynamic model [2]. Hence, it may be good at this stage to retain several options, which will be evaluated at the end of the problem-solving procedure.

5.1.1.1 Properties required

Concerning the **properties required**: which have the most impact on the final results? What simplifications can be afforded? The properties can be subdivided in two main families: single phase or equilibrium.

A. Single phase properties

Is there a need for a calculation method for single phase properties (volume, enthalpy, entropy…)? Simplifying assumptions may be either low pressure (gas properties are ideal gas; liquid properties can be taken at the vapour pressure) or ideal mixture (no mixing properties), as discussed in section 2.2.2 (p. 52) or in section 4.1.4 (p. 259).

It is important to mention that some equilibrium calculations may require single phase properties. Examples are *PH* calculations (as in a distillation column), where an enthalpy calculation is needed in additional to the distribution coefficients.

B. Equilibrium properties

Both for chemical and phase equilibrium calculations, a model must be provided for calculating fugacities (or chemical potentials) of each component in each phase.

In case phase equilibrium is considered, these fugacities are often transformed into distribution coefficients (section 2.2.3.1.B, p. 70):

$$K_i = \frac{x_i^{\alpha}}{x_i^{\beta}} = \frac{\varphi_i^{\beta}}{\varphi_i^{\alpha}} \tag{5.1}$$

where α and β are the phase labels and φ_i is the fugacity coefficient. For **vapour-liquid equilibrium**, following simplifying approximations can be envisaged:

- Distribution coefficients independent of composition: this is the case for ideal mixtures. Whether or not a mixture is ideal is discussed in section 3.4.2.1 (p. 173).
- Low or moderate pressure (below 1.5 MPa), in addition to ideal mixture, in which case the distribution coefficient is written as:

$$\ln\left(K_i P\right) = \alpha_i\left(T\right) + \beta_i\left(T\right)P \tag{5.2}$$

as discussed in the example 4.2 page 277 (expressions for $\alpha_i\left(T\right)$ and $\beta_i\left(T\right)$ are provided). At pressures below 0.5 MPa, equation (5.2) can be further simplified to yield Raoult's law:

$$K_i P = P_i^{\sigma}\left(T\right) \tag{5.3}$$

In case the mixture is not ideal, but low pressure conditions are validated, the use of the activity coefficient may be convenient, as it can help understand phase behaviour as discussed in section 3.4.1 (p. 160):

$$K_i = \frac{P_i^{\sigma}\left(T\right)}{P}\gamma_i\left(T,x\right) \tag{5.4}$$

In some cases, when a high dilution problem or a supercritical component is involved, it may be useful to write equation (5.4) using the asymetric convention, presented in section 2.2.3.1.B (p. 68), using the Henry constant for the solute, keeping (5.3) or (5.4) for the solvents:

$$K_i = \frac{H_i\left(T\right)}{P} \tag{5.5}$$

When **liquid-liquid equilibrium** is expected, the mixture is per definition not ideal. The assumption that is often made is that pressure has no effect on this type of equilibrium. Only activity coefficients are in principle enough for calculating these type of equilibria (see discussion section 3.4.1.3, p. 166).

When **equilibrium with a solid phase** is to be calculated, the freezing or crystallisation properties must also be known, as shown in section 3.4.4.1 (p. 222) or example 4.6 (p. 282).

If a **chemical equilibrium** is to be calculated, the discussion in section 2.2.4 (p. 86) and example 2.11 (p. 92), shows the importance of the accuracy of the formation properties.

Phase equilibrium may be required for different purposes, as discussed in section 2.2.3.2 (p. 81) (component separation according to volatility, component separation according to affinity, component separation using crystallization; phase boundary calculation or calculation of the relative amount of each phase): it is important that the process engineer know what his/her true problem is in order to further identify the key components in his/her mixture (see below).

5.1.1.2 Fluid composition

Concerning the **fluid composition**, the process engineer should be able to distinguish the type of mixture as well as the key components in his/her mixture. Chapter 3 is designed to help him/her in this search.

The type of mixture is needed both for identifying the adequate physical model as for identifying the best method for parameter identification:

- Three types of pure components have been identified (section 3.1.2, p. 121): database components, complex molecular components and pseudo-components. Each of these require a different approach for parameter identification.
- It is stressed, in section 3.4.1 (p. 160) that a component in a mixture may behave very differently from the same component as it is pure. This is probably the reason why so many different methods and mixing rules have been developed. All section 3.4 (p. 160) is devoted to the discussion of the main models that are available today.

Since the parameters are often determined by fitting on experimental data, section 3.3, (p. 148) discusses the methodology of parameter regression.

Yet, in view of the specific process problem that is addressed, all mixture components may not have the same importance. As an example, it can be understood that if the condition of water condensation from a natural gas must be calculated, the exact ratio of methane and ethane is probably of little importance. Instead, the water concentration must be known with high accuracy. This is why in section 3.5 (p. 225) the question of the "key components" is raised. Obviously, the answer to that question very much depends on the process problem. This question refers back, therefore, to the true property that must be investigated:

- **Phase boundary**: here, it is the composition of the nascent phase that will provide the answer. The fugacity or chemical potential of the majority (or perhaps only) component of this phase must be known with great precision within the bulk phase. This is often a challenge because the key component may in reality be a minority component in the bulk phase, and as a result behave in a strongly non-ideal manner.
- **Component separation**: in this case, the distribution coefficient of some components should be regarded as more important than others. Again, it can happen that one of the component to be separated from the others is a minority component (e.g. in case of severe specifications).

As a conclusion of this analysis, the process engineer must be able to identify the nature of the components, both taken individually and in a mixture. He or she must then evaluate whether key component(s) should be identified, either as individual component, or as binary, ternary, etc.

5.1.1.3 Representative phases

The fourth chapter of this book describes in detail the type of phases than can be encountered for a number of industrially relevant systems. Throughout the chapter, model recommendations are provided depending on the type of mixture and of problem considered. The conclusion (section 4.3, p. 325) of this chapter provides some general guidelines.

The applicability of a model may be restricted to some particular phase:

- **Vapour phases** require an equation of state (which can be very simple – the ideal gas – if only low pressure conditions are considered). The use of these type of equations is summarised in section 2.2.2.1 (p. 52) and a list of the most common such equations is found in section 3.4.3 (p. 189).
- **Liquid phases** can be described by either en equation of state or an activity coefficient model. The use of this latter type of models is summarised in section 2.2.2.2 (p. 60) and a list of the most common such models is found in section 3.4.2 (p. 171). The choice will be guided by the degree of non-ideality of the mixture, as explained in section 3.4.2.1 (p. 173).
- **Critical conditions** are found when two phases become identical. Obviously, if this is the region of interest, it is essential that the same model be used in the two phases: vapour-liquid critical points can only be computed by equations of state. Yet, the true properties close to such a critical point may require a specific approach, as the asymptotic behaviour close to this point can not be described by any analytic equation.
- **Solid, crystalline phases** necessitate a specific model, as shown in section 2.2.3.1.A (p. 67); section 3.4.4.1 (p. 222) and illustrated in section 4.2.2.3 (p. 282). They are calculated from pure component fusion properties (section 3.1.1.1.F, p. 105 and 3.1.1.2.B and C, p. 113), and possibly with an activity coefficient model.

5.1.2 Search for the most significant physical data

Once a list of models capable of describing the requested phenomena have been identified, a set of experimental data needs to be found for:

- adjusting the model parameters (discussed in section 3.3, p. 148),
- evaluating the quality of the model(s).

As is stressed over and over again, the experimental data will be the final referee on whether a model is good or not. However, it is important to select the correct type of data, and to make the right decisions if new data must be gathered. In case no data exist, it is important to be able to still provide some reasonable guidelines as to the trends that can be expected. This is why different types of data can be identified. They are listed here in decreasing order of pertinence:

1. Obviously, of most value are the **data that originate from a recognised laboratory** and that correspond to the actual physical property that must be calculated, in the process pressure and temperature range. If possible, these data should come with an evaluation of their experimental uncertainty. It is important to make sure that the data are obtained in true equilibrium conditions, are sampled correctly and analysed according to well-accepted methods. Each of these steps requires a good knowledge of the phenomena. This is not so much a question of date or equipment (very ancient data may be quite accurate), than of expertise of the researchers. From this point of view, data originating from a pilot plant

should be considered with great caution, as many phenomena may occur simultaneously (reaction, mechanical entrainment, non-equilibrium sampling) that are unrelated to the equilibrium properties that we are interested in. Evaluation of the quality of experimental data has been discussed in section 3.1.3 (p. 139) for pure components, and section 3.2 (p. 142) for mixtures.

2. When no direct experimental data exist in the process conditions, **unrelated data can be extrapolated**. This can be done, but should require a good knowledge of the physical trends. Examples are extrapolation of vapour pressures or solubilities according to well-known equations (for example using the corresponding states principle introduced in section 2.2.2.1.C, p. 57). It is also possible to consider data for other components of the same family, and to consider that their behaviour will be similarly. Finally, the use of thermodynamic relationships for calculating data of a different type (e.g. enthalpy of vapourisation using a vapour pressure curve) is also acceptable. All of these options have been further discussed in sections 3.1 (p. 102) and 3.2 (p. 142), where some methods for evaluating the quality of the data are proposed.

3. So-called **"pseudo-experimental" data** can be produced using predictive calculation methods. They have the advantage of offering the possibility to calculate data in conditions that are experimentally difficult to realise (unstable components, toxic components, extreme pressure or temperature conditions...). Yet, the quality of these data obviously greatly depends on the complexity of the predictive calculation tool:

- Probably the most promising approach for this purpose is the use of **molecular simulation** [3-5]. The panoply of potential models available in both Monte-Carlo (for equilibrium properties) as dynamic simulation (for both equilibrium and transport properties) is rapidly increasing and commercial tools [1] are becoming now increasingly user-friendly for non-expert use. Quantum-mechanical tools can also be used, as for example COSMO-RS [2] [6]. Their results should first be compared to true experimental data for validation, but they allow completing the databases in a reasonable amount of time for mixtures or molecules that are not too complex.

- In a more classical mode, **group contribution methods** may provide pseudo-experimental data that are of rather good quality as long as they are used for applications that have been validated previously. Today, many such tools exist, of which the most well-known is probably PSRK [7], but others have been developed for equation of state type calculations (for cubic equations: [8, 9], or for SAFT-type equations [10-13]) or for pure component properties [14-18]. They are further discussed in chapter 3.

---

1. From companies as Materials Design; Accelrys; Scienomics.
2. Available from COSMO-Logic.

5.1.3 Evaluation of the result

The data that have been identified should be separated in two sets. One will be used for parameter regression, the other for model validation. Sometimes (for predictive models, or when parameters are provided through a database), no regression is needed. In that case, all data can be used for validation. In any case, it is preferable to keep in the second category data that are close to the process conditions, and somewhat in extrapolation with respect to the data used for regression. This way, it will be possible to discriminate between the approaches that have been selected in step 1 here above (section 5.1.1, p. 350).

Data regression is an issue that can be discussed in great detail, but others have done this with much more authority [19], and we therefore refer to them directly. A short section is devoted to this topic in section 3.3 (p. 148) of this book. Let us simply stress at this stage the importance of a good match between the choice of objective function and parameters to regress. As an example, it seems pointless to regress parameters of a component that is highly diluted on bulk properties (a density, for example).

Once the different model/parameter sets have been developed, they can be **tested on the validation** set, in order to select the most appropriate combination for use in the process simulator. Most probably, the process engineer will select the set that yields the lowest overall deviation on the validation. However, probably more importantly, he/she should look at the trends of these deviations with respect to the process conditions. If a wrong trend exists in a domain where the process may run, the model should be abandoned, as its true behaviour cannot be ascertained. On the opposite, if no real trend exists, he/she can use the values of the deviations as an estimate of the uncertainty of the model, which he/she should incorporate in his/her design plan.

5.2 REVIEW OF MAJOR PROCESS PROBLEMS

It is impossible to provide an exhaustive review of all possible problems that may be encountered in the process industry. Yet, in order to provide some general guidelines, we discuss here a number of issues to be kept in mind for a list of process operations.

5.2.1 Flash separations

Properties given (see section 2.2.2, p. 52):

- Most often pressure and temperature are given.
- Sometimes, adiabatic conditions are provided in which case enthalpy and pressure are given.

Properties needed (see section 2.2.2, p. 52):

- Composition of the phases.
- Amount of the phases.

Components of interest:

- Database components (section 3.1.2.1, p. 121).
- Pseudo-components (section 3.1.2.3, p. 129): there is a possible need for lumping.
- Complex components (section 3.1.2.2, p. 124) possibly.

Mixture properties to be considered (section 3.5, p. 225):

- Make sure that the close-boiling binaries are correctly represented (risk for azeotropes).

Phases:

- Vapour – liquid.

Choice of model (depends on mixture):

- Cubic equation of state (must be corresponding states compatible for pseudo-components) (section 3.4.3, p. 189).
- PPR78.
- NRTL or UNIQUAC if polar components (no pseudo-components) (section 3.4.2, p. 171).
- UNIFAC in case of unknown interactions between polar components.

Watch (always validate with experimental data):

- Pure component vapour pressures (sections 3.1.1.1, p. 102 and 3.4.3, p. 189, for equations of state).
- Pure component enthalpies of vapourisation (section 3.1.1.1, p. 102), if duties are needed.
- Binary interaction parameters for binaries that are identified above (section 3.4.3, p. 189).

5.2.2 Simple distillation (continuous or batch)

Properties given (see section 2.2.2, p. 52):
- Isenthalpic flash at fixed pressure: the pressure is usually determined so that the distillate bubble temperature is close to ambient (or another convenient cooling temperature).

Properties needed (see section 2.2.2, p. 52):
- Temperature (through the flash calculation, see section 2.2.3.1, p. 63).
- Enthalpy (section 2.2, p. 43).
- Relative volatilities, or partition coefficient (VLE) (section 2.2.3.2, p. 81).

Components of interest:
- Database components (section 3.1.2.1, p. 121).
- Pseudo-components (section 3.1.2.3, p. 129).
- There is a possible need for lumping.

Mixture properties to be considered (section 3.5, p. 225):
- Most important is to describe correctly the volatility of all cuts.
- Almost ideal behaviour expected.
- If severe specifications for impurities are given, also evaluate binaries between this impurity and major components (infinite dilution properties, as stressed in section 3.5.2.3, p. 232).

Phases:
- Vapour – liquid.

Choice of model:
- Cubic equation of state (must be corresponding states compatible for pseudo-components) (section 3.4.3, p. 189).
- BK10 for heavy petroleum cuts at subatmospheric pressures.

Watch (always validate with experimental data):
- Pure component vapour pressures (sections 3.1.1.1, p. 102 and 3.4.3, p. 189 for equations of state).
- If a cubic EoS is used, the alpha function should be adapted in order to ensure that the vapour pressures are correctly represented.
- Remove water from the simulation (no decant): water always goes with the non-condensable products for hydrocarbon mixtures.

5.2.3 Close volatility distillation

This case must be considered when side-streams are taken on intermediate stages of the distillation column.

Properties given (see section 2.2.2, p. 52):

- Isenthalpic flash.

Properties needed (see section 2.2.2, p. 52):

- Temperature (through the flash calculation, see section 2.2.3.1, p. 63).
- Enthalpy (section 2.2, p. 43).
- Relative volatilities, or partition coefficient (VLE) (section 2.2.3.2, p. 81).

Components of interest:

- Database components (section 3.1.2.1, p. 121).

Mixture properties to be considered (section 3.5, p. 225):

- Only the azeotropes between key components are of importance (close to cut-point, as discussed in section 3.5.2 (p. 229).
- If severe specifications for impurities are given, also evaluate binaries between this impurity and major components (infinite dilution properties, as stressed in section 3.5.2.3, p. 232).

Phases:

- Vapour – liquid.

Choice of model:

- Cubic equation of state with good mixing rule (section 3.4.3, p. 189).
- PPR78.

Watch (always validate with experimental data):

- Pure component vapour pressures (section 3.1.1.1, p. 102 and 3.4.3, p. 189 for equations of state).
- Pure component enthalpies of vapourisation (section 3.1.1.1, p. 102).
- Binary interaction parameters for binaries that are identified above (section 3.4.3, p. 189).
- Remove water from the simulation (no decant): water always goes with the non-condensable products for hydrocarbon mixtures.

5.2.4 Extractive distillation

Properties given (see section 2.2.2, p. 52):
- Isenthalpic flash: the pressure is usually determined so that the distillate bubble temperature is close to ambient (or another convenient cooling temperature).

Properties needed (see section 2.2.2, p. 52):
- Temperature (through the flash calculation, see section 2.2.3.1, p. 63).
- Enthalpy (section 2.2, p. 43).
- Relative volatilities, or partition coefficient (VLE) (section 2.2.3.2, p. 81).

Components of interest:
- Database components (section 3.1.2.1, p. 121).

Mixture properties to be considered (section 3.5, p. 225):
- Make sure that the binaries between extraction solvent and key feed components (those to be separated) are well described.
- Ideally, ternary data should be checked between the key feed components at high dilution in the extraction solvent.
- If severe specifications for impurities are given, also evaluate binaries between this impurity and major components (infinite dilution properties, as stressed in section 3.5.2.3, p. 232).

Phases:
- Vapour – liquid.

Choice of model:
- Cubic equation of state on the condition of having an adequate mixing rule (G^E-based is probably best) (section 3.4.3, p. 189).
- NRTL or UNIQUAC (section 3.4.2, p. 171).
- Excess enthalpies may not be negligible, so include them explicitly in the model for enthalpy calculations.

Watch (always validate with experimental data):
- Pure component vapour pressures (sections 3.1.1.1, p. 102 and 3.4.3, p. 189 for equations of state).
- Pure component enthalpies of vapourisation (section 3.1.1.1, p. 102).
- Binary interaction parameters for binaries that are identified above (section 3.4.3, p. 189).

5.2.5 Azeotropic distillation

The azeotrope is driving the distillation.

Properties given (see section 2.2.2, p. 52):
- Isenthalpic flash: the pressure is usually determined so that the azeotrope is located correctly with respect to feed composition.

Properties needed (see section 2.2.2, p. 52):
- Temperature (through the flash calculation, see section 2.2.3.1, p. 63).
- Enthalpy (section 2.2, p. 43).
- Relative volatilities, or partition coefficient (VLE) (section 2.2.3.2, p. 81).

Components of interest:
- Database components (section 3.1.2.1, p. 121).

Mixture properties to be considered (section 3.5, p. 225):
- Make sure that the driving azeotrope is accurate throughout the temperature and composition range of the column.
- If severe specifications for impurities are given, also evaluate binaries between this impurity and major components (infinite dilution properties, as stressed in section 3.5.2.3, p. 232).

Phases:
- Vapour – liquid.

Choice of model:
- Cubic equation of state with adequate mixing rule.
- NRTL or UNIQUAC if polar components (no pseudo-components) (section 3.4.2, p. 171).
- UNIFAC or PSRK in case of unknown interactions between polar components.

Watch (always validate with experimental data):
- Pure component vapour pressures (sections 3.1.1.1, p. 102 and 3.4.3, p. 189 for equations of state).
- Pure component enthalpies of vapourisation (section 3.1.1.1, p. 102).
- Binary interaction parameters for binaries that are identified above (section 3.4.3, p. 189).

5.2.6 Liquid-liquid decant vessel

Non-idealities are strong in this case.

Properties given (section 2.2.2, p. 52):
- Pressure and temperature are given.

Properties needed (see section 2.2.2, p. 52):
- The phase quantities must be known with good accuracy.
- The phase compositions are not always crucial (depends on downstream operations).
- Density (or volume) will be required to ensure that the correct phase is tapped.

Components of interest:
- Database components (section 3.1.2.1, p. 121).
- Complex (group contribution) components (section 3.1.2.2, p. 124).

Mixture properties to be considered (section 3.5, p. 225):
- If only bulk quantities are needed, a thorough investigation of the mutual solubilities of the main components may be sufficient.
- If severe specifications for impurities are given, also evaluate binaries between this impurity and the two solvents (infinite dilution properties, as stressed in section 3.5.2.3, p. 232) (ternary data!).
- If the composition is needed: the presence of traces of co-solvent may modify dramatically the partitioning of minority components.

Phases:
- Liquid – liquid (sometimes including a vapour).

Choice of model:
- For water-hydrocarbon, the "decant" option may be sufficient.
- Cubic equation of state with complex mixing rule.
- Specific correlations may be best for mutual water / hydrocarbon mutual solubilities (see section 4.2.4.2, p. 297).
- When co-solvents may be present, or for other systems than water + hydrocarbons, specific NRTL or UNIQUAC parameters must be fitted (section 3.4.2, p. 171).
- UNIFAC or PSRK in case of unknown interactions between polar components.

Watch (always validate with experimental data):
- Accuracy strongly depends on data.
- Co-solvents may have dramatic effects.
- Watch for possible density inversions.

5.2.7 Stripping (vapour-liquid extraction)

The stripping medium is often steam, or otherwise an inert gas (nitrogen or hydrogen)

Properties given (see section 2.2.2, p. 52):

• Pressure and temperature.

Properties needed (see section 2.2.2, p. 52):

• Distribution coefficients of components to be stripped.
• Vapourisation enthalpy (section 2.2, p. 43).
• Vapourised fractions.

Components of interest:

• Database components (section 3.1.2.1, p. 121).

Mixture properties to be considered (section 3.5, p. 225):

• Henry constants (infinite dilution properties).

Phases:

• Vapour – liquid, low pressure.

Choice of model:

• Henry constant (see section 2.2.3.1, p. 63).
• An equation of state (e.g. cubic) can be used, on the condition that infinite dilution properties have been validated with binary interaction parameters (k_{ij}).

Watch (always validate with experimental data):

• Infinite dilution properties.
• In case of steam stripping, make sure no water decant option is selected!.
• The stripped stream is loaded with water, and must be dried downstream of the steam stripping.

5.2.8 Natural gas liquefaction

Properties given (section 2.2.2, p. 52):
- Pressure and temperature.

Properties needed (section 2.2.2, p. 52):
- Gas distribution coefficients.
- Enthalpy (section 2.2, p. 43).
- Phase boundaries.

Components of interest:
- Database components (section 3.1.2.1, p. 121).

Mixture properties to be considered (section 3.5, p. 225):
- Heavy components description.
- Interactions between light and heavy components.

Phases:
- Vapour – liquid.

Choice of model:
- Cubic equation of state (use k_{ij} for light + heavy interactions) (section 3.4.3, p. 189).
- Very accurate equations of state (see section 4.1.4.1, p. 260) for single phase gas properties.

Watch (always validate with experimental data):
- Small concentrations of heavy components may have large effect on the dew point.
- Due to low temperatures, heavy components, even in very low concentrations, may crystallise: see "flow assurance" (section 5.2.13, p. 368).

5.2.9 Gas treatment (physical)

Examples are gas dehydration or absorption of acid gases using a physical solvent.

Properties given (section 2.2.2, p. 52):

• Pressure and temperature.

Properties needed (see section 2.2.2, p. 52):

• Distribution coefficients of all components.
• Vapour fraction.
• Enthalpy (section 2.2, p. 43).

Components of interest:

• Database components (section 3.1.2.1, p. 121).

Mixture properties to be considered (section 3.5, p. 225):

• All binaries should be correctly described, but more in particular the binaries between the solvent and the components to be extracted (water, acid gases, ...).
• Infinite dilution properties in the solvent are also important.
• Check also for mixing enthalpies.

Phases:

• Vapour – liquid.

Choice of model:

• CPA is recommended in this case (section 3.4.3.5, p. 216).
• If not available, a modified version of any cubic Eos, that includes a complex mixing rule (at least two interaction parameters, or otherwise a G^E-type mixing rule, see section 3.4.3.4, p. 204).
• Excess enthalpies may not be negligible, so include it if possible in the enthalpy calculations.

Watch (always validate with experimental data):

• Solvent volatility: it will evaporate and end up in the lean gas.
• Solvent purity: any volatile impurity in the solvent will be directly found in the lean gas.

5.2.10 Liquid-liquid extraction

Properties given (section 2.2.2, p. 52):

- Pressure and temperature.

Properties needed (section 2.2.2, p. 52):

- Liquid-liquid distribution coefficients.

Components of interest:

- Database components (section 3.1.2.1, p. 121).
- Complex, group contribution components may be present (section 3.1.2.2, p. 124).

Mixture properties to be considered (section 3.5, p. 225):

- Ternary systems should be considered, where the two liquid phases are represented with their main components (solvents) and the partitioning of the component to be extracted is investigated.
- Infinite dilution properties in both phases are important.

Phases:

- Liquid – liquid.

Choice of model:

- Liquid-liquid extraction is very difficult to model accurately. Correlations based on experimental data are often used.
- Otherwise, activity coefficient models are most adapted (NRTL, UNIQUAC).

Watch (always validate with experimental data):

- The possible presence of a co-solvent may dramatically change the partitioning between the two phases (section 3.5.1.2, p. 228).

5.2.11 Supercritical extraction

An example is de-asphalting of heavy ends using a solvent close to its critical point. This is a two-step process: extraction in the dense solvent (see liquid-liquid extraction) and flash recovery from the light solvent (vapour-liquid equilibrium). Instead of heating for the recovery, the strong density change upon pressure drop for a supercritical solvent is used for controlling solvent power.

Properties given (see section 2.2.2, p. 52):

- Pressure and temperature.

Properties needed (section 2.2.2, p. 52):

- Liquid-liquid and vapour-liquid distribution coefficients of components to be extracted.
- Solubility of these components when changing pressure and temperature conditions.

Components of interest:

- Database components (section 3.1.2.1, p. 121).
- Pseudo-components (the detailed composition is often unknown): chemical affinity (determined through SARA-type analysis) is probably more important than the volatility (section 3.1.2.3, p. 129).
- It may be convenient to take representative molecules of each fraction of interest, and to describe them using group contribution.

Mixture properties to be considered (section 3.5, p. 225):

- No true predictive method exists (see also on asphaltene deposits in section 4.2.2.4, p. 286). Hence, it is recommended to collect data on representative fluids.

Phases:

- Liquid – liquid and vapour – liquid.

Choice of model:

- The model must be an equation of state, since the solvent power depends on density.
- A cubic EoS is convenient, as the location of the critical point is exact for the pure components. Yet, it probably needs complex mixing rules (G^E-type) validated on experimental data.
- In case all components can be represented by a specific molecule, the PSRK EoS can be used as a first approximation.

Watch (always validate with experimental data):

- Either vapour-liquid-liquid or vapour-liquid-solid equilibrium can be observed, depending on whether the fraction to be extracted is a solid or a liquid in process conditions.

5.2.12 Crystallisation

As for example in para-xylene or waxes separation.

Properties given (section 2.2.2, p. 52):

• Pressure and temperature.

Properties needed (see section 2.2.2, p. 52):

• Fusion properties of the component(s) to be crystallised.
• Activity (fugacity) of the same component(s) in the fluid phase.

Components of interest:

• Database components (section 3.1.2.1, p. 122).
• Sometimes complex molecules (properties calculated from the molecular structure, section 3.1.2.2, p. 124).

Mixture properties to be considered (section 3.5, p. 225):

• Interactions of bulk fluid (solvent) with components to be crystallised (can be
  validated with VLE or LLE data).

Phases:

• Fluid – solid.

Choice of model:

• Activity coefficient model (NRTL; UNIQUAC).
• In case of complex molecules, UNIFAC can be of help.
• If vapour phase or high pressure, an equation of state should be used: validate the
  interactions as indicated above.

Watch (always validate with experimental data):

• Solid solutions or coexistence of several solid phases may sometimes be encountered (e.g. for waxes, see section 4.2.2.3, p. 281).
• Crystallisation is often kinetically-driven, so a purely thermodynamic analysis
  may not be enough to describe the phenomena: supersaturation information may
  be needed.

5.2.13 Flow assurance

Example is the risk of hydrate formation in pipe transport (see section 4.2.4.3, p. 304). It may also be waxes (section 4.2.2.3, p. 281) or asphaltene (section 4.2.2.4, p. 286) precipitation. Scale formation resulting from salt precipitation is generally rather described with a chemical equilibrium.

Properties given (section 2.2.2, p. 52):
- Pressure and temperature.

Properties needed (section 2.2.2, p. 52):
- Phase boundaries: risk of condensation of a dense phase (often solid).
- Fusion properties of the component(s) to be crystallised.
- Fugacity of the same component(s) in the fluid phase.

Components of interest:
- Database components (section 3.1.2.1, p. 121).
- Sometimes complex molecules (properties calculated from the molecular structure).
- Pseudo-components may also be used (section 3.1.2.3, p. 129).

Mixture properties to be considered (section 3.5, p. 225):
- Interactions of bulk fluid (solvent) with components to be crystallised (can be validated with VLE or LLE data).

Phases:
- Fluid – solid.

Choice of model:
- A separate model must be defined for the incipient phase (solid or hydrate, as discussed in section 3.4.4, p. 222).
- Activity coefficient model (NRTL; UNIQUAC).
- In case of complex molecules, UNIFAC can be of help.
- If vapour phase or high pressure, an equation of state should be used, but the mixing rule should be good enough to validate the binaries as indicated above.
- For scales, a chemical equilibrium constant is used as discussed in section 2.2.4, p. 86.

Watch (always validate with experimental data):
- Crystallisation is often kinetically-driven, so a purely thermodynamic analysis may not be enough to describe the phenomena: supersaturation information may be needed.

5.2.14 Kinetically controlled reactions

The reaction is in that case not in thermodynamic equilibrium. A catalyst is generally used. Examples are hydrotreatments or the Fischer Tropsch reactor.

Properties given (section 2.2.2, p. 52):

• Pressure and temperature.

Properties needed (section 2.2.2, p. 52):

• Gas distribution coefficients (both reactants and products, etc.).
• Enthalpy, including enthalpy of reaction (section 2.2, p. 43).
• Vapourised fraction (for pressure drop calculation).

Components of interest:

• Gases (section 3.1.2.1, p. 121).
• Pseudo-components (section 3.1.2.3, p. 129): lumping may sometimes be needed. Lumping criteria for phase equilibrium and for chemical reactions may be different.

Mixture properties to be considered (section 3.5, p. 225):

• Dilution properties of gases in the liquid mixtures (Henry constants).
• Volatility (vapour pressure) of the liquid components.

Phases:

• Vapour – liquid.

Choice of model:

• Grayson Streed if light hydrocarbons (nC < 10) (section 4.2.3, p. 288).
• Cubic equation of state (section 3.4.3, p. 189): a binary interaction parameter between gases and heavy components is essential.
• If the molecules are well-defined, SAFT could also be used with benefit because of its good capacity to describe strongly asymmetric systems.

Watch (always validate with experimental data):

• Especially for heavy residues, the extrapolation of the vapour pressure or gas solubility is not always validated.
• During the reaction, H_2S, NH_3 or CO_2 gases may be formed, that may yield scale (ammonium sulfates): see "Flow Assurance".
• When water is present, it may condense downstream, forming very corrosive aqueous solutions. For describing this condensation, a specific electrolyte package is needed (e.g. e-NRTL or similar approaches).

5.2.15 Equilibrated reactions

As an example, the water-gas shift reaction which occurs in a single phase. Either the "Gibbs reactor" or a conventional equilibrium constant approach can be used (see section 2.2.4, p. 86).

Properties given (section 2.2.2, p. 52):

- Pressure and temperature.

Properties needed (section 2.2.2, p. 52):

- Equilibrium composition.
- Extent of reaction.
- Enthalpy, including enthalpy of reaction (section 2.2, p. 43).

Components of interest:

- Database components (section 3.1.2.1, p. 121).
- Complex components (section 3.1.2.2, p. 124).

Mixture properties to be considered (section 3.5, p. 225):

- In dense phases, the non-idealities of the reacting species must be correctly described: watch for highly diluted species.

Phases:

- Single phase (fluid).

Choice of model:

- Formation properties or, equivalently, equilibrium constants must be known (section 2.2.4, p. 86).
- A thermodynamic model may be needed for taking into account non-idealities due to density effects (unless the ideal gas assumption is acceptable): an equation of state or an activity coefficient model depending on the phase at hand.

Watch (always validate with experimental data):

- Impurities may be formed: all possible reaction products should be considered.
- Potential appearance of a new phase resulting from the chemical shift (liquid or solid).

5.2.16 Simultaneous phase equilibrium and chemical reactions

The most complex case occurs when chemical and phase equilibrium occur simultaneously. An example is gas treatment using amine solutions.

Properties given (section 2.2.2, p. 52):

- Pressure and temperature.

Properties needed (section 2.2.2, p. 52):

- All species distribution coefficients (solvents, reactants and products).
- Enthalpy, including enthalpy of reaction (section 2.2, p. 43).
- Vapourised fractions (for pressure drop calculation).

Components of interest:

- Database components (section 3.1.2.1, p. 121) including ionic species.

Mixture properties to be considered (section 3.5, p. 225):

- All binaries, and sometimes ternary interactions should be investigated (see discussion section 4.2.5.3, p. 310).
- Interactions between solvents and reacting species are essential.

Phases:

- Vapour – liquid.

Choice of model:

- Formation properties or, equivalently, chemical equilibrium constants must be known (section 2.2.4, p. 86).
- In case of electrolytic mixtures (e.g. amines), specific models are needed (section 3.4.2.5, p. 186).

Watch (always validate with experimental data):

- True (including chemical reactions) and apparent (global, i.e. not taking into account chemical shift) solubilities are different.
- Potential appearance of a new phase resulting from the chemical shift (liquid or solid).

Chapter 5 • Case Studies

REFERENCE LIST

[1] O'Connell, J. P., Gani, R., Mathias, P.M., Maurer, G., Olson, J.D. and Crafts P.A. "Thermodynamic Property Modeling for Chemical Process and Product Engineering: some Perspectives Industrial & Engineering Chemistry Research" **2009**, **48**, 4619-4637.

[2] de Hemptinne, J. C. and Behar, E. "Thermodynamic Modelling of Petroleum Fluids" *Oil & Gas Science and Technology-Revue de l'Institut Français du Pétrole* **2006**, **61**, 3, 303-317.

[3] Ungerer, P., Tavitian, B. and Boutin, A. *"Applications of Molecular Simulation in the Oil and Gas Industry"* 1st Ed.; Editions Technip: Paris, **2005**.

[4] Ungerer, P., Nieto-Draghi, C., Lachet, V., Wender, A., Di Lella, A., Boutin, A., Rousseau, B. and Fuchs, A. H. "Molecular Simulation Applied to Fluid Properties in the Oil and Gas Industry" *Molecular Simulation* **2007**, **33**, 4-5, 287-304.

[5] Ungerer, P., Nieto-Draghi, C., Rousseau, B., Ahunbay, G. and Lachet, V. "Molecular Simulation of the Thermophysical Properties of Fluids: From Understanding Toward Quantitative Predictions" *Journal of Molecular Liquids* **2007**, **134**, 1-3, 71-89.

[6] Klamt, A. *"COSMO-RS From Quantum Chemistry to Fluid Phase Thermodynamics and Drug Design"* 1st Ed.; Elsevier: Amsterdam, **2005**.

[7] Ahlers, J. and Gmehling, J. "Development of a Universal Group Contribution Equation of State. 2. Prediction of Vapor-Liquid Equilibria for Asymmetric Systems" *Industrial & Engineering Chemistry Research* **2002**, **41**, 14, 3489-3498.

[8] Jaubert, J. N., Vitu, S., Mutelet, F. and Corriou, J. P. "Extension of the PPR78 Model (Predictive 1978, Peng-Robinson EOS with Temperature Dependent k(ij) Calculated Through a Group Contribution Method) to Systems Containing Aromatic Compounds" *Fluid Phase Equilibria* **2005**, **237**, 1-2, 193-211.

[9] Vitu, S., Jaubert, J. N. and Mutelet, F. "Extension of the PPR78 Model (Predictive 1978, Peng-Robinson EOS with Temperature Dependent kij Calculated Through a Group Contribution Method) to Systems Containing Naphtenic Compounds" *Fluid Phase Equilibria* **2006**, **243**, 1-2, 9-28.

[10] Nguyen-Huynh, D., Passarello, J. P., Tobaly, P. and de Hemptinne, J. C. "Modeling Phase Equilibria of Asymmetric Mixtures Using a Group-Contribution SAFT (GC-SAFT) with a k(ij) Correlation Method Based on London's Theory. 1. Application to CO_2 + *n*-Alkane, Methane plus *n*-Alkane, and Ethane plus *n*-Alkane Systems" *Industrial & Engineering Chemistry Research* **2008**, **47**, 22, 8847-8858.

[11] Nguyen-Huynh, D., Tran, T. K. S., Tamouza, S., Passarello, J. P., Tobaly, P. and de Hemptinne, J. C. "Modeling Phase Equilibria of Asymmetric Mixtures Using a Group-Contribution SAFT (GC-SAFT) with a k(ij) Correlation Method Based on London's Theory. 2. Application to Binary Mixtures Containing Aromatic Hydrocarbons, *n*-Alkanes, CO_2, N_2, and H_2S" *Industrial & Engineering Chemistry Research* **2008**, **47**, 22, 8859-8868.

[12] Lymperiadis, A., Adjiman, C. S., Galindo, A. and Jackson, G. A. "Group Contribution Method for Associating Chain Molecules Based on the Statistical Associating Fluid Theory (SAFT-Gamma)" *Journal of Chemical Physics* **2007**, 127.

[13] Lymperiadis, A., Adjiman, C. S., Jackson, G. and Galindo, A. "A Generalisation of the SAFT-Gamma Group Contribution Method for Groups Comprising Multiple Spherical Segments" *Fluid Phase Equilibria* **2008**, 274, 85-104.

[14] Marrero, J. and Gani, R. "Group-Contribution-Based Estimation of Octanol/Water Partition Coefficient and Aqueous Solubility" *Industrial & Engineering Chemistry Research* **2002**, **41**, 25, 6623-6633.

[15] Marrero, J. and Gani, R. "Group-Contribution Based Estimation of Pure Component Properties" *Fluid Phase Equilibria* **2001**, 183, 183-208.

[16] Nannoolal, Y., Rarey J., Ramjugernath D. and Cordes, W. "Estimation of Pure Component Properties. Part 1. Estimation of the Normal Boiling Point of Non-Electrolyte Organic Compounds Via Group Contributions and Group Interactions" *Fluid Phase Equilibria* **2004**, 226, 45-63.

[17] Nannoolal, Y., Rarey J. and Ramjugernath D. "Estimation of Pure Component Properties. Part 2. Estimation of Critical Property Data by Group Contribution" *Fluid Phase Equilibria* **2007**, 252, 1-27.

[18] Nannoolal, Y., Rarey J. and Ramjugernath D. "Estimation of Pure Component Properties. Part 3. Estimation of the Vapor Pressure of Non-Electrolyte Organic Compounds Via Group Contributions and Group Interactions" *Fluid Phase Equilibria* **2008**, 269, 117-133.

[19] Englezos, P. and Kalogerakis, N. "Applied Parameter Estimation for Chemical Engineers" 1st Ed.; Marcel Dekker, Inc.: New York, Basel, **2001**.

Index

F

First principle 45
Flash 70
    calculation 70
    closed system 77
    isenthalpic 77
    isentropic 77
    isochoric 77
    $P\theta$ 72
    PT 72
    separation 356
    types 41
Flory-Huggins model 180, 182, 188
Flow assurance 368
Formation properties 91, 108
Free volume 181
Fugacity 52
    coefficient 56, 64, 191, 205
    liquid phase 64
    solid phase 67
    vapour phase 64
Fundamental thermodynamic relations 44, 46
Fusion 28, 67, 105, 113

G

Gas treatment 364
GC-Flory equation 182
G^E based mixing rules 214
Gel 270
Gibbs energy 11, 26, 46, 167
    excess 61
    formation 89, 108
    reaction 89, 91
Gibbs phase rule 27
Gibbs-Duhem criterion 145
Gibbs-Helmholtz equation 47, 62, 89
Glass 270
Grayson and Streed 224, 290
Group contribution 122, 125, 214, 303, 354
Group Contribution Association (GCA) model 219
Group Contribution Lattice Fluid (GCLF) model 220

H

Hard sphere 193
Hayden and O'Connell 200
Heat *(see also enthalpy)* 45
    vapourisation 109, 119
Heat capacity
    ideal gas 115
    isobaric 47, 254, 257
    isochoric 47
Helmholtz energy 26, 190
Henry
    constant 66, 129, 144, 224, 273
    law 69
Heteroatom 102, 316
Heteroazeotrope 35, 293, 296, 317
Heterogeneous approach 68, 70, 326
Homogeneous approach 68, 326
Huron and Vidal mixing rule 214
Hybrid model 218
Hydrate 223, 271, 304
Hydrogen bonding 102, 174, 317

I

Ice 304
Ideal gas 53, 115, 193
Ideal mixture 60, 69, 277
Impurity 232
Infinite dilution 65, 232
    properties 224
Interaction
    ion-dipole 174
    dispersive 175
    Van der Waals 175
Internal energy 11, 26, 45
    excess 62
Ionic liquid 325

J

Joule-Thomson coefficient 48, 258

www.ingramcontent.com/pod-product-compliance
Lightning Source LLC
Chambersburg PA
CBHW081040220326
41598CB00038B/6940